PROGRESS IN CHEMOMETRICS RESEARCH

PROGRESS IN CHEMOMETRICS RESEARCH

ALEXEY L. POMERANTSEV
EDITOR

Nova Science Publishers, Inc.
New York

Production Coordinator: Donna Dennis
Editorial Production Manager: Donna Dennis
Senior Production Editor: Susan Boriotti
Office Manager: Annette Hellinger
Graphics: Andrea Charles, Magdalena Nuñez and Annemarie VanDeWater
Editorial Production: Marius Andronie, Maya Columbus, Keti Datunashvili, Melissa Diaz, Andrew Kallio, Vladimir Klestov, Lorna Loperfido, Joann Overton and Rusudan Razmadze
Circulation: Ave Maria Gonzalez, Vera Popovic, Luis Aviles, Alexandra Columbus, Raymond Davis, Cathy DeGregory, Marlene Nuñez, Jeannie Pappas, Lauren Perl and Katie Sutherland

Library of Congress Cataloging-in-Publication Data
Available upon request

ISBN 1-59454-257-0

400 Oser Ave, Suite 1600
Hauppauge, New York 11788-3619
Tele. 631-231-7269 Fax 631-231-8175
e-mail: Novascience@earthlink.net
Web Site: http://www.novapublishers.com

Printed in the United States of America

CONTENTS

PREFACE

Chemometrics is what chemometricians do. Chemometricians are people who drink beer and steal ideas from statisticians.

These jocular definitions belong to Swedish scientist Swante Wold, who (together with Bruce Kowalski) introduced the term chemometrics in the early 1970s. Chemometrics is a new science that has no strict definition. The more official definition (by Massart) says that –

Chemometrics is the chemical discipline that uses mathematical, statistical and other methods employing formal logic:

- to design or select optimal measurement procedures and experiments, and
- to provide maximum relevant chemical information by analyzing chemical data.

However this definition is as ambiguous as the definition of geometry as the geodetic discipline that uses mathematical methods to measure distances and squares on earth. Being conceived as a branch of analytical chemistry, chemometrics now is a general approach. It extracts relevant information out of measured data, regardless of their origin: chemical, physical, biological, and *etc.* Chemometrics has been applied in different areas, and most successfully in multivariate calibration, pattern recognition, classification and discriminant analysis, multivariate modeling, and monitoring of processes. The main chemometric principle is a concept of hidden data structures that can be found using methods of multivariate data analysis. These are the well-known statistic tools such as partial least squares (PLS), soft independent modeling of class analogy (SIMCA), principal-component regression (PCR), wavelet analysis, and many others. Current activities of chemometricians fall into two main two categories: (1) development of new methods for manipulating multivariate data and (2) new applications of the known chemometric techniques in different areas such as environment control, food industry, agriculture, medicine, and engineering.

In this book we collected papers of scientists, who participated in the Third Winter School (now, Symposium) on Chemometrics (WSC-3) held in Russia, in Pushkinskie Gory village (near Pskov city), 2004. The conference is a regular event in the DRUSHBAMETRICS ("friendship-metrics") project, established in 1999 by the Danish scientist Kim H. Esbensen and his Russian colleagues Oxana Rodionova and Alexey Pomerantsev. The overall objective of DRUSHBAMETRICS is to provide young (and young-at-heart) scientists and specialists an opportunity to get acquainted with chemometric multivariate data analysis. An equally important goal is to facilitate international networking with colleagues from abroad. WSC has become a place mainly for presenting new ideas and fruitful discussions in a friendly informal atmosphere.

Participants of WSC-3

Scientists from Austria, Belarus, Canada, Denmark, Finland, Iran, Norway, Sweden, and Russia presented their work for this book. Among them there are the world famous specialists such as Rolf Ergon (Norway), Paul Geladi (Sweden), Kim Esbensen (Denmark), Kurt Varmuza (Austria), as well as the young researchers who make the first steps in chemometrics. We hope that this collection will open a series of the DRUSHBAMETRICS books dedicated to chemometrics and its applications.

Alexey L. Pomerantsev
Semenov Institute of Chemical Physics
Moscow, Russia

PART I. THEORY

In: Progress in Chemometrics Research
Editor: Alexey L. Pomerantsev, pp. 3-11
ISBN 1-59454-257-0

Chapter 1

ANALYSIS OF SIMULATED FRACTURE SURFACES USING AMT AND FRACTAL GEOMETRY METHODS

S. Kucheryavski, V. Polyakov and A. Govorov
Altai State University, Barnaul, Russia, svk@asu.ru

ABSTRACT

Conventional fractal analysis and new AMT (Angle Measure Technique) method are applied for investigation of deformation structure that appears in metals under plastic deformation. The problem of quantitative description of fracture surfaces is important for understanding the deformation processes and therefore for new materials development. In order to evaluate the ability of the methods to discriminate and to classify the surfaces regarding their complexity, both fractal analysis and AMT are applied to simulated surface images having different roughness and noise level. It is shown that the AMT method gives better results for the noised images and for images with a non self-similar structure.

Keywords: fractal analysis; angle measure technique; porous metals; deformation structure

INTRODUCTION

Investigation of deformation structure that occurs in the metal materials under plastic deformation and their fracture surface enables to understand the deformation processes and to obtain the most effective operative conditions. This structure, depending on the stage of deformation, could consist of dislocations, micro-, mezo- and macro-cracks as well as the fracture surfaces. Studying these structure objects we can determine the conditions of fracture and understand how deformation characteristics of materials depend on different parameters. Thus, investigation of the deformation structure is an actual task of material science.

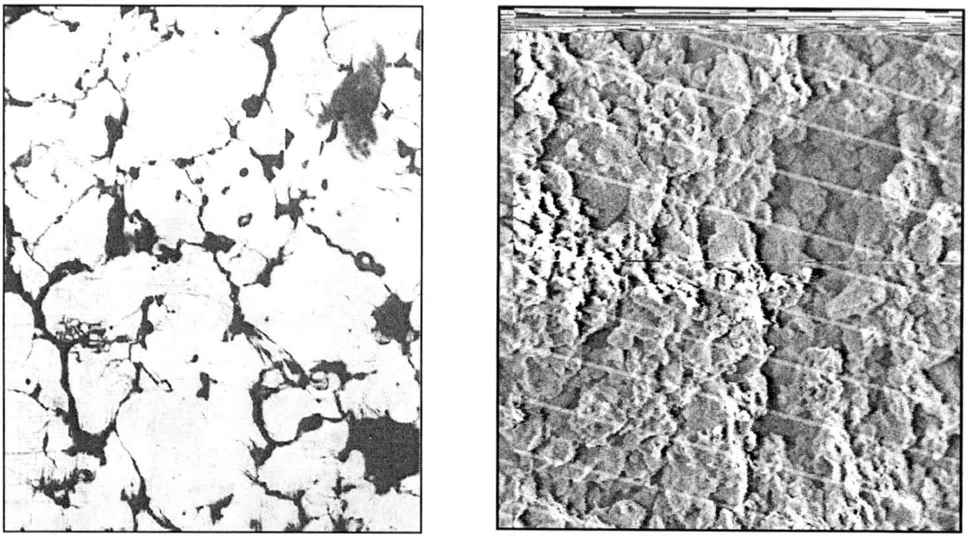

a) b)

Fig. 1. Micro cracks (a) and fracture surface (b) of deformed specimen of porous iron

The main difficulty in such investigations is obtaining the quantitative description of the fracture surfaces, which, usually, have a non-uniform structure. The problem is still more complicated when one studies complex multiphase materials with a very heterogeneous internal structure, such as, for example, porous materials (see Fig. 1). The known approaches to this problem can be divided in two groups:

- Statistical approaches (i.e. Mean Absolute Deviation).
- Non traditional approaches that deal with structure peculiarities at different scales

Fractal analysis is a well-known method that recently is widely used. The basic ideas of the fractal geometry methods were collected and published in the book of American scientist B. Mandelbrot [1]. The main concept of the fractal geometry is fractal dimension that can be considered as a quantitative measure of objects heterogeneity or their self-similarity. Mandelbrot defines fractal dimension as follows. Let us consider a set of points J in the space and choose some function, $h(\delta) = \gamma(d)\delta^d$, for example, square, cube, sphere, or segment. Here, $\gamma(d)$ is a geometric coefficient, which is equal to 1 for segments and squares, $\gamma = \pi / 4$ for circles, and $\gamma = \pi / 6$ for spheres. After that, we cover set J with the chosen functions, forming a measure $M_d = \sum h(\delta)$ of this set. As $\delta \to 0$, measure M_d, tends to zero, or to infinity, depending on the measure dimension, d. The Housdorf dimension of a set D is the critical value of d, for which M_d changes its limiting value from 0 to ∞:

$$M_d = \sum \gamma(d)\delta^d = \gamma(d)N(\delta)\delta^d \xrightarrow[\delta \to 0]{} \begin{cases} 0, d > D \\ \infty, d < D \end{cases}$$

The Hoursdorf dimension is a local property of the set. It can be shown, that for such objects as lines and planes, *d* is equal to 1 and 2, accordingly. However, for some objects, the value of *d* is not an integer. In this case, we would say that such objects have a fractal dimension. There are many algorithms that allow to obtain the fractal dimension in practice. The most simple and popular is the box-counting algorithm [2].

Thus, in simplified representation, one could interpret the fractal dimension as a measure of heterogeneity of a set of points in plane, or in space, or, if we deal with the surfaces, it is a measure of the surface roughness.

There are papers that report that fractal characteristics of deformation structure of porous metal materials depend on their physical and mechanical properties [3, 4]. These results show that in some cases there are correlations between fractal dimension of the materials fracture surface and the deformation stages and conditions. However, it was determined that the fractal dimension does not give good results, if surfaces are smooth ($D \sim 2.1\text{-}2.2$). Moreover, there are some cases, when we cannot calculate *D* with a satisfactory accuracy. This is because of non-linear dependence of the numbers of measure functions covering a set of points, which are used in the box-counting algorithm versus their scale factor. Moreover, fractal analysis fails when we deal with noisy surfaces images. In this connection, the main aim of this work is to apply new methods of quantitative analysis of the heterogeneous structures and to compare them with conventional fractal analysis.

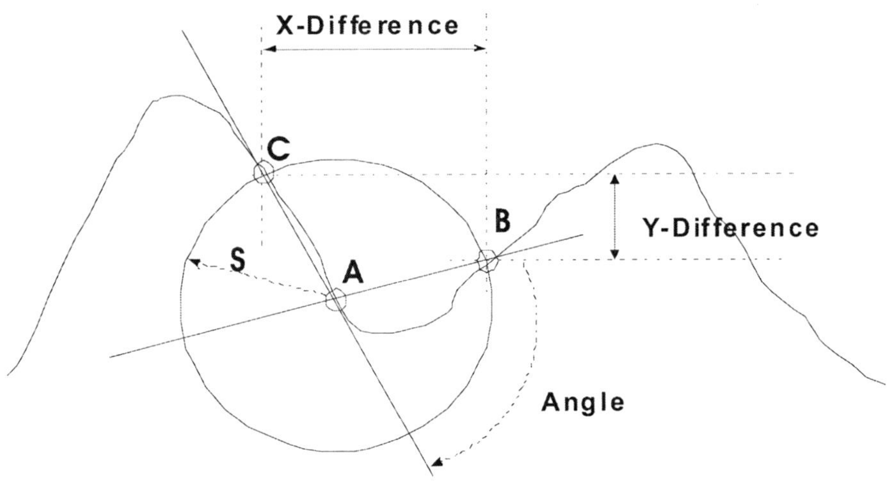

Fig. 2. The AMT-spectra obtaining

When we consider different methods of the structure analysis, we pay a special attention to their simplicity as well as to the closeness to ideology of fractal geometry. Finally, the AMT (Angle Measure Technique [5]) method was chosen for this purpose. One of the main reasons is that AMT has been developed for investigations of structures, for which the conventional fractal analysis failed.

The Angle Measure Technique is a rather new method for image processing and analysis purposes. It was developed by Robert Andrle as a substitute of fractal analysis for the purpose of complexity of geomorphic lines investigation. At present, AMT is widely applied in chemometrics as a general method for analysis of the 1D and the 2D complex signals [6].

The AMT algorithm is very simple. At the first stage, the rows of the 2D image matrix are unfolded into the 1D signal. Along this signal we choose a point (point *A* in Fig. 2) and choose initial value of scale factor, s_0. Afterward, for every point on the signal curve, we draw a circle, which has center in this point and radius *s*, calculate, and store the supplement to the angle CAB. The values of the X-difference and the Y-difference (see Fig. 2) are also obtained. Finally, we calculate the mean angle measure and the mean Y-difference by averaging values for all points. By repeating these calculations for each scale value s_i in the range $s_0 \leq s_i \leq s$, we get the AMT-spectra, which can be used for further analysis of complex spectra. Taking into account that these spectra consist of a large number of data, it is useful to apply methods of multivariate data analysis, such as PCA or PLS.

EXPERIMENTAL

To check the effectiveness of the AMT method for analysis of fracture surfaces, we apply this technique for classification of surfaces, which were simulated with and without noise. We made 225 stochastic fractal surfaces (*D* values are changed from 2.1 to 2.9) with the help of the Diamond-Square Algorithm [7]. The surfaces images have the 512×512 pixels size. The examples of such images are shown in Fig. 3.

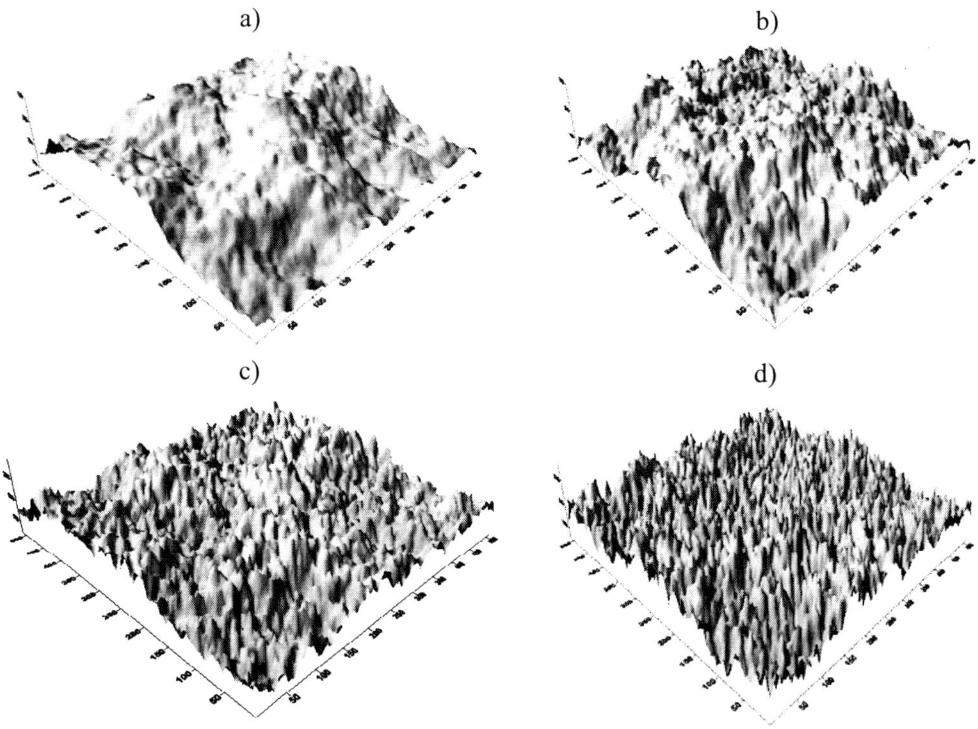

Fig. 3. Simulated stochastic fractal surfaces with fractal dimension D = 2.1 (a), 2.4 (b), 2.6 (c) and 2.9 (d)

The surface analysis includes two stages. At first we obtain the AMT spectra and the Y-difference spectra. At the next stage we perform the PCA analysis of these spectra for the surface classification. The results are presented below on the score plot in Fig. 4.

Two first components describe about 90% of residual variation. The correspondent score plot for the surfaces with value of *D* in the range of $2.1 < D < 2.9$ with step 0.1, is shown in Fig. 4. From this figure one can see that AMT discriminates samples with different values of *D*.

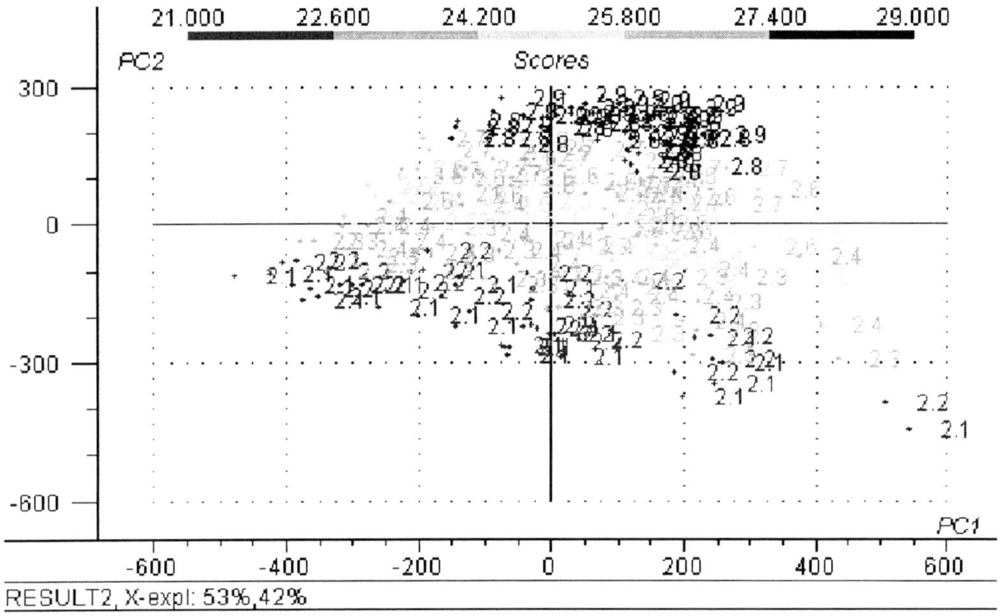

Fig. 4 Score plot of the AMT-spectra of samples with different D

The first component, PC1, most probably stands for such smooth distinction between surfaces, which cannot be described by the fractal dimension. PC2 corresponds to the fractal dimension, as the samples with larger values of *D* lie higher along this principal component. However, the clusters with neighboring *D* values are overlapped; thus we cannot make good discrimination of samples, which have similar fractal dimension.

The score plot constructed for the AMT spectrum of the samples with *D* equals 2.1 and 2.9 is shown in Fig. 5. As we can see, application of PCA discrimination to the AMT-spectra of samples with very different *D* values gives a clear-cut distinction. In this case, PC1 corresponds to *D* and PC2 describes the difference in smooth properties of the surfaces. Thus, we can make the conclusion, that AMT is also sensitive to the fractal properties of objects.

The real surfaces that are obtained with microscopes, various video and photo instruments differ from the simulated surfaces, first of all by noise presence. In the second part of this work we consider how noise in images affects discrimination ability both in fractal analysis and in the AMT method. For this purpose, the Gauss noise with various magnitudes is added to the simulated surfaces images. Afterwards, the fractal dimension is calculated and the AMT-spectra are obtained for these noisy surfaces. The examples of such noisy surfaces, together with their initial and calculated fractal dimensions are shown in Fig. 6. Obviously, that noise makes the surfaces more heterogeneous, and thus calculated fractal dimension does not correspond to the initial one. In this case, we cannot perform classification using fractal analysis.

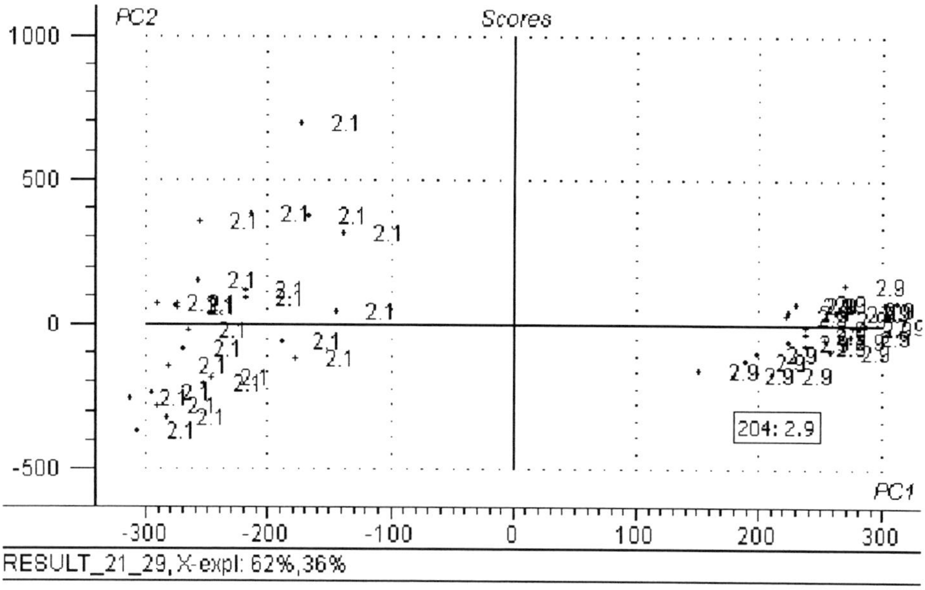

Fig. 5. Score plot of the AMT-spectra of samples with D equals 2.1 and 2.9.

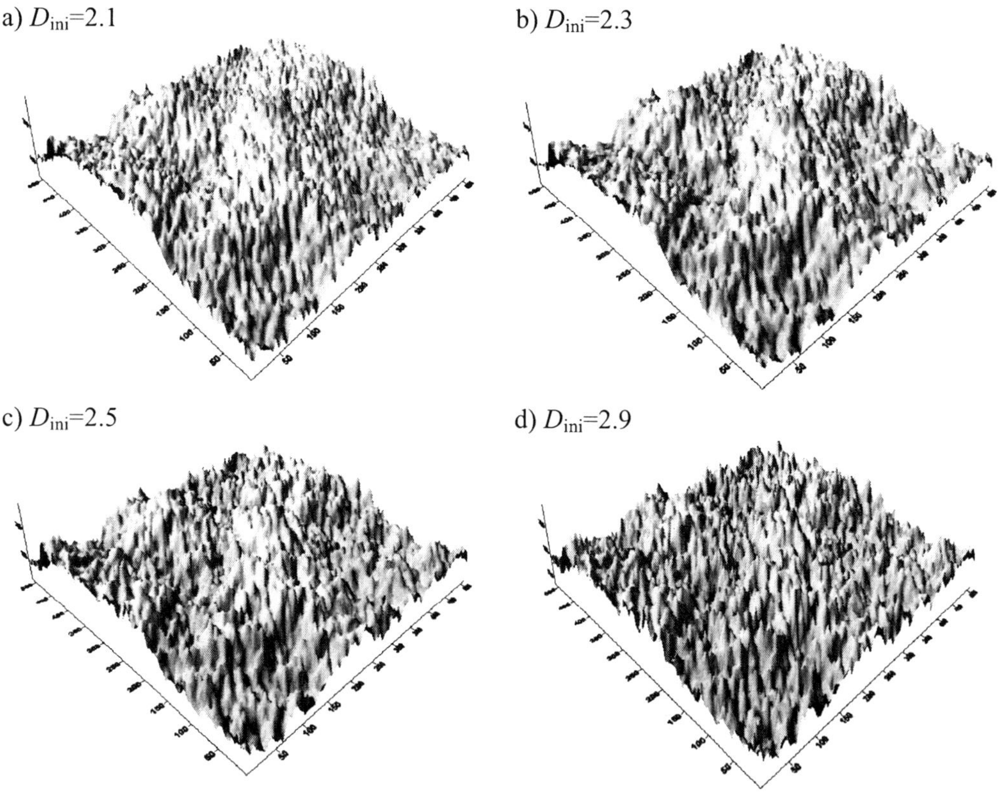

Fig. 6. Samples of noised surface with their initial (D_{ini}) and calculated (D_{cal}) fractal dimensions

PCA score plot of the AMT spectra of the Gaussian noised samples with initial fractal dimension *D* from 2.1 to 2.9 is presented on Fig. 7.

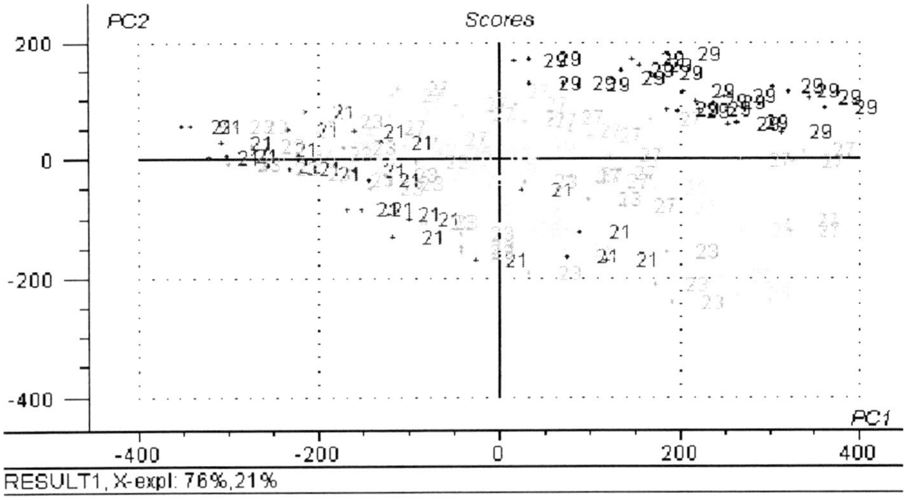

Fig. 7.Scoreplot (PC1 vs. PC2) of AMT-spectra of surfaces with noise

First component describes 93% of the residual variation. However, from the score plot one can see that PC2 is also important. As well as for the noiseless samples, the AMT method discriminates the surfaces with different fractal dimension, although clusters that correspond to the close values of fractal dimensions are even more overlapped.

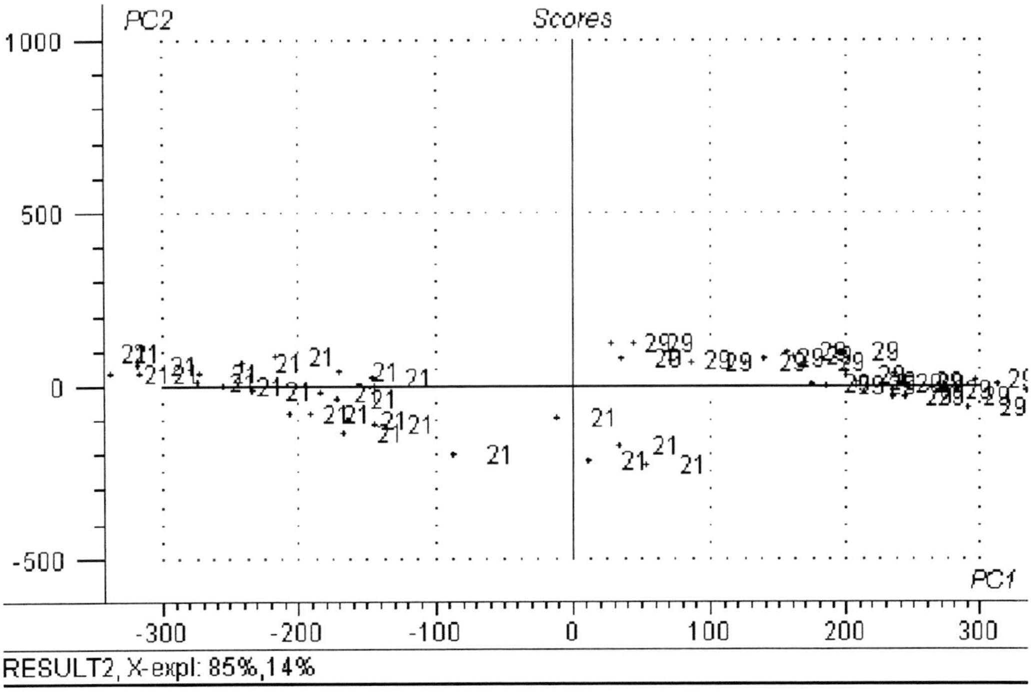

Fig. 8. Score plots of the AMT spectra of noised surfaces with initial fractal dimension *D*=2.1 and *D*=2.9. First scale of PC1/PC2 axis

Fig. 8 and 9 show the score plots for samples with $D = 2.1$ and $D = 2.9$. The plots represent the same data and they differ in the scales of PC1/PC2 axes, only. It is obviously that addition of noise leads to stretching along the PC1 axis. Nevertheless, from the second plot we can see that changing of axis scale gives us a clear-cut distinction between these two clouds of points.

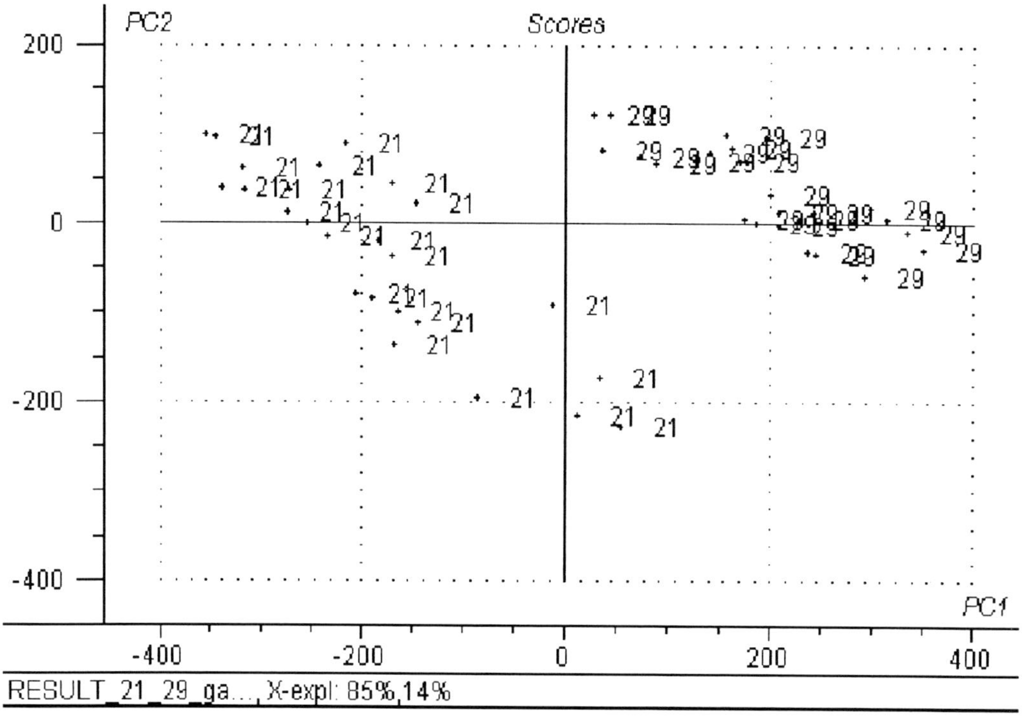

Fig. 9. Same data as in Fig. 8 with new scale of PC1/PC2 axis

Conclusions

Summarizing the above-presented results, we can conclude that the AMT method can classify objects with different fractal dimension, especially for $D>2.3$. Moreover, this method gives good results when we consider the noisy images, for which the conventional fractal analysis fails.

References

[1] B. Mandelbrot, *Fractals: Form, Chance, and Dimension*. Freeman. 1977.

[2] B. Mandelbrot, *The Fractal Geometry of Nature*. Freeman, 1983.

[3] V. Polyakov, S. Kucheryavski, *Let. to J. of Tech. Phys.*, V. 27, 14, 42-46, (2001).

[4] V. Polyakov, S. Kucheryavski, A. Egorov, *Proc. Europ. Conf. "Euromat-99"*, Vol. 4, 7-10, (2000).

[5] R. Andrle, *Math. Geol.*, 16, 83-79, (1996).

[6] K.H. Esbensen, K, Kvaal, K.H. Hjelmen, *J. Chemom.*, 10, 569-590, (1996).
[7] G.S.P. Miller,*Comp. Graph.*, 20, 39-48, (1986).

In: Progress in Chemometrics Research
Editor: Alexey L. Pomerantsev, pp. 13-19

ISBN 1-59454-257-0

Chapter 2

INFORMATIVE SCORE-LOADING PLOTS FOR MULTI-RESPONSE PROCESS MONITORING

R. Ergon
Telemark University College, Porsgrunn, Norway, rolf.ergon@hit.no

ABSTRACT

An ordinary principal component regression (PCR) latent variables model, using any number of components, can be reduced to a model with as many components as the number of response variables. This is done by projection onto a subspace spanned by the PCR estimators for the responses. By defining new orthogonal loading vectors in that subspace, a new latent variables model with exact correspondence between loadings and scores is obtained. For the special case with two response variables, the result of this can be shown in a single score-loading biplot, and in this plot the contributions to the score from the different regressor variables can be shown by use of contribution vectors. A confidence ellipse defining upper control limits for score movements may also be included. The confidence ellipse and the contribution vectors give both fault detection and fault diagnosis.

Keywords: PCR latent variables model, model reduction, score-loading correspondence, contribution vectors

INTRODUCTION

The projection based multivariate data methods of principal component regression (PCR) and partial least regression (PLSR) are well established in the field of process monitoring [1,2]. The problem is then to predict primary response variables from a number of measured secondary process variables. These response variables are typically product qualities that cannot be measured online, and due to many and/or correlated process measurements the regression problem is often ill-posed.

The strength of PCR and PLSR as compared with e.g. ridge regression, is the visualization possibilities given by use of score and loading plots, and this is absolutely essential in process monitoring applications, where not only predictions of the responses but also fault detection and diagnosis are of interest. The interpretation of these plots are, however, often made difficult by the fact that more than two principal or PLS components are necessary in order to obtain good predictions. Thus many plots have to be monitored simultaneously, and in order to see which process variables that are involved in detected process upsets, intricate contribution plots have to be designed and interpreted [1,2].

The basic insight behind the present paper is that the number of components for prediction and visualization need not be the same. Even though prediction may often require many components, visualization should be based on as few components as possible. A reduction of the number of components is the major result of the often rather cumbersome orthogonal signal correction (OSC) methods, e.g. OPLS [3]. Recently it has also been shown that this reduction can be achieved directly by use of especially chosen projection subspaces [4], and the present paper is based on that specific insight.

I will here limit the treatment to PCR, and only very shortly discuss the corresponding PLSR solutions. First I will show that the number of PCR components can be reduced to the number of responses, with unchanged prediction ability. I will then focus on the case with two responses, i.e. two final components, and thus one final score plot and one final loading plot only. Based on correspondence properties [5], these plots may be combined in a score-loading biplot, and the contribution to scores from the different regressor variables can also be shown in the same plot by use of contribution vectors. A confidence ellipse defining upper control limits for score movements may also be included. The confidence ellipse and the contribution vectors give both fault detection and fault diagnosis. The theoretical development is finally illustrated by use of data from a mineral processing plant.

THEORY

Principal Component Regression

PCR for m response variables is based on the latent variables model

$$Y = \hat{T}\hat{Q}^T + \hat{F} \tag{1}$$

$$X = \hat{T}\hat{P}^T + \hat{E} \tag{2}$$

where $X = [x_1 \quad x_2 \quad \ldots \quad x_p]$ and $Y = [y_1 \quad y_2 \quad \ldots \quad y_m]$ are $N \times p$ and $N \times m$ matrices with calibration data, and where Eq. (2) is a principal component analysis (PCA) factorization with $\hat{P}^T\hat{P} = I$. The number of PCA components is here determined by validation, preferably by use of new independent test data. From Eq. (1) and Eq. (2) follow the least squares solutions $\hat{Q}^T = (\hat{T}^T\hat{T})^{-1}\hat{T}^TY$ and $\hat{T} = X\hat{P}$, and thus $\hat{Y} = X\hat{P}\hat{Q}^T = X\hat{P}(\hat{P}^TX^TX\hat{P})^{-1}\hat{P}^TX^TY = X\hat{B}$. The estimator for prediction of Y from new x-data is thus

$$\hat{B} = [\hat{b}_1 \quad \hat{b}_2 \quad \dots \quad \hat{b}_m] = \hat{P}(\hat{P}^T X^T X\hat{P})^{-1}\hat{P}^T X^T Y \tag{3}$$

Compression into Two Components

The m estimators $\hat{b}_1$ to $\hat{b}_m$ define an m-dimensional subspace in the variable space spanned by $\hat{p}_1$ to $\hat{p}_p$, as shown in Fig. 1 for the case of $m = 2$.

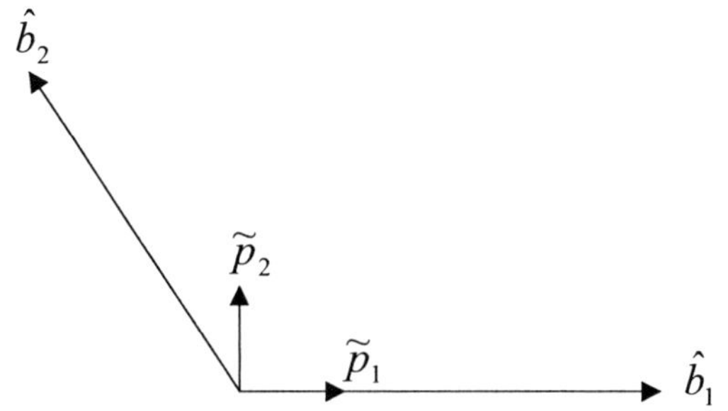

Fig. 1 New loading vectors in subspace spanned by columns in $\hat{B}$.

In that subspace we may define a new orthonormal loading matrix $\tilde{P} = [\tilde{p}_1 \quad \tilde{p}_2 \quad \dots \quad \tilde{p}_m]$. Using $\tilde{P}$ we may also define a new non-orthogonal score matrix $\tilde{T} = X\tilde{P}$, and a new latent variables model

$$Y = \tilde{T}\tilde{Q}^T + \tilde{F} \tag{4}$$

$$X = \tilde{T}\tilde{P}^T + \tilde{E} \tag{5}$$

In the same way as we found the estimator (3), we now find the estimator

$$\tilde{B} = \tilde{P}(\tilde{P}^T X^T X\tilde{P})^{-1}\tilde{P}^T X^T Y \tag{6}$$

Since $\tilde{P}$ is a linear combination of $\hat{B}$, $\tilde{P} = \hat{B}M$, where M is invertible, and since it is well known that $\hat{B}$ fits into its own expression (3) instead of $\hat{P}$, we find

$$\tilde{B} = \hat{B}M(M^T\hat{B}^T X^T X\hat{B}M)^{-1}M^T\hat{B}^T X^T Y = \hat{B}(\hat{B}^T X^T X\hat{B})^{-1}\hat{B}^T X^T Y = \hat{B} \tag{7}$$

i.e. the new estimator $\tilde{B}$ is exactly the same as the original estimator $\hat{B}$. From this also follows that the predictions are the same.

Score-Loading Correspondence and Contribution Vectors

Since $\widetilde{T} = X\widetilde{P}$ there is a total score-loading correspondence [5], i.e. a deviation $\Delta x^T = [0 \;\; \ldots \;\; 0 \;\; \Delta x_{ij} \;\; 0 \;\; \ldots \;\; 0]$ caused by a single variable results in a score movement

$$[\Delta\widetilde{t}_{i1} \quad \Delta\widetilde{t}_{i2}] = \Delta x_{ij}[\widetilde{p}_{i1} \quad \widetilde{p}_{i1}] \tag{8}$$

From this follows that the size of the contribution is determined by the value Δx_{ij} of the variable (including sign), while the direction is determined by the direction of the corresponding loadings in $[\widetilde{p}_1 \quad \widetilde{p}_2]$. This may be visualized by use of contribution vectors in the score-loading biplot, as illustrated in the industrial data example in Section 3.

Score Statistics

A comparison of a new sample with the calibration data set can be done using the Hotelling's T^2 statistic [6] based on the estimated score covariance matrix, which with centered data and use of the model (4,5) is

$$S = \frac{1}{N-1}\widetilde{T}^T\widetilde{T} \tag{9}$$

With centered data the T^2 statistic for a new observation is

$$T^2 = \widetilde{t}_{new}^T S^{-1}\widetilde{t}_{new} \tag{10}$$

The upper control limit for a new score based on N past multivariate data is

$$T_{UCL}^2 = \frac{(N^2-1)2}{N(N-2)}F_\alpha(2, N-2) \tag{11}$$

where $F_\alpha(2, N-2)$ is the upper 100α % critical point of the F distribution with $(2, N-2)$ degrees of freedom [1,2]. With two components only, the upper control limit for normal score movements can be shown as a confidence ellipse in the score-loading plot, as illustrated in the industrial data example in Section 3. The confidence ellipse and the contribution vectors thus give both fault detection and fault diagnosis.

INDUSTRIAL DATA EXAMPLE

The following example uses multivariate regression data from a mineral processing plant [7] (the 'cleaner' data, originally published in [8]). The problem considered here is to predict two given response variables from twelve known process variables. For the purpose of finding an initial PCR factorization, samples 1 to 120 in the data sets xce and yce [7] were used for modeling, while samples 151 to 240 in the same data sets were used for validation. The data were centered and scaled to unit variance (autoscaled), and the result was a PCR model with A = 10 components, resulting in validation root mean square errors RMSEP = 0.21 and RMSEP = 0.28 for the two responses respectively (as compared with RMSEP = 1 for A = 0). The PCA factorization (2) was finally compressed into the two-component factorization (5), resulting in the new loading matrix $\widetilde{P}$.

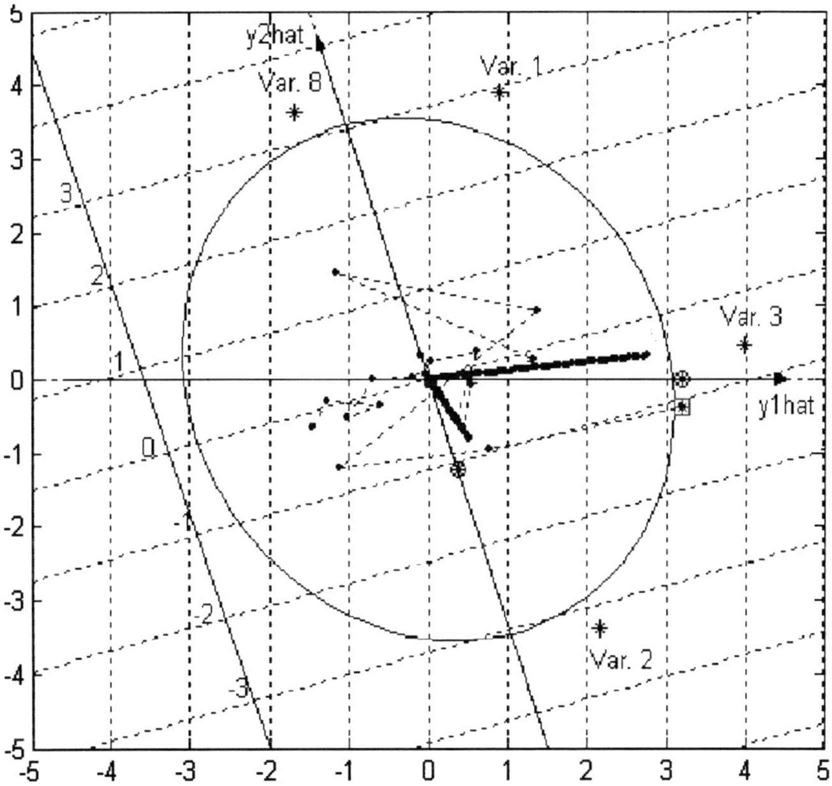

Fig. 2 Score-loading plot with variable contribution vectors and confidence ellipse

The score-loading-contribution plot described in Section 2 is shown in Fig. 2. The directions of the loadings are here shown by stars, while the variable contributions are shown by radial vector lines out from the origin. In the projection plane according to Fig. 1, only variables number 1, 2, 3 and 8 give significant contributions to the score, while the rest of the twelve variables have very little to say. The confidence ellipse is computed according Eq. (12), with α = 0.01. The score for the last test sample is marked by a square, while older scores are shown by dots connected by dashed lines. The last score in Fig. 2 is found slightly outside the confidence ellipse, and the obvious reason for this is a large positive value of variable number 3, and also a positive value of variable number 2. The corresponding predictions are marked by star filled circles on the axes for y_1 and y_2.

For a somewhat later sample the plot is shown in Fig. 3. Here variable number 1 has a positive value, while variables number 2, 3 and 8 have negative values. Note that the sum of the contribution vectors gives the score position (exactly when all the twelve contribution vectors are included).

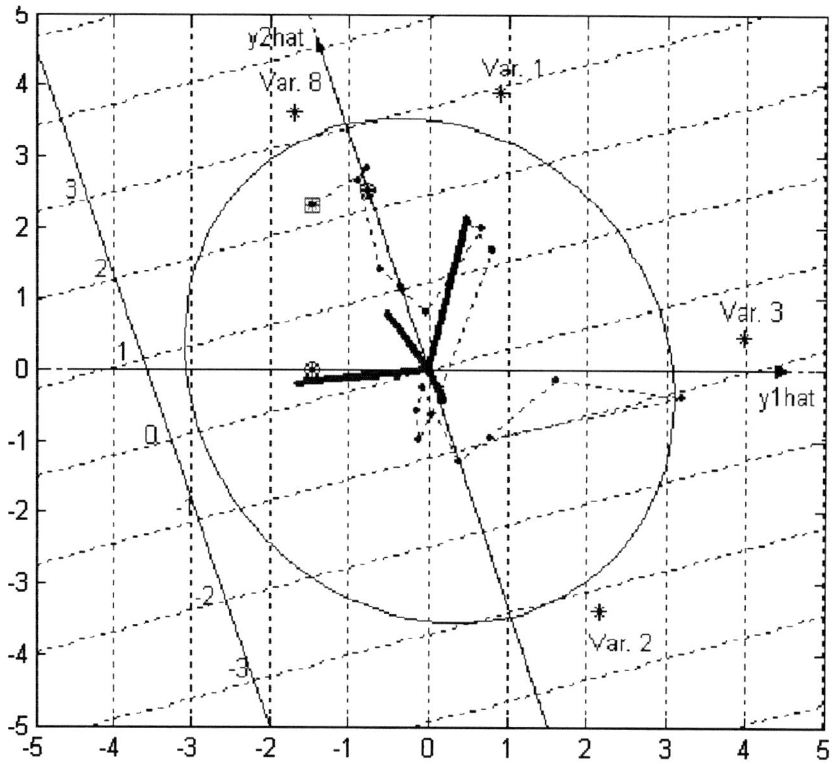

Fig. 3 Score-loading plot with variable contribution vectors and confidence ellipse

Discussion and Conclusions

For the case with two response variables, projection onto a plane spanned by the two PCR estimators and use of two orthogonal loading vectors in that plane, result in score and loading plots with exact correspondence. In the score-loading biplot the position of the score will thus be exactly determined by contribution vectors for the regressor variables. For the general case with *m* response variables, the corresponding projection subspace is spanned by the *m* estimators.

With use of PLSR the corresponding results will be only approximate. The reason for this is that the estimators then will be located in different subspaces.

REFERENCES

[1] T. Kourti, J.F. MacGregor, *Journal of Quality Technology*, 28, 409-428, (1996)

[2] L.H. Chiang, E.L. Russel, R.D. Braatz, *Fault Detection and Diagnosis in Industrial Systems*, Springer, London, 2001

[3] J. Trygg, S. Wold, *J. Chemometrics*, 16, 119-128, (2002)

[4] R. Ergon, *J. Chemometrics*, 17, 303-312, (2003)

[5] R. Ergon, *J. Chemometrics*, 16, 368-373, (2002)

[6] R.A. Johnson, D.W. Wichern, *Applied Multivariate Statistical Analysis*, Prentice-Hall, New Jersey, 1992

[7] Höskuldsson, *Prediction Methods in Science and Technology, Vol. 1 Basic Theory*, Thor Publishing, Copenhagen, 1996

[8] D. Hodouin, J.F. MacGregor, M. Hou, M. Franklin, *Can. Inst. Mining Bull.* 86, 23-34 (1993)

In: Progress in Chemometrics Research
Editor: Alexey L. Pomerantsev, pp. 21-36
ISBN 1-59454-257-0

Chapter 3

WAVELETS IN COMPRESSING SPECTRAL DATA

Satu-Pia Reinikainen
Lappeenranta University of Technology,
Lappeenranta, Finland, satu-pia.reinikainen@lut.fi

ABSTRACT

Wavelet transformation has been developed during the last few years to improve the shortcomings of the Fourier transform technique. The major problem of Fourier transform is the complementarily of time and frequency resolution, i.e., when increasing the frequency resolution the transformation length increases and time resolution decreases considerably. In this paper some main aspects, related to the *wavelet transform* or *wavelet analysis:* such as denoising, smoothing, and compression multivariate data analysis. 1-D and 2-D spectral data is considered and illustrated by using, e.g., NIR and NMR data.

Keywords: Wavelet transformation, PLS, spectral data

INTRODUCTION

Nowadays need for compression methods are increasing due to the ever increasing production of large data sets with thousands of variables. There is more computation power and storage space needed to calculate or handle the large data sets. Wavelet methods provide powerful tools for analyzing, encoding, compressing, reconstructing, and modeling large spectral data, and images. Wavelet analysis may offer added insight in data analysis situations where Fourier techniques have previously been used. It is well known from Fourier theory that a signal can be expressed as the sum of a, possibly infinite, series of sines and cosines. This sum is also referred to as a Fourier expansion. The disadvantage of the Fourier expansion is that, unlike wavelet transformations, the FT's basic functions are continuous over time and are, therefore, not suitable for representing time-related information of the transformed signal. Nevertheless, the Fourier transform is the most important signal analysis technique today. The *wavelet transform* or *wavelet analysis* is probably the most recent solution to overcome the shortcomings of the Fourier transform. In wavelet analysis the use of a fully scalable modulated window solves the time scale problem. The window is shifted along the signal and for every position the spectrum is calculated. Then this process is repeated many times with a slightly shorter (or longer) window for every new cycle. In the end the result will be a collection of time-frequency representations of the signal, all with different resolutions.

The use of the wavelet transform has tended to focus on the compression and noise reduction of separate signals. Wavelet analysis and compression, followed by PCA or PLS and their generalizations, form a simple and practical tool, allowing computations to be made using a reduced set of variables, without any loss of information. Results can be interpreted in the original variable space, using the inverse wavelet transform on resultant model parameters such as loadings. Additionally, the time-frequency nature of many types of signals can facilitate separation of different features, thus increasing the detection limit and improving fault detection. Some common aspects, related to denoising, smoothing, and compression of data in applying wavelet transformations and multivariate data analysis in 1-D and 2-D spectral data is considered and illustrated by using, e.g., NIR and NMR data. These spectra are typically redundant by nature, and therefore suitable for wavelet compression (and simultaneous denoising).

Overview of Wavelet Transformation

Wavelets are functions that satisfy certain mathematical requirements and are used in representing data or other functions. This idea is not new. The basic idea of representing a signal as the sum of analyzing functions was presented already in 1807 by Joseph Fourier. Joseph Fourier discovered that he could superpose sines and cosines to represent other functions.

Wavelets are mathematical functions that cut up data into different frequency components, and then study each component with a resolution matched to its scale. They have advantages over traditional Fourier methods in analyzing physical situations where the signal contains discontinuities and sharp spikes. Wavelets were developed independently in the fields of mathematics, quantum physics, electrical engineering, and seismic geology. Interchanges between these fields during the last ten years have led to many new wavelet applications such as image compression, turbulence, machine vision, radar, and earthquake prediction.

For many decades, scientists have wanted more appropriate functions than the sines and cosines, which comprise the bases of Fourier analysis. By their definition, these functions are non-local (and stretch out to infinity). Unlike sines and cosines, the wavelets functions have a very important property, along the fact that they are local in both time and frequency, which make the functions versatile and useful. Typical wavelet functions look like small oscillation waves (Fig. 1). In wavelet analysis, the scale plays a special role, since the functions have ability to analyze a signal according to scale, i.e., inverse frequency. Wavelet algorithms process data at different scales or resolutions (Fig. 2). The size of the analyzing window varies with different scales. A wide window is used for low frequencies and a narrow window

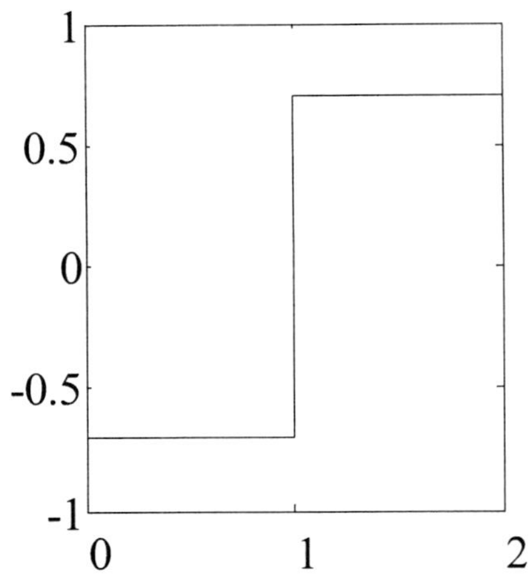

The Haar filter (which could be considered a Daubechies-2) was the first wavelet, though not called as such, and is discontinuous.

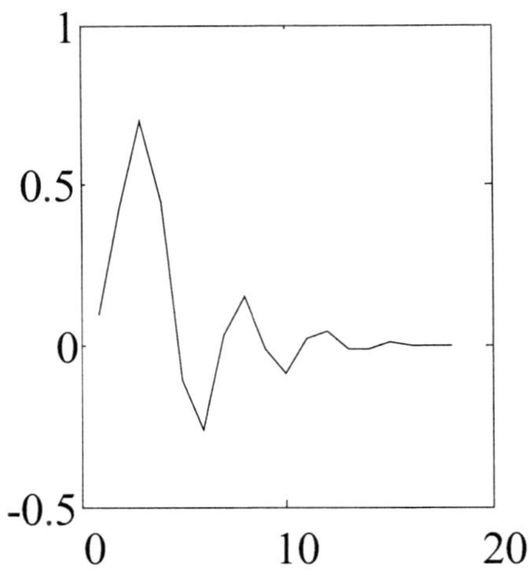

The Beylkin filter places roots for the frequency response function close to the Nyquist frequency on the real axis.

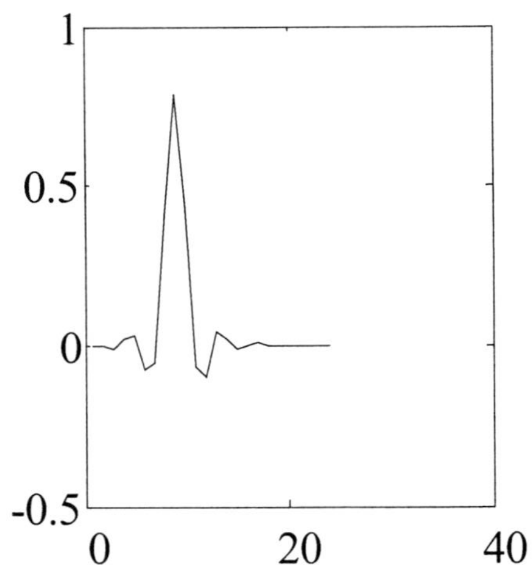

The Coiflet filters are designed to give both the mother and father wavelets 2Par vanishing moments; here Par may be one of 1,2,3,**4** or 5.

Figure 1 continued on next page

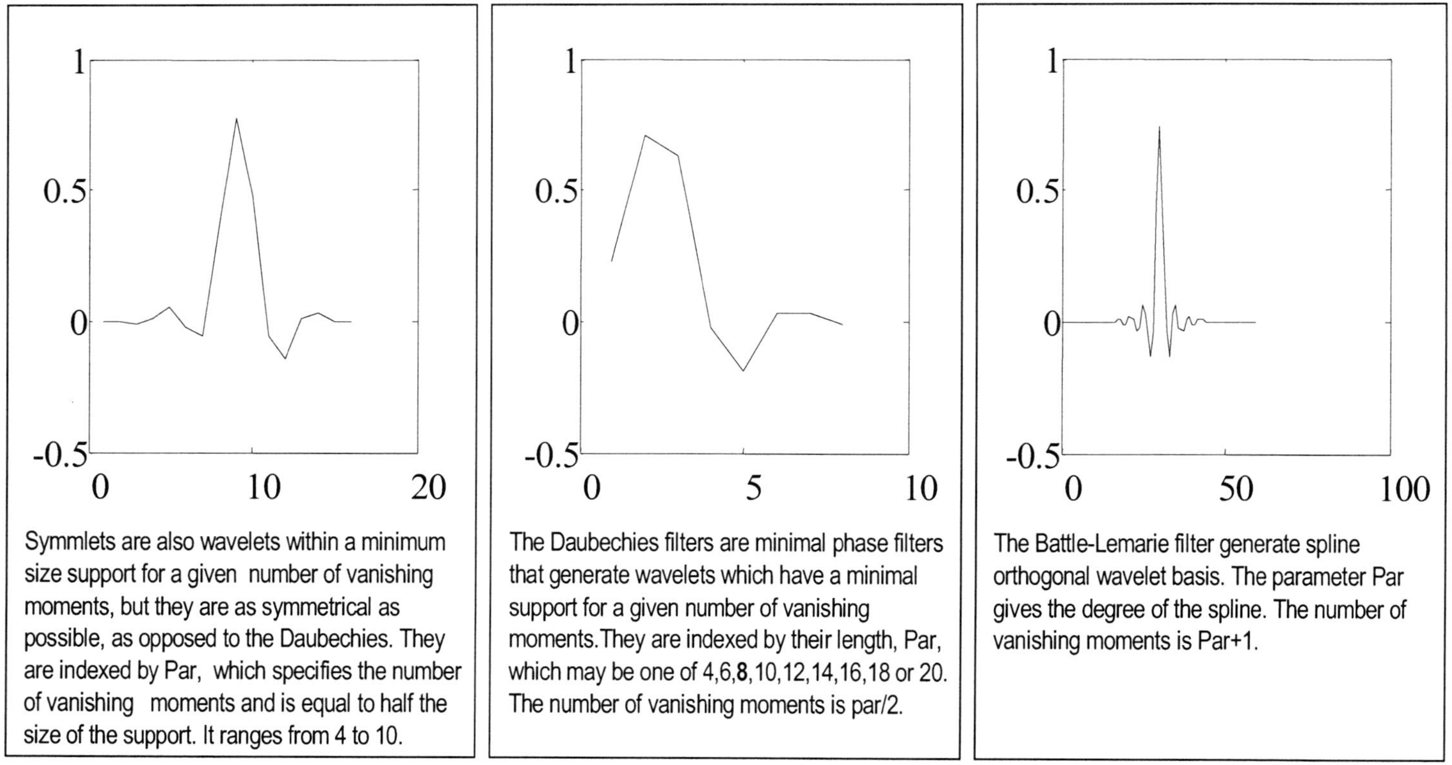

Figure 1. Several different families of wavelets: Haar, Beylkin, Coiflet(4), Symmlet(8), Daubechies(8) and Battle-Lernarie(5) The number next to the wavelet name represents the number of vanishing moments (A stringent mathematical definition related to the number of wavelet coefficients) for the subclass of wavelet in Fig.

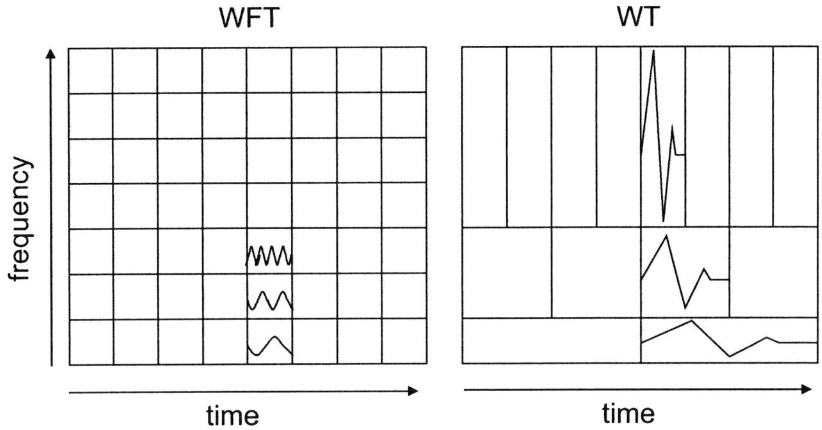

Figure 2. Windowed Fourier Transformation (WTF): the window is simply a square wave; Wavelet Transformation: the window size (time-frequency) is changing:. When the scale decreases, the time support is reduced, but the frequency spread increases and covers an interval that is shifted towards high frequencies. Transformation can also be made in time of frequency domain.

for high frequencies. When a signal is transformed with a wide window, more general features are detected. Similarly, when a signal is transformed with a small window, small features will be found. The result in wavelet analysis is to see both the forest and the trees, so to speak. Figure 2 shows the preliminary idea in changing the window widths. There exist continuous and discrete time wavelets. However, all computations are performed over discrete signals, with conjugate mirror filters.

Wavelet Transformation in Brief

The signal to be compressed has to be of length 2^n (n is integer). The wavelet transform analyzes signals locally without prejudice to scale. The signal is transformed with so called "mother function", which has a certain scale, i.e., window width. The mother function is then stretched or compressed to create other scales as shown in Fig. 2. The mother wavelet is local in time (or in wavelength) and frequency. For a signal length 2^n the procedure is performed n times, creating n levels of different scales (Fig. 3).

The wavelet filter produces a detailed part, which is called wavelet coefficients. At each step a scaling filter ("father function") creates an approximate description of the signal, which is used to represent the signal at the next level. When the highest scale is reached, only one scaling coefficient is produced, and it is related to the mean value of the original signal. It is possible to reverse the steps and reconstruct the original signal exactly, using the scaling coefficient and the wavelet coefficients. The size of the transformed signal is the same as the original signal. The wavelet transform and its associated wavelet functions are defined below.

$$\left(f(t), \psi_{b,l}(t)\right) = \int_{-\infty}^{\infty} f(t), \psi_{b,l}(t) dt$$

$$\psi_{b,l}(t) = 2^{-b/2} \psi_{b,l}(2^{-b/2} t - l)$$

where the signal is *f(t)* and the wavelet function is $\psi_{b,l}(t)$, *b* is the scale and *l* is the time shift. Each scale (stretched version) of the wavelet covers a separate band of frequencies. For detailed information see references [1-7]. The literature of wavelets is huge and several excellent books on wavelets can be found. There exist also several toolboxes, e.g., for Matlab, which can be found via Internet.

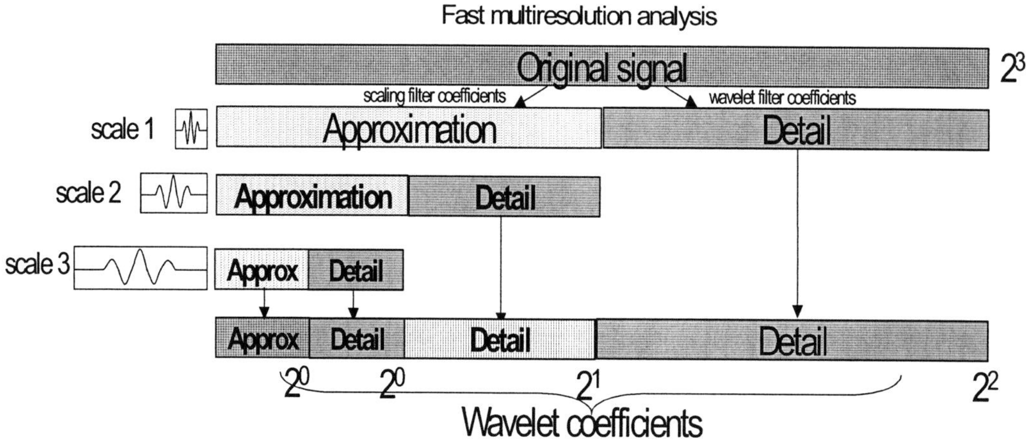

Figure 3. Overview of the wavelet transform and multi-resolution analysis scheme proposed by Mallat [8]. On each step signal (or approximation part from previous scale) is filtered with scaling and wavelet filter. The scaling filter produces the approximation part to be used in next scale. The wavelet filter produces detail part at the present scale. At last scale only two values are presented: normalized average value of the signal and the detailed part. These values are combined with the other wavelet coefficients into a vector, which has same length as the original signal.

Wavelet Basis Functions

A wavelet is a localized pulse (wave) that decreases to zero, and has an integral of zero. The wavelet functions should also have compact support, some degree of smoothness and, preferably, be orthogonal. Two popular classes of basis functions are the orthogonal and biorthogonal wavelets. The orthogonal wavelets provide simplicity, speed, absence of redundancy and perfect reconstruction [1]. They are asymmetric and do not provide a linear phase, except in the case of the simplest orthogonal wavelet, the Haar wavelet. Daubechies and Symmlet are two of the orthogonal wavelet families and each consists of several wavelet functions, which differ in their compactness and smoothness. The biorthogonal wavelets also provide perfect reconstruction [1]. As the name implies, instead of one set of lowpass and highpass filters, biorthogonal wavelets use two sets of wavelets, one set for decomposing and another set for reconstructing the signal. The biorthogonal wavelets can be symmetrical and have a linear phase. The total error cannot be computed from the wavelet coefficients without reconstructing the whole signal and the redundant coefficients are harder to interpret.

APPLICATIONS AND EXAMPLES

Denoising, Smoothing

Wavelet transformations offer a valuable tool for denoising of data by extracting the signal from the noise. The main idea is that coefficients smaller than a calculated or predetermined threshold are eliminated and the signal is reconstructed. Two common approaches to denoising are hard and soft thresholding [1]. The hard thresholding is "keep or kill" method. A threshold value (z) is set, and all coefficients below that are set to zero. Soft thresholding is "shrink or kill" method, where all values below the threshold are set to zero. The threshold value is subtracted from those larger than it. Common denoising methods include VISU [9], SURE [10], HYBRID [11], SOFT [10] and denoising by cross validation [12]. Figure 4 shows an example of denoising of a part of ^{13}C NMR spectrum.

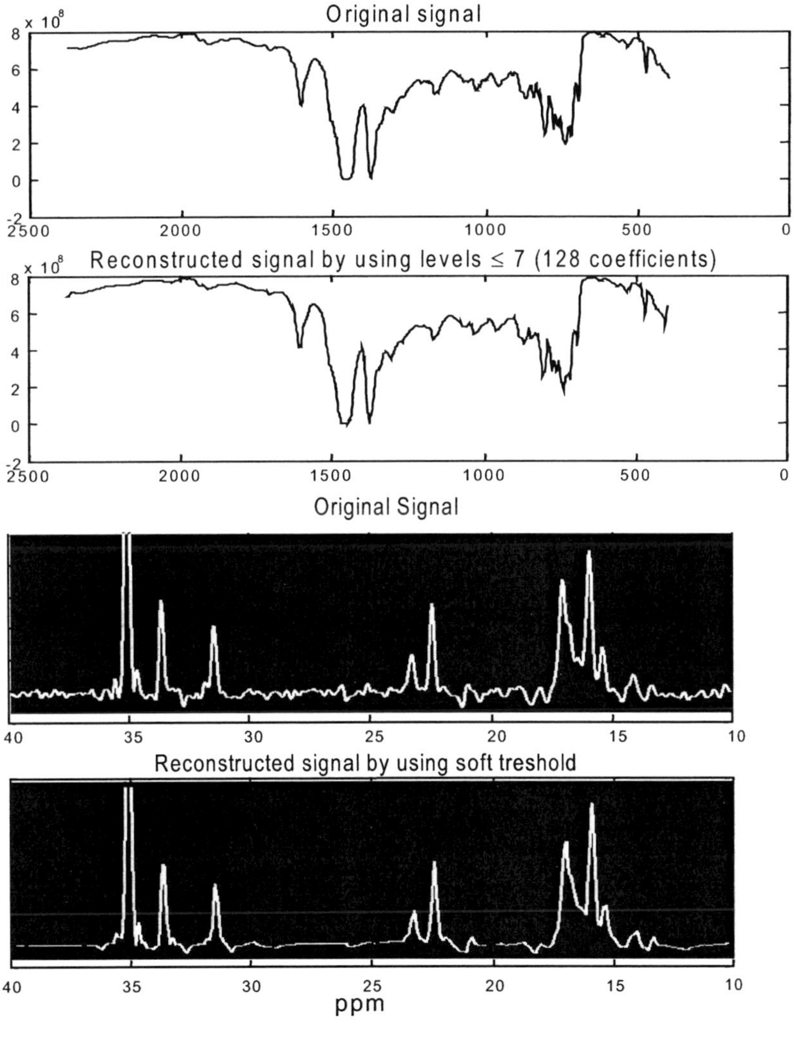

Figure 4. Original and reconstructed NIR and NMR spectrum; compression can be gained, when number of wavelet coefficients is reduced. Spectra are transformed with Daubechies (4).

Feature Extraction, Compression and Multivariate Models

Wavelet transformation can be done for both 1-D and 2-D signals (or images). 1-D signals can be analyzed and compressed row-wise or column-wise. With row-wise transformation of individual spectrum (e.g., each sample is transformed separately) the spectra profile can be studied and compressed.

With column-wise analysis in time series the time related variation can be studied and compressed. 2-D signals, such as 2D-NMR spectra, can be compressed with 2-D wavelet functions. In the 2D case both dimensions of 2D-spectrum should be set equal to 2^n. 2-D spectra can also be compressed as one-dimensional spectra (row-wide), and the analysis and compression can be continued with N-mode multivariate methods (such as 2-D PCA or PLS models) [13].

Feature studies can be made by using, so-called, power spectrum [12] and/or with multivariate methods (wavelet transformation is used as pre-treatment method). Compression of data is achieved by eliminating coefficients, that are smaller than some threshold value, or by using some other criterion, e.g., in PLS models based on variance spectra of wavelet coefficients [2].

EXAMPLES

Example 1: Wavelet Transformation and Compression of Spectral Data

Wavelet transformations have been found to be suitable in compression of, e.g., IR and NMR spectra. A good overview of the IR modeling procedure is given in Ref. [2]. Daubechies wavelets are the most commonly used wavelet functions with IR data. Namely the functions with 4-8 vanishing moments are used. The results are often found to be similar, when Beylkin, Coiflet, Symmlet, or Daubechies functions are used. However, when selecting the wavelet function, some attention to computation time may be paid (different functions requires different amount of computation). In general, when the number of vanishing moments increases the number of required computations increases. Fig. 4 shows example of a reproduced IR spectra. For wavelet transformation the spectra should be of length 2^n. If this is not the case, several methods exist for extrapolating the signal. Linear padding and zero padding are two such methods [1]. Padding with zeros is not recommended for most chemical signals, because artifacts are easily introduced. The method of linear padding is easily implemented and works reasonably well. It takes advantage of the 'vanishing moments' property of the wavelets.

Contrary to popular belief, wavelets themselves do not compress the data: Their job is to make compression possible. By using all wavelet coefficients the spectra can be completely reconstructed and no compression of data is achieved. The wavelet coefficients can be used to interpret individual spectra, however. The compression of data is achieved, when a part of the wavelet coefficients is removed. Spectra, such as NMR or IR spectra, have typically a lot of redundant variation and they are ideal for data compression. In several studies wavelet transformation has been used as pretreatment method before multivariate analysis. The Example 2 introduces the main principles. It has been reported that, e.g., by retaining 3-10 % of wavelet coefficients the relevant information of a spectra can be stored [2]. Selection of wavelet coefficients can be based on general variable selection methods, or as reported in Refs [1,2] on a variance spectrum of wavelet coefficients. Variable selection has also been done by using "larges scales" as selection criterion, where the wavelet scales with largest variance is selected.

Example 2: PLS Model for Density (^{13}C NMR Spectra)

In general PLS method has been found essential, when spectral data is modeled. The theory of PLS is well presented in the literature. Therefore it is not presented here. The aim of this example is introduce a simple and useful procedure, in which wavelets are applied as pretreatment method of PLS analysis. There exist several more advanced and sophisticated methods, which can be found from the literature and articles. In this case study a data set containing 25 calibration ^{13}CNMR spectra and 15 test spectra is transformed with Symmlet (8) function. The NMR data originate from an oil refinery. The spectral information was used to model density of the samples (diesel).

Main Steps in Analysis

1. Signal should be length of 2^n (n is integer)
   An informative range of size 2^{10} is selected from the NMR spectra.
2. Select the mother function and make the wavelet transformation separately for each spectrum.
   Symmlet (8) function was selected (several other functions, such as Daubechies and Coiflet, would give almost similar results.)
3. Select the number of wavelet coefficients compression is achieved by eliminating the coefficients. Particularly, when IR data is compressed the highest frequency often contains white noise (level 2^{n-1}). Variance spectrum can be used (initial guess) [2]:

$$Variance_k = \frac{\sum (x_{nk} - \bar{x}_k)^2}{N-1}$$

$$k = 1,2,...,K \text{(column index)}$$

$$n = 1,2,...,N \text{ (row index)}$$

Wavelet coefficients of NIR and NMR spectra differ from each another, and the compression level of NMR spectra is often lower (Fig. 6). Also the traditional variable selection criteria can be used in variable selection. A collection of the criteria is presented in Ref [14]. A variance spectrum is shown in Fig. 5. Fig 6 shows typical wavelet coefficients, when full transformation is applied to NIR and NMR data.

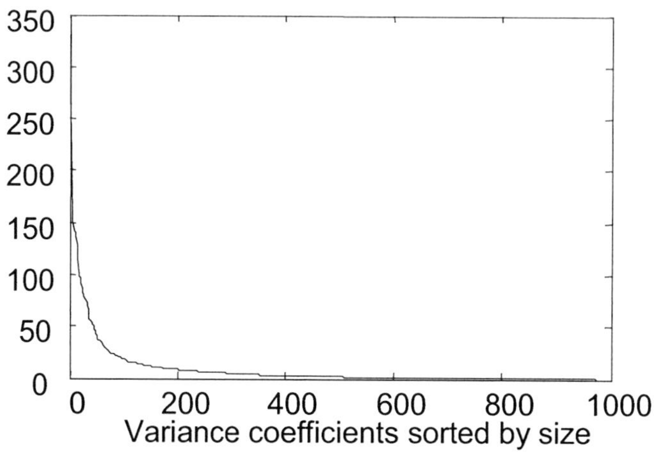

Figure 5. Variance spectrum: Size distribution plot of variance coefficients of the wavelet coefficient matrix.

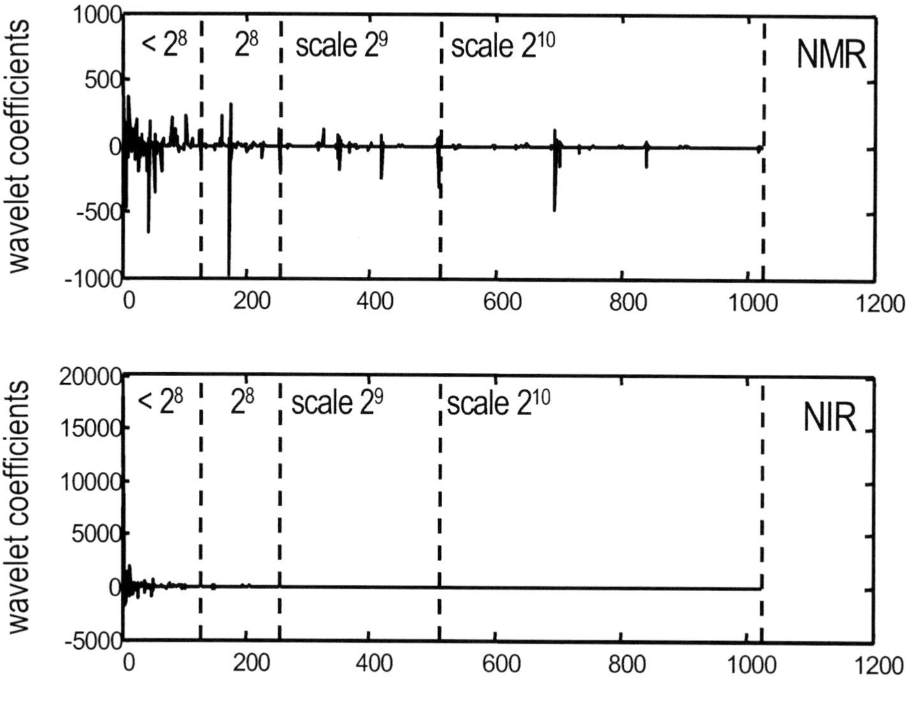

Figure 6. Examples of wavelet coefficients on different scales. NIR coefficients on largest scales (29, 210) can be removed.

4. Extract the selected columns from wavelet coefficient matrix into X matrix. Variable center the wavelet coefficients, and compute/validate PLS model (use independent test set): repeat steps 3 and 4 until model is satisfactory or no improvement is achieved.
   The goodness of PLS model can be checked with, e.g., coefficient of determination (for calibration set, for cross validation set, and for test set):

$$R^2(Y) = 1 - \sum(\hat{y} - y)^2 / \sum y^2$$

Also root mean square error (RMSEP) is another very often-used measure:

$$RMSEP = \sqrt{\frac{\sum(\hat{y} - y)^2}{N}}$$

R^2 and RMSEP values for the final PLS model is shown in Table 1, in which also the indices for a model based on original spectra are presented.

Table 1 Modeling results for the original PLS model and for the compressed PLS model (Example 2: NMR data).

	Original data (centered NMR data)			
No. of variables	No. of LVs	R^2(Y)	RMSEP	Q^2(Y)
1024	7	97.9	2.53	95.2
	Compressed data (centered wavelet coefficients) Criterion: maximum variance coefficient (Fig. 5)			
No. of variables	No. of LVs	R^2(Y)	RMSEP	Q^2(Y)
50	7	96.1	3.32	92.2
100	7	96.9	2.95	94.8
150	**7**	**97.1**	**2.88**	**95.0**
200	7	97.1	2.84	94.5
300	7	97.3	2.76	94.5
	Compressed data (centered wavelet coefficients) Criterion: maximum squared covariance, cov(x,y)			
No. of variables	No. of LVs	R^2(Y)	RMSEP	Q^2(Y)
100	7	97.6	2.61	95.0

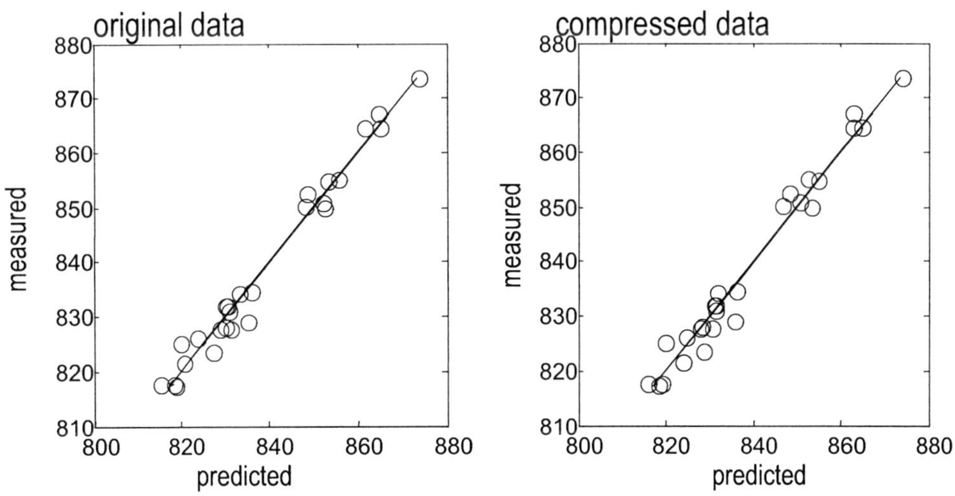

Figure 7. Measured vs. predicted densities (Example 2).

5. To compute the original loadings, place the coefficients in their original position in wavelet vector, fill all other positions with zeros, and perform inverse wavelet transformation. Some main figures are shown as examples in Figs 7-8.

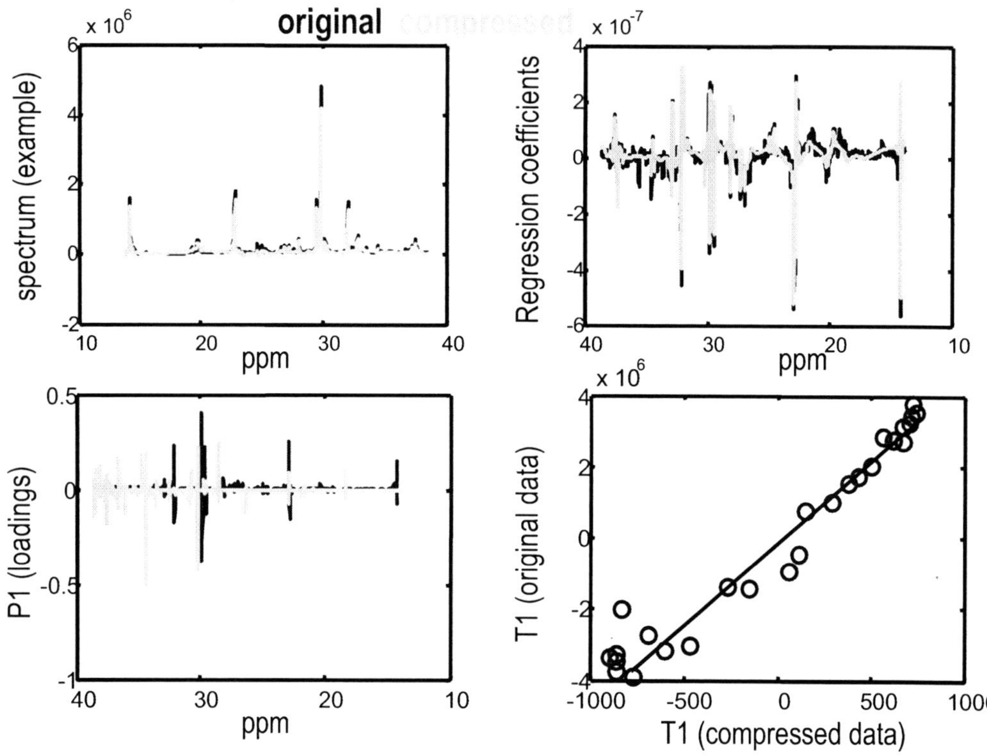

Figure 8. Measured vs. predicted densities (Example 2)

Example 3: 2-Way PLS Models for Density (^{13}C-^{1}H NMR Spectra)

NMR can be used to determine molecular structure. The idea of 2D-NMR was presented in 1970s'. In this example ^{13}C-^{1}H NMR spectra of small oil sample set are related to density of the samples. The aim of the example is to introduce some considerations in compression of 2D-data sets for multivariate analysis, and the theory of 2D-NMR is not introduced. However, for successful data analysis the knowledge of chemistry and NMR theory is essential. Conventional NMR spectra (one-dimensional spectra, Example 2) are plots of intensity vs. frequency; while in two-dimensional spectroscopy intensity is plotted as a function of two frequencies. Fig 9 shows a 2D spectrum as an example. Spectra were arranged on three-dimensional **X** matrix, in which the third dimension were samples (no autocorrelation between samples).

N-mode PLS based on PARAFAC type of model was applied to model relationship between the 2D spectra and density of samples [15]. The tridimensional model is not fitted in a least squares sense but, in accordance with the philosophy of PLS, it seeks to describe the covariance of the dependent and the independent variables. As a results two loading matrices ($\mathbf{W}_j$, $\mathbf{W}_k$) and one scores matrix (**T**) is constructed from **X** block. The **Y** block is divided into loadings ($\mathbf{Q}_l$, $\mathbf{Q}_m$) and scores ($\mathbf{U}$). And the relationship between these blocks is described with $\mathbf{B}$: $\mathbf{U} = \mathbf{TB} + \mathbf{E}_u$.

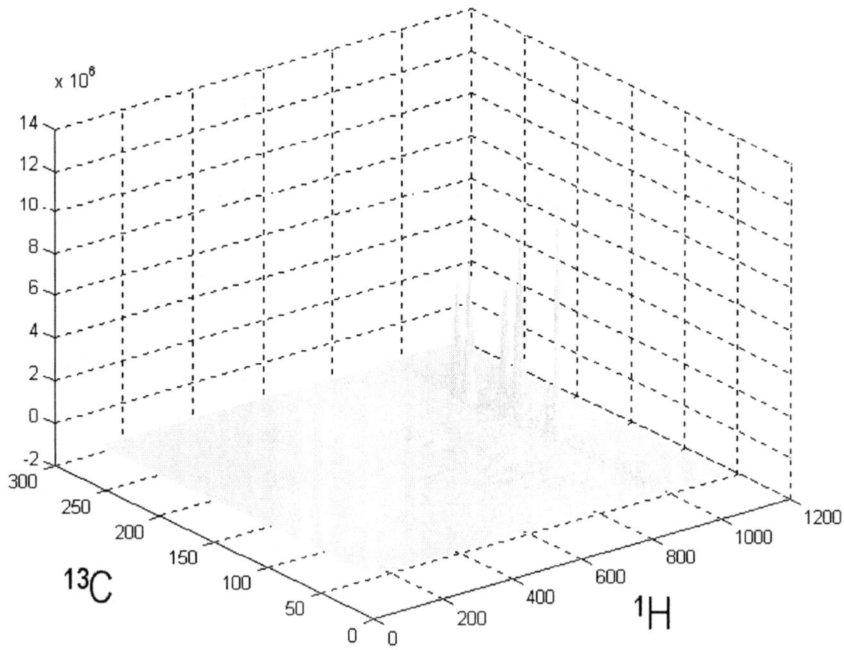

Figure 9. 2D NMR spectrum (Example 3).

Figure 10 illustrates the modeling procedure.

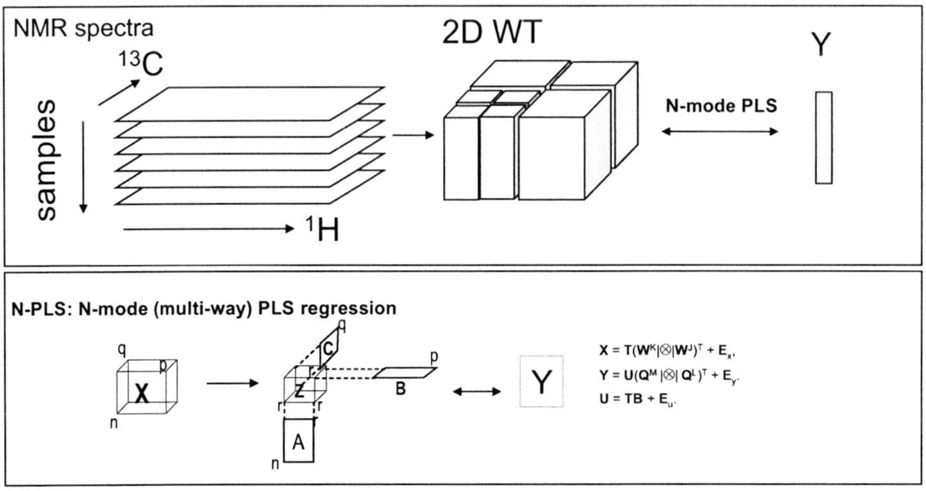

Figure 10. 2D wavelet transformation was applied to NMR spectra and the WT coefficients (of sublevels) were modeled and analyzed with N-mode PLS models.

The wavelet transformation can be done simultaneously in both column and row direction. In that case the transformation is called 2D transformation. The Symmlet (8) transformation was applied to the 2D NMR data.

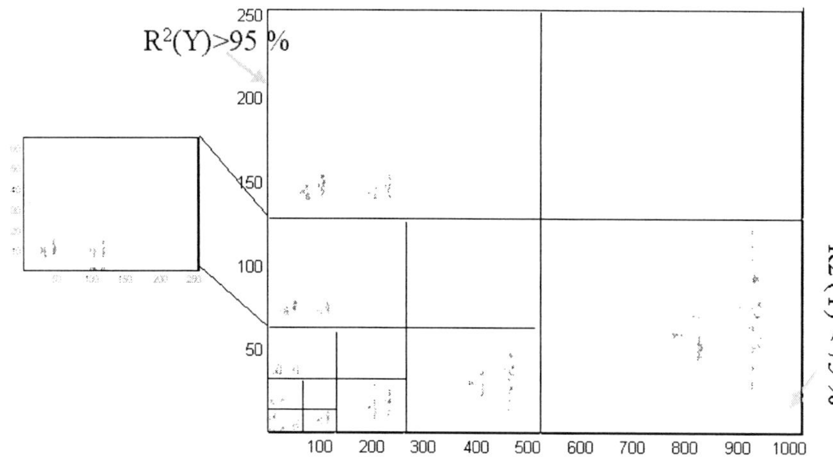

Figure 11. Variance of wavelet coefficients (Example 3)

There exists several ways to describe the transformation, Fig. 11 shows the variation of wavelet coefficients on different scales. In this example different scales (or levels) were studied separately and it was found that the certain scales ("subplots") tend to be very similar. N-way PLS was applied to these "subplots" and it was found, that by using certain "subplots" shown in Fig. 11 over 95 % of the variance of density of oil samples could be explained. The other "subplots", with dissimilar structure, were found to contain dissimilar information, which were not informative in means of density estimates. By retaining the wavelet coefficients from the "best" subplot the 2D NMR spectra were reconstructed and the loadings from N-mode PLS model were associated to the original NMR dimensions by using inverse wavelet transformation. Main results are shown in Fig. 12.

The purpose of analysis was to interpret the data and find a relationship between density and spectral data. The results were found to be very similar to the ones made by using whole spectra. However, the computation time was decreased dramatically. Depending on the purpose of a model, selection of the subplots or coefficients must be done carefully. With wrong and careless selection the result will became unreliable and misleading. It is often necessary to include several "subplots" into the model. If all the "subplots" are included no compression is achieved. However, the multivariate nature of data can be studied in details.

The study was continued with residuals. Residuals in multivariate analyses can be used for various purposes. In this example the residual matrix of excluded wavelet coefficients, e.g., can be simply used to study the excluded variation of NMR data. The residuals of test set can detect dissimilarities between calibration and test set (either residuals of wavelets or reconstructed spectra together with distance to model criterion can be used).

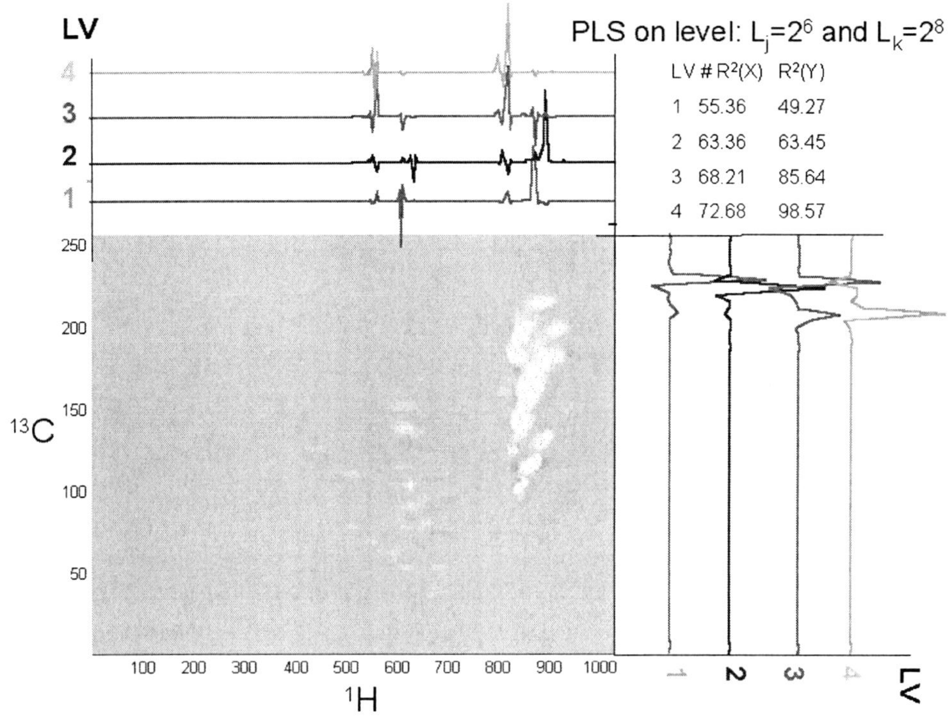

Figure 12. 2D NMR spectra were compressed with WT and the WT coefficients were modeled with N-way PLS. The results were studied and re-associated to the original data. Loadings of N-mode PLS model and an example spectra.

Conclusions

With the foundations of wavelet theory securely in place, the field has grown rapidly over the last decade. Although wavelets are best known for image compression, the application fields will continue increasing. Wavelets offer a valuable tool for handling any large data set and chemometricians, among other researchers and engineers, have found ever increasing creative applications.

References

[1] Trygg J., Wavelets in Chemometrics-compression, denoising and feature extraction, *Homepage of Chemometrics*, Editorial, May 2002.

[2] Trygg J., Wold S., *Chemom. Int. Lab. Syst.* 42 (1998) 209-220.

[3] Hubbard B. *The world according to wavelets*, A K Peters: Wellesley, Massachusetts, 1995;

[4] Alsberg B K, Woodward A M, Kell D B. *Chemom. Intell. Lab. Syst.,* 37:(1997) 215-239.

[5] Walczak B, Massart D L. *Trac Trends Anal. Chem*., 16, (1997) 451-463.

[6] Walczak B, vanden Bogaert B, Massart D L., 68 (1996): 1742-1747.

[7] Jetter K, Depczynski U, Molt K, Niemoller A.. *Anal. Chim. Acta*, 420 (2000): 169-180.

[8] Mallat S G.A., *IEEE Trans. Pattern Anal. Mach. Intell.,* 11 (1989): 674-693.
[9] Donoho D L, Johnstone I M, Kerkyacharian G, Picard D. *J. of the Royal Stat. Soc,* Series B, 57 (1995): 301-370.
[10] Donoho D L, Johnstone I M. *J. Am. Stat. Assoc.*, 90 (1995): 1200..
[11] Alsberg B K, Woodward A M, Winson M K, Rowland J, Kell D B. *Analyst,* 122 (1997): 645-652.
[12] Nason G P. *J. R. Statist. Soc.* B, 58 (1996): 463-479.
[13] Trygg, J., N. Kettaneh Wold, et al., *J. Chemometrics* 15 (2001)): 299-319.
[14] Hösskulsson A., *Prediction Methods in Science and Technology*, Poland (1996)
[15] R. Bro, *J. Chemomometrics*, 10 (1996) 47

In: Progress in Chemometrics Research
Editor: Alexey L. Pomerantsev, pp. 37-41
ISBN 1-59454-257-0

Chapter 4

APPLICATION OF WAVELET NEURAL NETWORKS IN MULTIVARIATE DATA ANALYSIS

T. Khayamian, M. Esteki and A. Abbasi
Isfahan University of Technology, Iran

ABSTRACT

A wavelet neural network (WNN) model is proposed in quantitative structure property relationship (QSPR) to predict critical micelle concentration (cmc) for a variety of nonionic surfactants.

Keywords: QSPR, Wavelet, Neural Networks

INTRODUCTION

In 1995 Huibers et al. [1] proposed a reliable relationship between the three topological descriptors and cmc (Eq. 1), which can be effectively predict cmc for a diverse set of nonionic surfactants by using the following equation.

$$\mathrm{Log}_{10}\,\mathrm{cmc} = -\,(1.8 \pm 0.16) - (0.567 \pm 0.009)\,\mathrm{KH0} + (1.054 \pm 0.048)\,\mathrm{AIC2} + +(7.5 \pm 1.0)\,\mathrm{RNNO} \qquad (1)$$

In this equation the Kier & Hall index of 0 order for the hydrophobic fragment, KH0, represents the size of the hydrophobic fragment and contains group contributions from all non-hydrogen atoms in the fragment. The average information content of order 2 for the hydrophobic fragment (AIC2) represents the complexity of the hydrophobic fragment, which increases with the structural diversity of the fragment and the corresponding increase in cmc. The last descriptor, the relative number of nitrogen and oxygen atoms (RNNO), represents the size of hydrophilic fragments and accounts for the increase in cmc with increasing the number of ethylene oxide group.

Wavelet Neural Networks

Wavelet has been used in processing analytical data such as compression, denoising, and variable selection [2]. In this work, a special application of wavelet as basis functions in neural networks method is described.

Function learning, function approximation or developing models from observed data is a fundamental work in many fields such as statistical analysis, signal processing and artificial intelligence [3]. The approximated function may be expressed as:

$$F(x) = \sum w_i \psi_i (x, k_i)$$

Where ψ_i $(i = 1, \infty)$ represents the set of basis functions and w_i and k_i, the corresponding set of adjustable weights [4].

Artificial neural network (ANN) is a popular tool in function learning due to its ability to learn rather complicated functions [3]. However, ANN has some limitations such as settle in local minima of the error surface and convergence too slowly [3]. Recently radial basis function neural networks (RBFNNs) is an alternative to multilayer perceptron (MLP). RBFNNs is characterized by a number of radial basis function units in the hidden layer. An RBF networks can represent any function that is the space spanned by the family of basis functions [3].

Wavelet neural network is another novel approach towards the learning function [4]. Wavelet networks, which combine the wavelet theory and feed-forward neural networks, utilize wavelets as the basis function to construct a network. Wavelet function as the same as the radial basis function is a local function. Although RBF is also a local function, but it does not have the spatial- spectral zooming property of the wavelet function, and therefore it cannot represent the local spatial–spectral characteristics of the function [4]. The wavelet neural network (WNN) shows surprising effectiveness in solving the conventional problems of poor convergence or even divergence encountered in other kinds of neural networks [5]. The WNN consists of three layers: input layer, hidden layer and output layer. In this work, a Morlet mother function is used as node activation function for the hidden layer.

Data and Methodology

Data Set

The data set consists of 73 molecules selected from ref. [1]. These molecules have a wide range of structural features of the surfactants.

Descriptor Calculations

The procedure for calculation of topological descriptors has been described in Ref [1]. The quantum chemical descriptors, the POL, and HOMO were obtained using the Hyperchem software (semi-empirical AM1 method). The optimization of the molecular structures was also carried out by AM1 Hamiltonian, before calculation of the descriptors is performed.

Numerical values of these two descriptors were calculated for 73 nonionic surfactants. All of the computations were carried out on a Pentium 4, 1.5 GHz PC computer. The WNN algorithm was written in MATLAB (Math Works, version 6.1) by the authors.

RESULTS AND DISCUSSION

In addition to the three topological descriptors, KH0, AIC2, and RNNO that they have used by Huibers et al [1] a number of other descriptors such as total energy, dipole moment, surface of molecule, lowest unoccupied molecular orbital (LUMO) energy, highest occupied molecular orbital (HOMO) energy, hydrophobe surface and its volume, hydrophil surface and its volume and volume of the molecule have also been examined in the MLR model. It was realized that the two quantum chemical descriptors, the POL and the HOMO, are able to improve the performance of the model. The polarizability descriptor is the relative tendency of the electron cloud of an atom or a molecule to be distorted from its normal shape by the presence of nearby ion or dipole, that is, by an external electric field. The cmc for more polarizable molecules appears at the lower concentration.

The HOMO is another new descriptor, which was introduced in this work.

WAVELET NEURAL NETWORK

A wavelet neural network model was developed using the five descriptors as inputs. The same data set consisting of 73 molecules were randomly divided into three groups, training set, prediction set and the test set which consisted of 52 and 16, 5 compounds respectively.

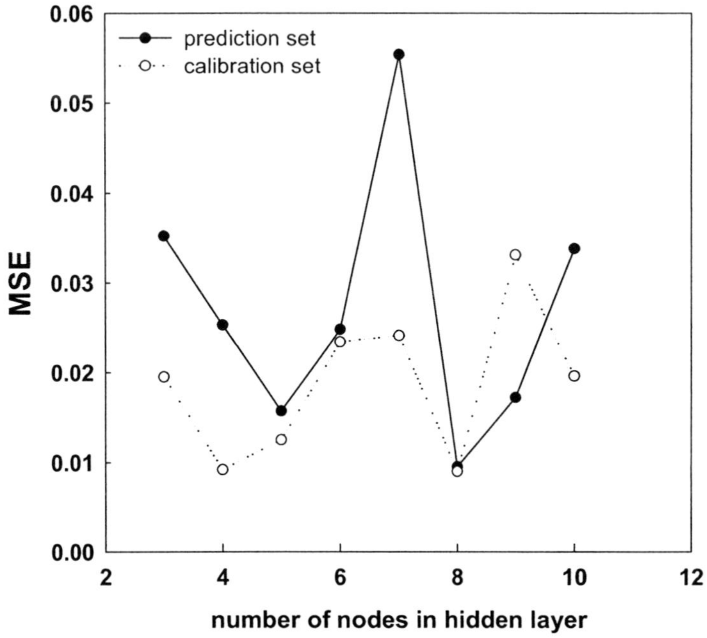

Fig. 1. Variation of MSE vs number of nodes in hidden layer

The network architecture consists of five neurons in input layer corresponding to the three topological, KH0, AIC2, RNNO and two quantum chemical, POL and HOMO, descriptors. The number of neurons in the hidden layer is unknown and need to be optimized. The output layer has one neuron that predicts cmc by the model. In addition to the number of the neurons in the hidden layer, the WNN parameters consist of the learning rate, the momentum and the number of iterations should also be optimized. In order to determine the optimum number of neurons in the hidden layer, the mean square error (MSE) against different number of neurons in the hidden layer was plotted for the training set and prediction set. The optimum number of nodes in the hidden layer is selected based on the minimum MSE value for the prediction set to prevent over-fitting of the model. Fig.1 shows this plot, from which it can be seen that the MSE error has its minimum value when the number of the neurons in the hidden layer is eight. This figure also shows that the behavior of MSE against the number of neurons in the hidden layer is approximately the same for the training set and the prediction set. The MSE against different numerical values for learning rate and momentum and also the MSE error against different number of iterations for calibration and prediction sets are plotted. The optimum numerical values for the learning rate, the momentum and the number of iterations are 0.02 and 0.9, 7000 respectively.

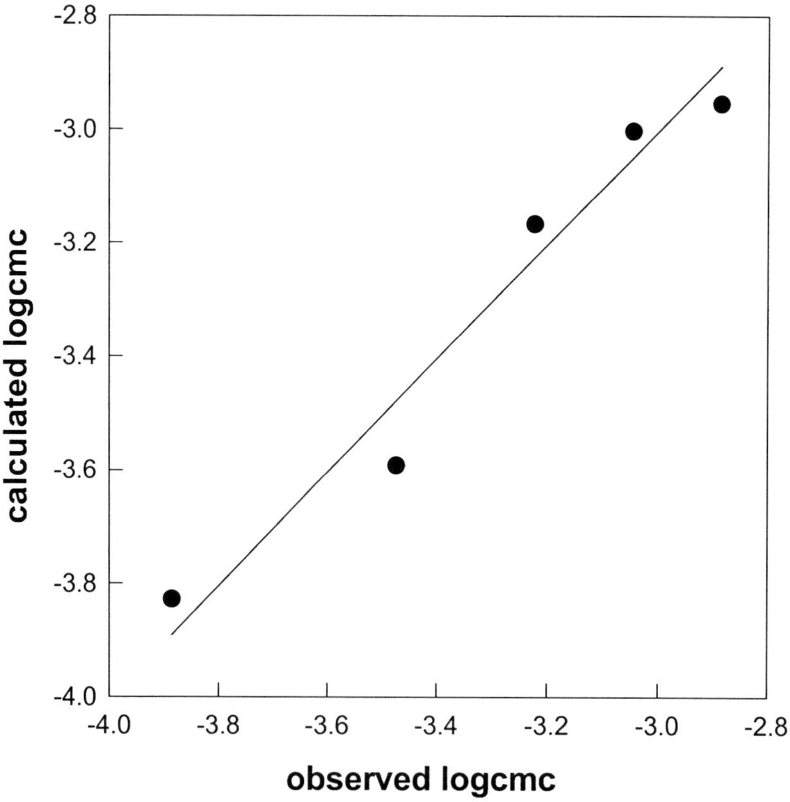

Fig. 2. Scatter plot of the calculated vs experimental logcmc (test set)

In order to evaluate the performance of the model, the scatter plot of the observed logcmc versus calculated log cmc is shown in Fig. 2. Additionally the relative errors were calculated for the test set. These values are tabulated in Table 1.

Table 1. Relative error percentage for the test set.

Experimental logcmc	Calculated logcmc	% relative error
-3.046	-3.003	-1.4
-3.886	-3.828	-1.5
-3.474	-3.491	0.5
-3.222	-3.167	1.7
-2.886	-2.954	2.4

The average relative error for test set is 1.5%.

CONCLUSIONS

A wavelet neural network model was conducted to predict cmc of nonionic surfactants using three topological and two new quantum chemical descriptors. The capability of the model was evaluated by plotting calculated values of log(cmc) vs. experimental values for the test set. The results indicate good performance of the model.

REFERENCES

[1] P.D.T Huibers, V.S. Lobanov, A.R. Katritzky, D.O. Shah, M. Karelson, *Lagmuir* 12 (1996) 1462.

[2] Daubechies, I., *Ten Lectures on Wavelet*, SIAM Press, Philadelphia, 1992

[3] Zhang,J.,Walter,G.G.,Miao,Y.,Lee,W.,1995,*IEEE Trans.*, Signal processing ,43,1485.

[4] Motard,R., Joseph,B., *Wavelet Applications in Chemical Engineering*, Kluwer Academic Publishers, Boston, 1994

[5] Zhang,X.,Qi,J.,Zhang,R.,Liu,M.,Hu,Z.,Xue,H.,Fan,B.,2001,*Comput.Chem.*,25, 125

In: Progress in Chemometrics Research
Editor: Alexey L. Pomerantsev, pp. 43-64
ISBN 1-59454-257-0

Chapter 5

PRINCIPLES OF SIMPLE INTERVAL CALCULATIONS

O. Ye. Rodionova and A. L. Pomerantsev
Institute of Chemical Physics, Moscow, Russia, rcs@chph.ras.ru

ABSTRACT

Simple Interval Calculation (SIC) is proposed for linear modelling in general, and for the prediction of interval estimation in particular. The roots of the method, based on ideas of Kantorovich, are to apply linear programming to the data analysis. The SIC approach is based on a single assumption that the error is finite. For prediction modelling, this is shown to lead to results that can be represented in a convenient interval form, and that account for all uncertainties present (X measurement errors, Y measurement errors). The SIC-interval is opposed to the traditional confidence interval estimators based upon theoretical error distributional model assumptions, which rarely hold for practical data analysis of technological and natural systems. The SIC-approach is also used for object status classification, i.e. a relative position of each sample regarding the calibration set. Several important theorems are outlined which allow determination of object status for both the calibration set - as well as the test set samples To illustrate different issues of the method there is used a specially simulated example.

Keywords: Prediction uncertainties, multivariate calibration, interval estimations, object status classification, linear programming, SIC-method.

Linear regression is one of the oldest and most studied methods used for data analysis in general and for *multivariate calibration* (MVC) in particular. The standard MVC procedure has been reviewed in many places; for example see [1], [2] and the literature references listed herein. MVC always consists of the two steps: modelling ('training') and prediction. The crucial part of the whole procedure is estimating the uncertainty of predicted values. It is obvious that the solution will be incomplete without that last step. The traditional approach is based on estimating of the covariance matrix of the regression parameters [3]. This matrix is used to calculate the variance of the predicted response values and then to compute confidence intervals [4], [5]. It is known that this method is reliable only if the standard assumptions about the error distribution are valid.

An approach that enjoys great popularity involves statistical simulation techniques. Thus in [6]-[8], for example, a jack-knife method is used for the estimation of the uncertainty of the pertinent model regression parameters. These techniques produce very reliable results [9] but they are very time-consuming.

Another approach, very popular among practitioners, is based on the comparison of the predicted values with those determined by a reference method. Cross-validation and test set validation are the two main methods used for this purpose. There is a huge literature devoted to these methods discussing their pros and cons in detail, e.g. [10], [11]. This approach works perfectly as long as new samples, sometimes called ‘normal samples’, lay ‘inside the model’. In practice it may be very difficult to classify new samples.

In this paper we suggest an unusual approach to the problem of uncertainty evaluation. This is the method of ‘Simple Interval Calculation’ (SIC), which gives the prediction directly in a convenient interval form. First it must be emphasized that the SIC-method has nothing in common with the ‘interval mathematics’ [12]. The current approach assesses the uncertainty of predicted values in such a way that each point of the resulting interval has equal ‘likeliness’. The SIC-approach does not use an(y) objective function (e.g. sum of squares) for the estimate search. In conventional regression analysis these estimates are the values of unknown parameters, which agree with the experimental data *in the best way*. In the proposed method any parameter value that *does not contradict* experimental data is accepted as a feasible estimate. No doubt such an approach makes sense only under certain assumptions.

The main (in fact, *the only*) assumption of the SIC-approach is that all errors are *limited*, which would appear to be reasonable in many practical applications. On the other hand to distinguish confidently truncated normal distribution and conventional normal distribution we need very large sample sets. For example, when truncated on 2σ we need 1000 sample, on 3σ – 10000 samples. Our goal is to use only the information that is really included in experimental data set and to estimate the uncertainty in prediction on the base on the modelling errors and errors really presented in the measurements. We do not use any particular features of the error distribution except the finiteness of the error range. However if somebody prefers to use error distributions with infinite range (e.g. normal), the SIC approach may be applied as well. In such cases one should set some value of confidence probability and calculate error bounds with it. Then predicted intervals could be implemented with the same *confidence* probability.

The history of the approach presented in this article is interesting. Apparently, *Kantorovich* [13] was the first who proposed this idea, but at that time it was neither accepted nor used. Some twenty years later, a number of applications devoted to the identification of parameters of hard physicochemical models were carried out using the idea of the SIC method [14], [15]. In 1986 one of the present authors ‘re-invented’ this method and published this in [16]. However, we do not know about any articles that deal with the theoretical aspects of the SIC-method and/or its validation. We propose that the Simple Interval Calculation may be a useful addition to the current repertoire of methods in multivariate data analysis.

The major goal of this paper is to present the main issues of the SIC-approach and to show that prediction intervals calculated by the SIC-method have fairly sensible size. We also try to show the powerful possibilities of the SIC-approach in object status classification.

SIC BASICS PRINCIPLES

Region of Possible Values

Let us consider a linear model

$$\eta = \Xi\alpha \tag{1}$$

where $\boldsymbol{\eta}$ is the n–dimensional vector containing the true response values; $\boldsymbol{\alpha}$ is the corresponding p-dimensional vector containing the true parameter values; Ξ is the $(n \times p)$–matrix containing the true predictor values. We assume that in the course of experiments we obtain both the manifest response values $\boldsymbol{y}$ and predictor values $\mathbf{X}$ with some errors.

$$\boldsymbol{y} = \mathbf{X}\boldsymbol{\alpha}+\boldsymbol{\varepsilon}. \tag{2}$$

Definition 1. The error finiteness means that there exists a maximum error deviation (MED) of the error ε, which equals β, i.e.

$$\exists\beta > 0 \quad \text{Prob}\{|\varepsilon| > \beta\} = 0, \text{ and for any } 0<b<\beta \text{ Prob}\{|\varepsilon| > b\} > 0 \tag{3}$$

where Prob{•} denotes the probability that an event occurs.

Symmetry of ε is used here for simplicity but this assumption is not essential for the method. We consider β to be common for all objects, i.e. we assume error homoscedasticity (which however is not critical).

Known β

First we assume that β is known. We call a pair $(\boldsymbol{x}_i, y_i)$, i=1,…,n a calibration object. Here the vector $\boldsymbol{x}_i^t$ is the i-th row in $\mathbf{X}$ matrix and corresponds to response value y_i. According to our main assumption Eq. (3) for each calibration object i=*1*,…,n we can write

$$y_i^- \le \boldsymbol{x}_i^t\boldsymbol{\alpha} \le y_i^+, \quad y_i^- = y_i - \beta, \qquad y_i^+ = y_i + \beta \tag{4}$$

Naturally we do not know the 'true' $\boldsymbol{\alpha}$ values but we can consider all vectors $\boldsymbol{a}$, which agree with Eq. (4). It is obvious that such vectors $\boldsymbol{a}$ for a given i, form a strip $S(\boldsymbol{x}_i, y_i)$ in the space of parameters R^p. The position and the width of the strip depend on the calibration object values $(\boldsymbol{x}_i, y_i)$. An example of such strips can be seen in Fig.1, where we have two unknown parameters, i.e. p=2 and five samples in the system, i.e. n=5.

Let us consider all calibration samples and their corresponding strips. It is obvious that any vector $\boldsymbol{a}$ satisfies all inequalities Eq. (4) simultaneously if and only if it belongs to all strips $S(\boldsymbol{x}_i, y_i)$ (Dark grey polyhedron in Fig. 1.)

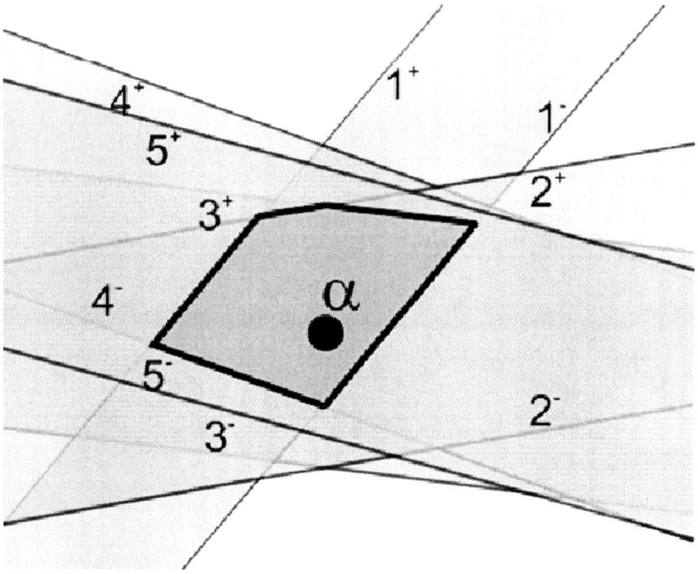

Fig 1. Strips in the parameter space, p=2Typical shape of RPV A (polyhedron), □ (●)- the 'true' value.

Definition 2 A *region of possible values* (RPV) *A* for parameter $\boldsymbol{\alpha}$ from Eq.(2) is a set in parameter space determined by the intersection of all strips, i.e.

$$A = \bigcap_{i=1}^{n} S(\boldsymbol{x}_i, y_i) \tag{5}$$

Region *A* is a *convex polyhedron* delineated by the boundaries of intersecting strips. This is a *random set* because the RPV is constructed using random values $\boldsymbol{y}$.

Remark. Naturally, randomness of *A* means that it depends on the vector $\boldsymbol{y}$ and varies when response values are changed. Formally, randomness of $A(\boldsymbol{y})$ means that for any $\boldsymbol{a} \in R^p$ we can calculate the measure $\{\boldsymbol{y} \in R^n : \boldsymbol{a} \in A(\boldsymbol{y})\}$ and thus calculate the probability $\text{Prob}\{\boldsymbol{a} \in A\}$. This is of course just a general definition [18] for the probability that confidence area *A* covers some point $\boldsymbol{a}$.

The RPV Properties

If the assumed value of MED is greater than or equal to the 'true' value β, the RPV *A* has the following properties for any linear model Eq. (2).

Property 1. The RPV is unbiased.

Remark. In the confidence interval theory [18] this means that the probability of covering a false value of the unknown parameter is not greater than the probability of covering the true value.

From the RPV definition it follows that the true value $\boldsymbol{\alpha}$ always belongs to RPV:

$$\text{Prob}\{\boldsymbol{\alpha} \in A\} = 1 \tag{6}$$

In particular, this means that if A consists only of one element $A=\{\boldsymbol{a}\}$, this element is the true parameter value, i.e. $\boldsymbol{a} = \boldsymbol{\alpha}$. It is interesting that in the SIC-analysis such a situation is possible even when the number of samples is finite ($n < \infty$) in contrast with the traditional statistical approach.

Property 2. The region A is bounded if and only if rank$\mathbf{X} = p$.

This means that if we work with a multicollinear system, we have to apply some regularisation procedure, for example a projection method, to overcome the problem of multicollinearity. We will not discuss this issue here because of the lack of space but it should be mentioned that the SIC method gives a rise to new possibilities for finding a solution for such a problem.

Property 3. The region A is consistent, i.e.

$$\text{Prob}\{A \cap \boldsymbol{\alpha}\} = 1 \quad \text{as} \quad n \to \infty \tag{7}$$

under the usual traditional 'weak' conditions [17]

$$\lambda_p \to \infty \quad \text{as} \quad n \to \infty \tag{8}$$

as for the OLS estimate.

This means that if the number of calibration samples increases, A shrinks to the true value.

Property 4. The RPV is formed not by all objects from the calibration set, but only by so-called *boundary objects*.

Therefore, if we exclude all objects from the calibration set except boundary ones, the RPV will not change. In Fig. 1 all objects except sample 5 are boundary.

Unknown β

Ordinarily, the true value of MED β is unknown and some estimate b is used instead of β. It is clear that the RPV A depends on b and that $A(b)$ is extended monotonically with increasing b

$$b_1 > b_2 \quad \Rightarrow \quad A(b_1) \supset A(b_2) \tag{9}$$

Furthermore,

$$A(0) = \varnothing, \quad A(\infty) \neq \varnothing \tag{10}$$

From Eqs.(9)-(10) it follows that there exists a *minimum* b such that $A(b) \neq \varnothing$. This minimum value can be taken as an estimator for the unknown parameter β

$$b_{\min} = \min\{b, \quad A(b) \neq \varnothing\} \tag{11}$$

Estimate in Eq.(11) is a consistent but biased, and it is the low limit of all possible β values. $b_{\min}$ is a useful characteristic of a calibration set and a calibration model (more about that below), but we need to estimate the upper limit too.

Using the following statistical approach, we try to construct the estimator b such that $b \geq \beta$ and b is as close to β as possible.

It is evident that the quality of b should depend on two issues:

1. The number of samples in the calibration set. The more objects we have, the closer is b to β.
2. The form of error distribution. Better to say, not exactly the form but the heaviness of the tails of error distribution. The lighter the tails of error distribution the worse is the estimate.

Let us consider $\hat{\boldsymbol{a}}$ that is some point (regression) estimate of parameter $\boldsymbol{\alpha}$, residuals $\boldsymbol{e} = \boldsymbol{y} - \boldsymbol{X}\hat{\boldsymbol{a}}$, and value

$$b_{\text{reg}}=\max(\ |e_1|,\ldots,|e_n|\ ). \tag{12}$$

Statistical simulations on the wide range of different finite distributions and different number of samples in the calibration set (see Appendix) show that enhanced estimator b_{SIC}

$$b_{\text{SIC}}=b_{\text{reg}}\ C(n,w) \tag{13}$$

is a reliable upper limit for β. Empirical function $C(n,w)$ (see Eq.(24)) depends on n that is the number of objects in the calibration set, and on the *variance coefficient* w,

$$w = \frac{1}{b_{\text{reg}}}\sqrt{\frac{1}{n}\sum_{i=1}^{n} e_i^2}\ , \tag{14}$$

which is a normalized factor that characterizes the heaviness of tails of error distribution.

So, for unknown β we can write, that

$$b_{\min} \leq \beta \leq b_{\text{SIC}} \tag{15}$$

and use these two values to characterize the accuracy of our calibration model.

Remark. Such a statistical technique is not the intrinsic part of the SIC method; this is only one of possible ways of β estimation. The details are presented in Appendix 1.

Model Accuracy

Values $b_{\min}$ and b_{SIC} (Eq. (15)) fully characterize the model accuracy. We can conclude that:

1. Any prior value of β is valid, only if it is greater than or equal to $b_{\min}$.
2. Aplication of the SIC-method with parameter b_{SIC} guarantees that 'true' response value lies inside the prediction interval for any *calibration* object.
3. In the worst case the half-width of the prediction interval for calibration objects is less than or equal to b_{SIC}
4. Both β estimates , i.e. $b_{\min}$ (defined by Eq. (11)) and b_{SIC} (defined by Eq. (13)), are consistent. This means that for any MED value from the interval Eq. (15) Properties 2-4 are true and Propertie 1 is true assimptotically. For any MED value that is greate than or equal to β Property 1 is true.

Predicting the Response

Consider a response prediction for any new vector $\boldsymbol{x}$ using the model in Eq.(2). If parameter $\boldsymbol{a}$ varies over RPV A, it is clear that predicted value $\hat{y} = (\boldsymbol{x}^t, \boldsymbol{a})$ belongs to am interval

$$V = [v^-, v^+] \tag{16}$$

where

$$v^- = \min_{\boldsymbol{a} \in A}(\boldsymbol{x}^t \boldsymbol{a}), \qquad v^+ = \max_{\boldsymbol{a} \in A}(\boldsymbol{x}^t \boldsymbol{a}) \tag{17}$$

In such a manner we can find prediction interval for any new object, using calibration set or RPV for a given model. Thus, we do not need to construct RPV explicitly, but obtain prediction in a simple interval form.

The solutions of Eq.(17) may be found with the help of linear programming methods [19]. These methods help us to find an extreme of bilinear form on the convex set. It is known that an extreme is achieved at one of the vertices of this set. The Simplex method helps us to move from one vertex to another in the direction of maximum change (increase or decrease) of the particular form. Being a standard numerical analysis technique it is not considered further in here.

From consistency of the RPV, it follows that V is a consistent interval estimate of the true response value y, i,e.

$$\text{Prob}\{y \in V\} = 1 \tag{18}$$

Thus, the *prediction interval* V (Eq. (16)) is a solution of prediction problem by the SIC-method.

Actually, we use b_{SIC} estimate instead of unknown, true MED value β. Therefore, there is a chance that b_{SIC} is less than β, and Eq. (18) is not true. However, this can be treated as negligibly small effect in practice. In Appendix 2 we prove this claim by calculating confidence (covering) probability for the SIC prediction interval and show that being on practical standpoint (0.9999=1) we can consider b_{sic} as the acceptable evaluation of unknown β.

SIC Object Status Theory and Quality of Prediction

When new objects are similar to the calibration objects, the accuracy of prediction is almost the same as the model accuracy. Otherwise the accuracy will vary greatly. To characterize this similarity and dissimilarity we introduce the SIC-object status theory.

The central idea is that since each sample forms a strip $S(\boldsymbol{x}, y)$ in the parameter space, so position of the strip in relation to the RPV $A(b)$ designates a sample (or an object) status. Let us consider an object $(\boldsymbol{x},y)$ and a strip $S(\boldsymbol{x},y)$ that is connected with this object (Eq.(4)) and analyse what will happen with PRV A if we add this new object to the initial calibration set. It is obvious, that in any case RPV cannot increase.

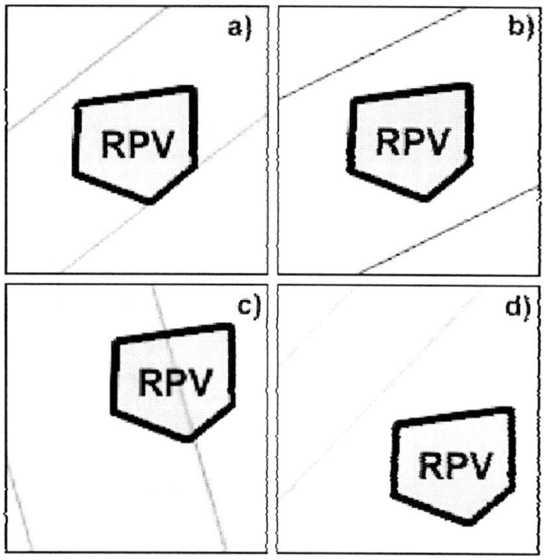

Fig. 2. Possible positions of a new sample strip w.r.t. given RPV (calibration set) in parameter space.

Definition 3. A sample $(\boldsymbol{x}, y)$ is called an *insider* if $A \subset S(\boldsymbol{x}, y)$

In other words $\left|\boldsymbol{x}'\boldsymbol{a} - y\right| \le b$ for *any* parameter $\boldsymbol{a}$ from A (Fig 2a, 2b). Each sample from the calibration set is an insider by definition.

Definition 4. A sample $(\boldsymbol{x}_i, y_i)$ from *calibration set* is called a *boundary sample* (Fig 2a), if there exists parameter $\boldsymbol{a}$ from RPV A such that $\left|\boldsymbol{x}_i^t\boldsymbol{a} - y_i\right| = b$

Boundary samples form the *borders* of RPV and they are thus of critical importance. Obviously RPV will not change if we leave only this type of samples in the model. Hence all other calibration set samples *may* be considered redundant, or *insignificant*, in this specific SIC sense.

Definition 5. A sample ($\boldsymbol{x}$, y) is called an *outsider* if $A \cap S(\boldsymbol{x}, y) \neq A$. In other words $|x'\boldsymbol{a} - y| > b$ for some parameters $\boldsymbol{a}$ from A (Fig 2c, 2d). We may conclude that all samples *except insiders* are outsiders. However we can make a useful further distinction among outsiders.

Definition 6. A sample ($\boldsymbol{x}$, y) is called an *outlier* if $A \cap S(\boldsymbol{x}, y) = \varnothing$. In other words $|\boldsymbol{x}'\boldsymbol{a} - y| > b$ for *any* parameter $\boldsymbol{a}$ from A (Fig 2d).

Definition 7. A sample ($\boldsymbol{x}$, y) is called an *absolute outsider* if $A \cap S(\boldsymbol{x}, y) \neq A$ for *any* y. It is to a great advantage that the status of an absolute outsider may be found out regardless its reference y value.

To understand the meaning of this classification let us consider what happens with a given RPV A in a case we *add* a new sample to the calibration set. If the sample is an *insider*, the RPV will not change. If the sample is an *outsider*, but not an *outlier*, the RPV decreases and the new sample becomes a boundary sample. If a new sample is an *outlier*, the RPV *disappears*, i.e. becomes empty. Fig 2 shows some of the possible relative positions of strips and RPV.

Simplest Example

The complete status of a sample (outsider, insider, ...) depends on the (predictor) vector $\boldsymbol{x}$ *as well as* its response value y. Below we illustrate the concept of mapping SIC sample status with the simplest possible one-dimensional model $y=xa+\varepsilon$ (Fig 3). The horizontal axis represents $\boldsymbol{x}$ and the vertical axis represents y.

The grey-shaded parallelogram in the centre of Fig 3 corresponds to the domain of the *insiders*. Calibration samples (three open dots ○ with error bars shown here) belong to this region. Two black dots ● – also with error bars – represent *boundary* samples, which lie on the border *between* the insiders and outsiders. *Outsiders* reside in the complementary space dichotomously outside the central insider area (parallelogram). *Outliers* occupy the upper-left and bottom-right areas respectively. Line ***1*** represents the 'true' model $y=x\alpha$, while boxes **2** represent the SIC prediction intervals $V(\boldsymbol{x})$.

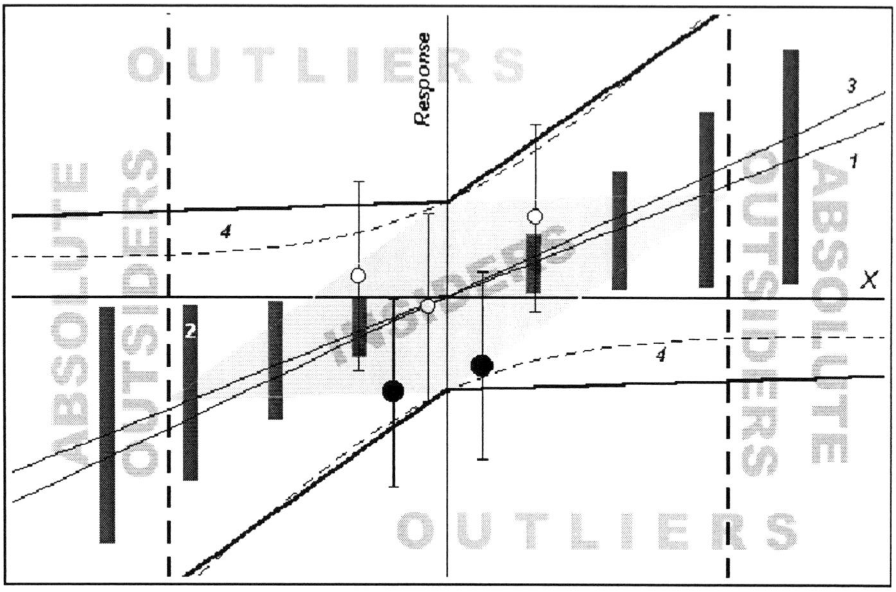

Fig. 3. One-dimensional model. Status classification plot. Response y vs. X-predictor

It can be seen that the farther away a particular object is from the origin, the wider is its SIC prediction interval. In order to compare the present approach with traditional regression techniques we also show the regression estimate of the model (line *3*) as well as its conventional 0.90 confidence limits (dashed curves *4*). It can be seen that these are very close to the SIC outlier borders, but not identical. Let us consider a sample ($\mathbf{x}$, y) and look what happens when its predictor vector $\mathbf{x}$ is fixed, while the response value y varies from $-\infty$ to $+\infty$. Ordinary its status changes in the following order: *outlier – outsider – insider – outsider – outlier*. The situation is visually accessible in Fig. 3. However, some vectors $\mathbf{x}$ cannot be insiders *for any value of y*. Such samples/objects/predictors may be termed *absolute outsiders*. The regions of absolute outsiders occupy the area to the right from the right-dashed vertical line and to the left from the left-dashed vertical line in Fig 3.

SIC Object Status Map

In the n-dimensional case the picture will of course be more complicated. For example the region of insiders will consist of two symmetrical pyramids with a common base, all together they form a convex polyhedron in the X-space. The apexes of this double pyramid can all be projected down to the origin of X-space. Luckily in the general case we do not need to construct such a *complexion* to understand the status of a new sample, because the SIC method offers explicit way for the object status classification, i.e. an object status map. Now it is necessary to introduce additional notions.

Definition 8. For any object ($\mathbf{x}$,y), value $r(\mathbf{x},y)$ is called (normalized) *SIC-residual*

$$r(\mathbf{x}, y) = \frac{1}{b}\left[y - \frac{v^{+}(\mathbf{x}) + v^{-}(\mathbf{x})}{2} \right] \tag{21}$$

where v^+ and v^- are the bounds of prediction interval (Eq. (17)). The SIC-residual may be calculated for each calibration object and for the objects from the test set, i.e. for all objects where the response is measured.

Definition 9. For any predictor $\boldsymbol{x}$, value $h(\boldsymbol{x})$ is called *SIC-leverage*

$$h(\boldsymbol{x}) = \frac{1}{b}\left[\frac{v^+(\boldsymbol{x}) - v^-(\boldsymbol{x})}{2}\right] \tag{22}$$

It is possible to calculate the SIC-leverage for *any* object. By definition $h(\boldsymbol{x})\geq 0$ for any $\boldsymbol{x}$, and $h(\boldsymbol{x})\leq 1$ for any calibration sample.

Last two definitions help to formalize the object statues classification. Taking into account Eq. (17) and Def. 3 – Def. 7, it may be proved that:

Statement 1. An object $(\boldsymbol{x},y)$ is an insider. iff $|r(\boldsymbol{x},y)| \leq 1-h(\boldsymbol{x})$;
Statement 2. The calibration object $(\boldsymbol{x}_i,y_i)$ is boundary, iff $|r(\boldsymbol{x}_i,y_i)| = 1-h(\boldsymbol{x}_i)$;
Statement 3. An object $(\boldsymbol{x},y)$ is an outlier iff $|r(\boldsymbol{x},y)| > 1+h(\boldsymbol{x})$
Statement 4. If for the given predictor $\boldsymbol{x}$ $h(\boldsymbol{x}) > 1$, then for any y, object $(\boldsymbol{x},y)$ is an absolute outsider and vice verse.

This last statement eliminates a special class of absolute outsiders from a class of all outsiders. They differ from the calibration samples greatly and may be found without response measurements. From Eq.(22) it follows that $h(\lambda\boldsymbol{x})=|\lambda|h(\boldsymbol{x})$, so for any calibration $\boldsymbol{x}_i$, vector $\lambda\boldsymbol{x}_i$ is an absolute outsider iff $|\lambda| > 1/h(\boldsymbol{x}_i)$. As a result, for any given calibration set we can easily construct the region in the predictors' space that is occupied by the absolute outsiders. This can be done using the following statement.

Statement 5. Let D be a set in the X space defined as a linear combination of weighted calibration predictors $\boldsymbol{x}_i$

$$\boldsymbol{x} = \sum_{i=1}^{n} \frac{\lambda_i}{h(\boldsymbol{x}_i)} \boldsymbol{x}_i, \quad \sum_{i=1}^{n} |\lambda_i| \leq 1$$

Then all *absolute outsiders* are to be found exclusively outside this region D.

This area is shown in Fig. 5a by the polygon. In multidimensional X-space we cannot construct such an area explicitly, therefore we use another representation. It is the *object status map* where object classification based on Statements 1-4 may be presented graphically (Fig.4).

Due to this classification, the end-user may analyse the quality of prediction as follows.

1. If an object is an insider (belongs to area 1, Fig.4), the half width of the prediction interval is less than or equal to b, the 'true' response value lies inside the prediction interval, the size of the concrete prediction interval characterises the uncertainty for the given object. Insider is a perfect prediction, as its accuracy is not worse, than calibration accuracy.

2. If an object is a boundary sample (lies on the border of area 1, Fig. 4), it has a special meaning in calibration set. This is very significant sample that produces (together with other boundary objects) the RPV area, i.e. the model.
3. If an object is an outsider (lies out of area 1), it is a doubtful sample. In this case prediction accuracy is worse then calibration, but this can be caused by some error in reference value y.
4. If predictor $\mathbf{x}$ lies in the area of absolute outsiders, the predictor vector of this sample greatly differs from predictors of calibration objects and its prediction will always be worse, than calibration. It is very important that this status can be detected for new samples with unknown response values.

In conclusion we can mention that triangular structure of insiders area in the status map looks like the influence plot [20]. However, the proposed status map has an evident advantage being presented explicitly and numerically

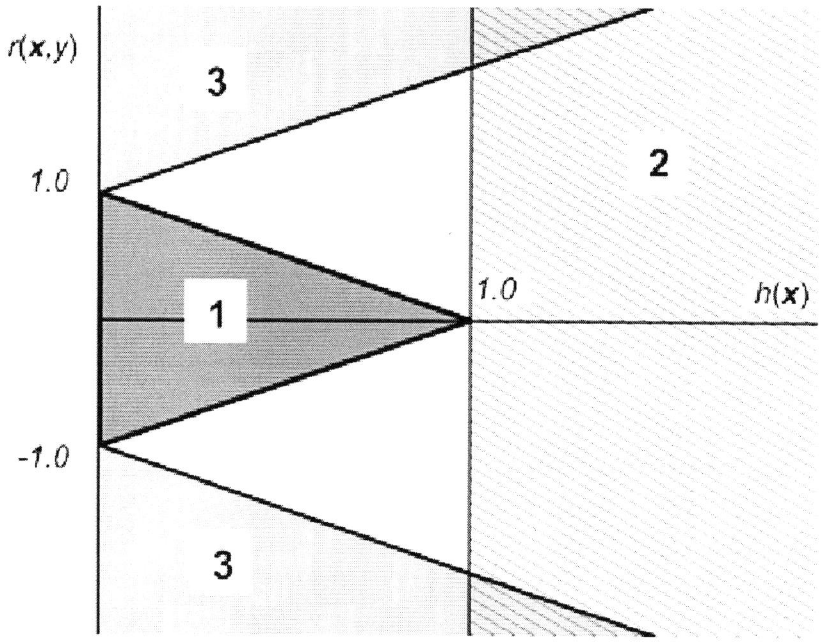

Figure 4. Object status map. SIC-residual vs. SIC-leverage

EXPERIMENTAL

Often, when we work with real data, we have a rank deficient problem. In these cases, the RPV A is unlimited, the MED estimate tends to zero and the width of prediction interval tends to infinity. To overcome these difficulties we may apply a standard projection technique [20], and after that we can use the SIC-method. We have applied the SIC-method together with projection techniques to the various real-world data [21] and received very interesting results. In this paper we will not discuss this aspect of the problem, since we would like to stop on the main features of the advocate method itself.

All SIC-calculations were made with newly implemented software, which, in addition to the SIC features, includes the following base algorithms: NIPALS algorithm [20] for the pertinent matrix decompositions, a standard Simplex algorithm [22] for optimisation, as well as the special procedures for preliminary data analysis that reduce initial problems to a form suitable for efficient application of the Simplex procedure.

Data

For illustration purposes we choose a model with two parameters (i.e. p=2) and a small calibration set in order to draw the RPV explicitly and to use a simulated data with known 'true' values. This example is derived from the real-world problem [21]. Calibration data set consists of 11 objects and a test set consists of 4 objects. In the model $\boldsymbol{y}=\mathbf{X}\boldsymbol{a}+\boldsymbol{\varepsilon}$, the 'true' parameter values are a_1=0.0305, a_2=-0.2073. Matrix **X** is the 11×2 orthogonal matrix, errors ε were simulated by the uniform distribution on interval [-0.15, +0.15], so the 'true' MED value is equal to 0.15. Predictors for calibration (dots) and test (squares) samples are shown in Fig. 5a.

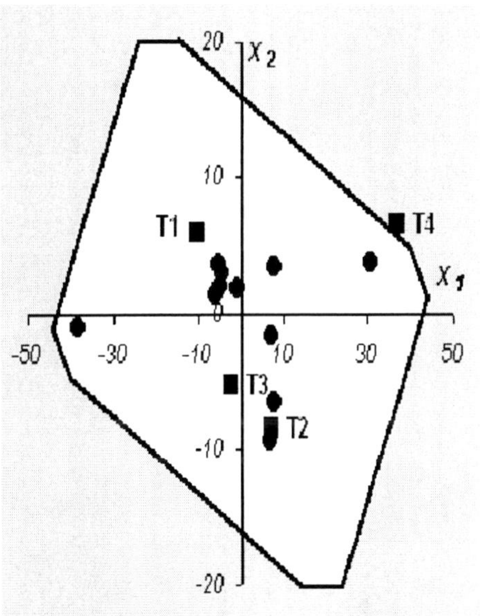

Fig.5a. Simulated example. Predictors (X-space). (●) calibration samples, (■) test samples, (—) border of the absolute outsiders region.

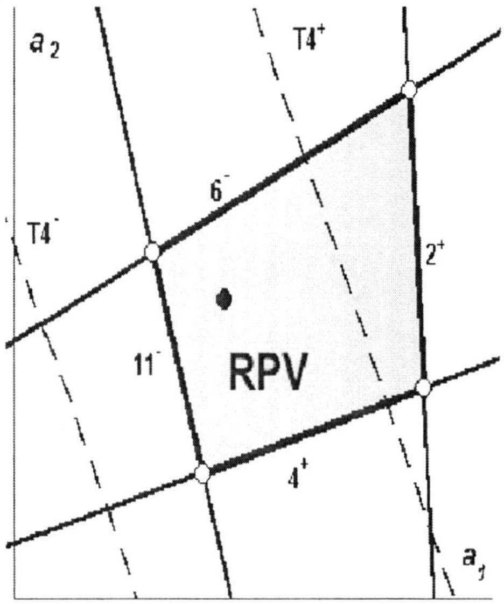

Fig.5b. Simulated example. Parameter space RPV formed by calibration samples 2,4,6,11 Dashed lines $T4^+$ and $T4^-$ represent the absolute outsider in the test set.

Calibration

First of all we have to estimate the MED value. Using Eq.(11) we get $b_{\min}$=0.115. This means that for the given data set we cannot apply the SIC-method with b<0.115, because the RPV will be empty. This is the underestimated value as it was supposed above (See Section 2.3). To increase b we use Eqs. ((12)-(14)) and calculations reflected in Appendix and have received that w=0.430, C=1.12 and b_{SIC}=0.182. This value was used in all subsequent calculations in this example. Also, we can conclude that for the given calibration set

$$0.115 \le \beta \le 0.182.$$

The 'true' value (β=0.15) lies in this interval. The dimension of parameter space equals two; so we can draw the RPV here explicitly for illustration purposes. As it was mentioned above we do not need to build it in a general case. Only four calibration samples (2,4,6,11) form RPV (See Fig 5b). They are the boundary objects. Strips that are formed by other calibration objects are so wide that RPV lies completely within these strips. To understand the status of each calibration object without explicit RPV construction we can look at the object status map (Fig. 6b). All calibration objects (dots) lie inside the central triangle, and only four of them (2,4,6,11) are on the border of the insider area.

Now we can characterise the accuracy of calibration by the SIC method. The maximum error, i.e. the upper bound is equal to 0.182. Due to Statement 5 we can construct the border for absolute outsiders in the X-space (see Fig 5a). That will help us to analyse the status of any new sample regardless its reference value.

Prediction

Now we calculate SIC-prediction intervals for the test set (see Fig 6a). SIC prediction results (grey bars) may be compared with the reference values (black dots). They also may be matched up to the 'true' values (white dots), which are shown with the error bars equal to the 'true' error deviation.

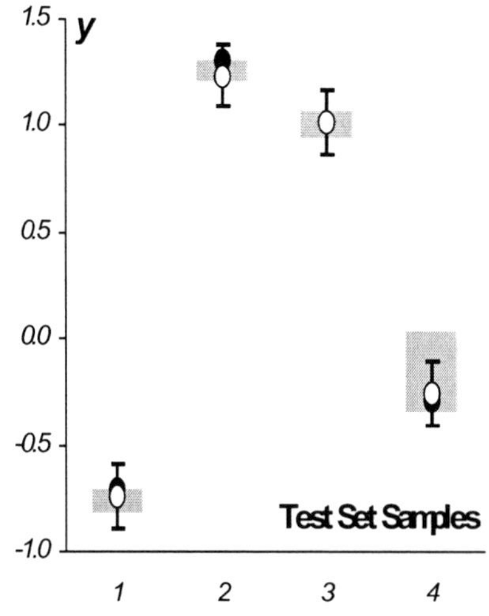

Fig.6a. Simulated example. Prediction for the test set. (○) true responses with 'true' error bars (●) reference response values, (▒) SIC-prediction intervals

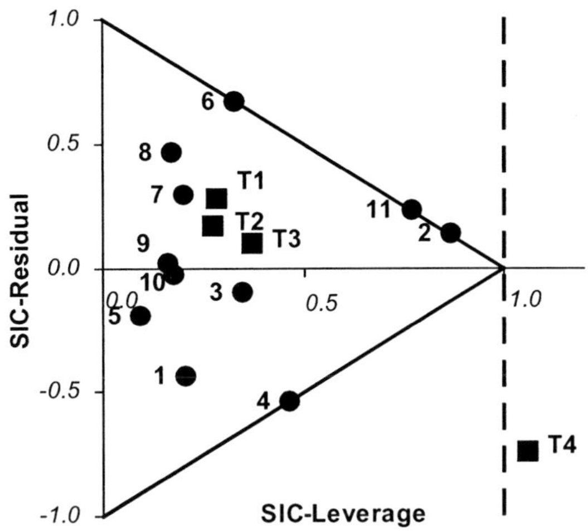

Fig.6b. Simulated example. Object status map. (●) calibration samples, (■) test samples

For each new object we get an individual prediction interval, i.e. we estimate the uncertainty in prediction, for every new sample. To analyse the prediction results it is helpful to consider Fig.6a and Fig. 6b simultaneously. Test objects are indicated as squares in Fig 6b. Samples T1-T3 are insiders; they lie in the insiders' area (compare with area 1 in Fig.4.). For all such samples the half width of the prediction intervals is less than b_{SIC}. Object T4 has another status. Though 'true' response value lies inside the SIC prediction interval (Fig.6a), this interval is too large (greater than $2b_{SIC}$.). This sample is an absolute outsider (compare with area 2 in Fig. 4 and see Def. 7). As described in Section 3.1, if we add such a sample to the calibration set the RPV decreases (dashed lines in Fig. 5b).

Figs. 6a and 6b may be constructed regardless the dimensionality of the initial problem and they are the standard output for the SIC-method. Figs. 5a and 5b are shown here for illustration purposes and they were specially built for this example.

Example Conclusions

We use calibration set to estimate the MED β. We cannot estimate β precisely here, as our calibration set is *too small*. We prefer to take b_{SIC} that is greater than β as at this case we guarantee that 'true' parameter values lies inside the RPV (black dot in Fig 5b). As expected the 'true' response values lie inside the SIC-prediction intervals (Fig. 6a). In the situation with absolute outsiders, SIC-intervals immediately signal about the abnormal samples. Though the 'true' value lie inside the SIC-interval, this interval is too large. We can notice the abnormal behaviour of absolute outsider on all four figures (Figs. 5.a –6.b). In Fig 5.a the absolute outsider is to be found out of the border of absolute outsiders in the X space. In Fig 5.b absolute outsider form the strip in the parameter space that crosses the RPV. In Fig.6a we observe the abnormal large interval for such a sample and in Fig.6b we also can explicitly find absolute outsider.

This example also demonstrates how the SIC approach answers to the main questions that are of great importance for the end-user.

1. The MED β estimation determines the accuracy of modelling, and the worst-case prediction precision for the samples that are similar to calibration samples.
2. The SIC prediction intervals calculate the uncertainty of each individual new sample.
3. The position of each object on the object status map helps to determine whether this or that object is similar to the calibration set or not, i.e. to determine the sensible range of application for the constructed model.

If we have to guarantee the accuracy of prediction then we can state that the half-width of the prediction interval is less than or equal to b_{SIC} for the objects from the test set that are insiders (Fig 2, area 1). We also can state that the constructed model cannot be applied to the absolute outsiders (Fig 2, area 2). Such samples differ seriously in their predictors' structure. It is very valuable that we do not need to know the response values for these samples.

CONCLUSIONS

We propose that the presented SIC-approach for prediction interval construction may be useful for many practical applications within multivariate data analysis and possesses some advantages over traditional regression methods. The SIC-method

- does not depend on the error distribution and can be applied for any kind of errors – in $\mathbf{y}$, in $\mathbf{X}$ and modelling
- gives the result in the interval form with all the uncertainties included
- helps to circumscribe the framework of model applicability

The SIC-method uses no extra parameters, which cannot be evaluated by the data set and have to be set a priori The main (in fact, *the only*) assumption of the SIC-approach, that all errors are *limited,* we consider being the merit of this approach. The validity of the SIC-method with respect to the above-mentioned assumptions is beyond all doubt. Arguments that justify the SIC-approach are so simple and evident that they should not raise objections.

The specific calculation aspects of the SIC-method are rather simple, since they are founded on the well-designed algorithms for linear programming.

APPENDIX 1. CONSTRUCTION OF THE CORRECTION FUNCTION $C(N,W)$

Correction function C is constructed as the upper confidence bound for the MED-estimator $b=b_{reg}$ (Eq. (12)). To make the bound more effective we take into account that the 'heavier' the tails of error distribution, the greater the estimate b and at that same time deviation is smaller. This heaviness may be characterized by the variance coefficient w (Eq. (14)). However variates b and w are not independent. They have a negative correlation that is noticeable for the small sample sets ($n<50$). This correlation competes with a positive causality. In practice function $C(n,w)$ is built using statistical simulation. In brief, the procedure is as follows.

1. We consider different number of objects (n) in calibration set, n= 10, 20, 50, 75, 100, 250.
2. For each data set we use different finite error distributions on the interval $[-\beta,+\beta]$, with β= 1. Each distribution is derived from the normal distribution $N(0, K^2)$, with K =0.3, 0.5, 1.0, 1.5, 2.0, 2.5, 3.0; totally 7 different functions. When parameter K=0.3 we get nearly uniform distribution and when K =3 we get a normal distribution truncated on 3σ. Thus, we analyse rather a wide range of distributions existed in real life. Normal distribution is taken only for simplicity, but it is not critical. Ordinarily we do not know the exact form of error distribution, especially for small sample sets.
3. For each n and for each K we make 500 series of data simulation, i.e. each calibration set is simulated 500 times. At each step we calculate the maximum error deviation b (Eq.(12)) and the variance coefficient w (Eq.(14)).
4. As in real life we do not know the value of parameter K, so we agglomerate the results of all simulations for a given n. Totally there are 3500 pairs of (b, w).

Our aim is to correct b value in such a way that 95 % of new estimates will be greater than or equal to 1 (i.e. 'true' value). Direct calculation shows that for each number of calibration samples n the right tail of conditional distribution of b at given w, may be modelled as the normal distribution with parameters

$$\mathrm{E}(b \mid w) = M(n, w) = a_1(n)w + a_2(n) \text{ and } \mathrm{D}(b \mid w) = D(n) \tag{23}$$

where a_1 and a_2 are empirical parameters, which depend on n.

This claim is illustrated in Fig.7 where we present sampling probabilities P_c versus normal probabilities P.

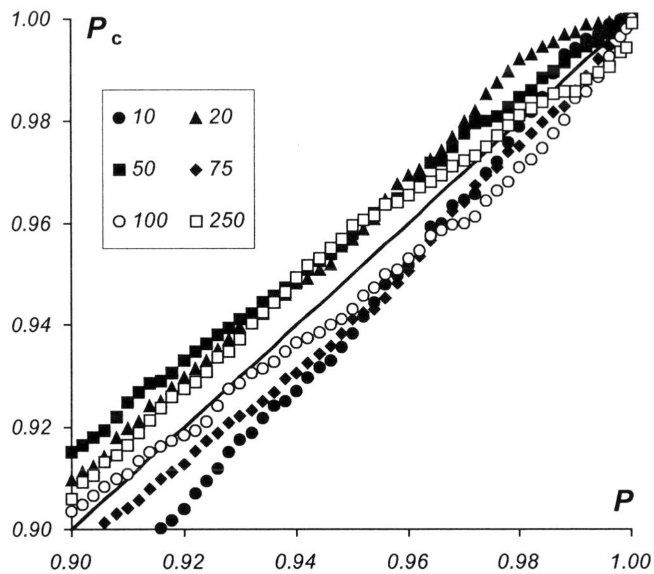

Fig 7. Sampling probabilities P_c (dots) versus theoretical probabilities P calculated by the normal distribution with parameters Eq (23)

From Fig. 7 it can be seen that in the range $0.90 < P < 1$ we have rather good approximation of the theoretical values. Therefore function C may be presented as

$$C(n,w) = (M - x_P D)^{-1}, \tag{24}$$

where x_P is the normal quantile value for a probability P=0.95. Parameters a_1, a_2, and D, which are used in Eq. (23), depend on the sample size n. Their correspondent empirical functions are shown below.

$$a_1(n) = \frac{1 - 172.812n^{-1.5287}}{1 + 172.812n^{-1.1626}}$$

$$a_2(n) = 0.9572\left[1 - \exp\left(-17.8528/\sqrt{n}\right)\right] \tag{25}$$

$$D(n) = 0.2011\left[1 - \exp\left(-73.4113/n\right)\right]$$

While designing these functions, we took into account the following evident properties

$$a_1(1) \approx -1,\ a_1(\infty) = 1,\ a_2(\infty) = 0,\ D(\infty) = 0$$

Fig 8 shows the dependence of $C(n,w)$ versus variance coefficient w for different sample sizes (n=10, 20,..., 250). Values calculated with the help of simulations (dots) agree well with those calculated with the help of correction function (lines) defined by Eq. (23). The growth of the correction function at n=10 at n=20 can be explained by the negative correlation between b and w.

As a result we yield the following algorithm for b_{SIC} determination for the given calibration set that consists of n samples:

i) calculation of b_{reg} (Eq. (12)) and w (Eq. (14)) with the help of least square regression method (or any projection technique – PCR or PLS – while required)
ii) calculation of the correction coefficient $C(n,w)$ as the function defined in Eq.(24) (with respect to Eq. (25))
iii) determination of $b_{SIC} = b_{reg}\, C(n,w)$

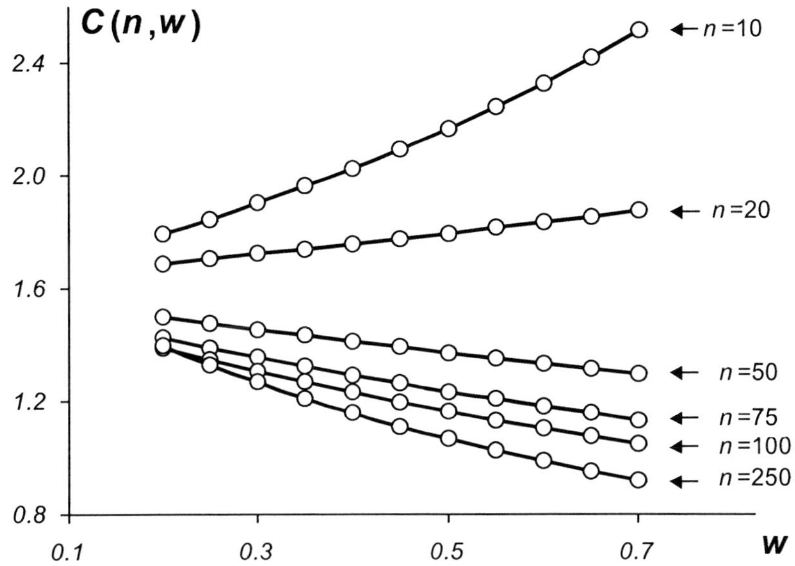

Fig. 8. Correction function $C(n,w)$ for different number of samples (n) in the calibration set. Dots are simulated values and curves are calculated values

APPENDIX 2. CALCULATION OF COVERING PROBABILITY

In this Appendix we show how to calculate the actual covering probability for the SIC prediction interval. In Appendix 1 we constructed the SIC estimate b_{SIC} (Eq. (13)) that gives the upper 0.95 confidence limit for unknown MED value β

$$\mathrm{Prob}\{b_{\mathrm{SIC}} > \beta\} = 0.95$$

One might think that this 0.95 confidence probability at the same time is the covering probability for the SIC prediction interval. This is quite wrong!

Let us consider an event that interval $V(b_{\mathrm{SIC}})$ (Eq. (16)) constructed with b_{SIC}, does not cover the true response value y. Actually, this event is a coincidence of two independent events

1. Estimate b_{SIC} is less than β
2. True y value lies on the edge of $V(\beta)$ in a gap between $V(\beta)$ and $V(b_{\mathrm{SIC}})$

To calculate the probabilities of these events we introduce a random variable

$$z = b_{\mathrm{SIC}}/\beta,$$

that changes in interval [0,1] when $0 \le b_{\mathrm{SIC}} \le \beta$. From Appendix 1 it follows that for a given value of w, estimate $b=b_{\mathrm{reg}}$ Eq (12) could be considered as a normally $N(M, D^2)$ distributed with parameters M and D given in Eq. (25). These parameters depend on the variance coefficient w (Eq. (14)) and on the number of objects (n) in calibration set. Therefore, estimate $b_{SIC}=bC(n,w)$ is also might be treated as the normal distributed, so the probability of the first event equals

$$P_1(z) = \Pr\{b_{SIC} < \beta\} = \Pr\{z < 1\} \approx \Phi\left(\frac{z - m_z}{\sigma_z}\right) \tag{26}$$

where Φ is the cumulative normal distribution function, $\sigma_z = DC$, $m_z=MC$. In the result, we can easily calculate value $P_1(z)$ for any given parameters w and n.

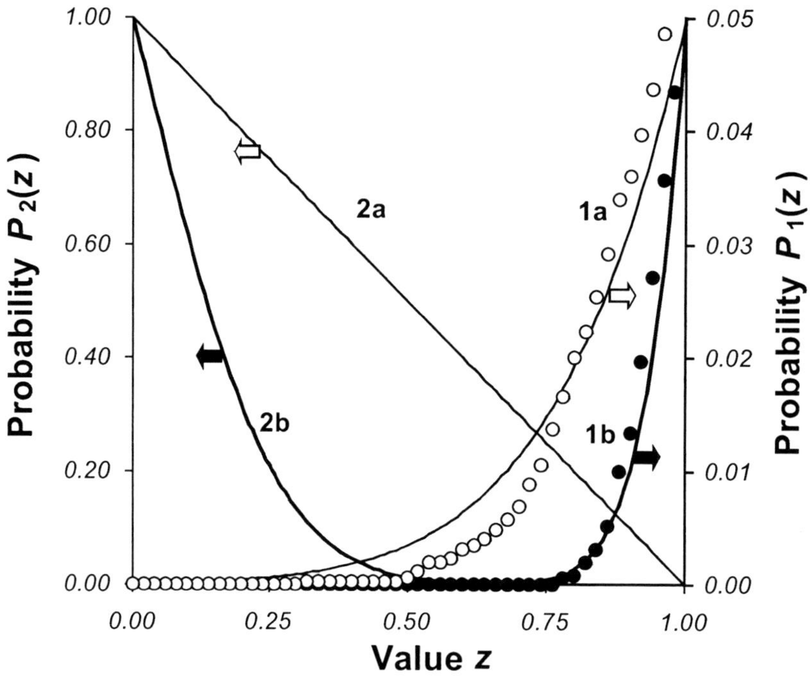

Fig. 9. Probabilities $P_1(z)$ and $P_2(z)$ for w=0.6, n=10 (a) and for n=100, w=0.2, (b) andcorresponding sampling probabilities P_2 for n=10 (open dots ○) and for n=100 (closed dots •)

To evaluate the probability of the second event $P_2(z)$ we have to take into account that this probability has the following boundary values: $P_2(0)=1$ and $P_2(1)=0$. Actually, this probability shows how prediction interval covers the true value. Let us consider the likelihood to find the true value at some small part Δ of the interval. For conventional confidence theory this likelihood is higher when Δ is situated in the middle of the interval in comparison with the case n when Δ lies at the interval edge. Therefore, we can suppose that probability $P_2(z)$ depends on the form of error distribution. Applying the form parameter K, which was introduced in Appendix 1, we can write an approximate formula

$$P_2(z) = 2\frac{\Phi(K) - \Phi(Kz)}{2\Phi(K) - 1} \tag{27}$$

It is clear that there is unique correspondence between the form parameter K and variance coefficient w. However, it is not presented here owing to complicity.

In Fig.9 we show two extreme cases of error distributions: the uniform distribution (i.e. K=0.3, w=0.6) for n=10 and the 5σ–truncated normal distribution (K =5, w=0.2) for n=100. Probabilities $P_1(z)$ (right Y axes) are represented by curve 1a (first case) and 1b (second case). Corresponding probabilities $P_2(z)$ (left Y axes) are shown by curves 2a (first case) and 2b (second case). In Fig. 9 we also plot the sampling probabilities, calculated with simulated data for n=10 (open dots) and for n=100 (closed dots). This data demonstrates that approximation for probability $P_1(z)$ proposed in Eq. (26), gives a close approach to the simulated values.

Now, we have to multiply probabilities (26) and (27) and integrate this over z from 0 to 1. In the result, we obtain that in the worst case (w=0.6, n=10) the covering probability is 0.9983, and in the best case (w=0.2, n=100) the probability is more than 0.9999999. This shows that being on practical standpoint (0.9999 ≡ 1) we can consider inequalities in Eq. (15) as the true limits for unknown β.

REFERENCES

[1] Brown P.J. *J. Royal Statistical Society,* B, **44**, 287-321, (1982).

[2] Hoskuldsson A. *Prediction Methods in science and Technology*. Thor Publishing, Copenhagen, Denmark, (1996)

[3] Seber G.A.F. *Linear Regression Analysis*. Wiley and sons, NY (1977)

[4] Donaldson J.R., Schnabel R.B. *Technometrics,* **29,** 67– 82, (1987)

[5] Faber K., Kowalski B. *J. Chemometrics*, **11**,181-238 , (1997)

[6] Efron,B. *Ann. Statist*., **7**, 1-26, (1979)

[7] Efron B., Tibshirani R. *Statistical Science*, **1**, 54-75, (1986)

[8] Martens H., Dijksterhuis G.B, Byrne D.V., *J. Chemometrics* , **14**, 441-462, (2000).

[9] Pomerantsev A.L. *Chemom. Intell. Lab. Systems*, **49**, 41-48, (1999)

[10] Efron B., Gong G. *The Amer. Statistician*, **37**, 36-48, (1983)

[11] Naes T, Isaksson T, Fearn T, Davies T. *Multivariate Calibration and Classification*. Christerer, UK, (2002)

[12] Moore B.E. *Interval analysis*, N.Y., Prentice-Hall. (1966),

[13] Kantorovich L.V. *Siberian Math Journal*, **3**, 701-709 (1962). (in Russian)

[14] Kokovin G.A., Titov V.A., Titov A.A., Spivak S.I., in *Mathematics in chemical thermodynamics*, Nauka, Siberian Department, Novosibirsk, 50-58. (1980) (in Russian)

[15] Votshinin A.P., Bochkov A.F., Sotirov. G.R., *Industrial Laboratory*, **56**, 76-95(1990)

[16] Anisimov V.M., Pomerantsev A.L., Novoradovsky A.G., Karpukhin O.N.. *J. Appl. Spectros.*, **46**, 117-122. (1987)

[17] Eicker F *Ann. Math. Stat*., **34**, 447-456, (1963).

[18] Lehmann E.L. *Testing Statistical Hypotheses*, Wiley, New York, (1960)

[19] Dantzig G. *Linear Programming and Extensions*, Princeton University Press, Princeton, New Jersey. (1963)

[20] Martens H., Naes T. *Multivariate Calibration*. Wiley, New York, (1998) .

[21] Pomerantsev A.L., Rodionova O.Ye. In *Aging of polymers, polymer blends and polymer composites*, Nova science Publishers, NY, v.**2**, 19-29, (2002)

[22] Rodionova O.Ye., Pomerantsev A.L. *This issue*, 187-203
[23] Taha H. *Operations Research. An Introduction, 3-d Ed.* Macmillan Publishing Co., New York, v.**1**, (1982)

PART II. CHEMISTRY

In: Progress in Chemometrics Research
Editor: Alexey L. Pomerantsev, pp. 67-87
ISBN 1-59454-257-0

Chapter 6

GLOBAL AND LOCAL CHEMOMETRIC MODELS OF SPECTRA-STRUCTURE RELATIONSHIPS

K. Varmuza
Vienna University of Technology, Austria; kvarmuza@email.tuwien.ac.at

ABSTRACT

The recognition of chemical structure information from spectral data challenged chemometricians since the pioneering time of this discipline. A lot of work has gone into the development of spectral classifiers based on multivariate classification methods. Actually, such classifiers are capable to detect with low error rates the presence or absence of some substructures automatically from IR or MS data. Together with the known molecular formula and using appropriate software for isomer generation, a systematic structure elucidation of organic compounds is possible in some cases. However, because of the complexity of spectra-structure relationships, the high diversity of chemical structures, and the large amount of data this global approach is limited in practical applications. Routinely used for the identification of unknowns are spectra similarity searches in spectroscopic databases. Result is a hitlist which contains the reference compounds exhibiting spectra that are most similar to the spectrum of the unknown. A method - based on binary substructure descriptors - has been developed for evaluating the structural similarity between the unknown and the hitlist compounds. This method has been applied to improve search methods in IR and MS databases with the aim to yield subsets of compounds that contain useful chemical structure information about the unknown, even in cases the unknown is not in the database. Spectral and structural data from the hitlist compounds can be used to create local models by PCA or PLS. The resulting scatter plots support the spectroscopist in spectra interpretation and finding useful relationships between spectra and chemical structures.

Keywords: chemical structure elucidation, similarity of spectra and structures, IR, MS, PCA and PLS mapping

INTRODUCTION

Structure elucidation of organic chemical compounds is a fundamental task in chemistry; it is mainly based on the evaluation and interpretation of spectroscopic data. Nuclear magnetic resonance (NMR) spectroscopy often provides best information about chemical structures of un-known substances [1]. However, NMR is not sensitive enough for trace analysis. For the identification of organic compounds in complex mixtures at the nanogram level and below, mass spectrometry (MS) and - to a lesser extent - infrared spectroscopy (IR) are applied [2,3]. Therefore, MS (in particular with electron impact ionization) and IR are indispensable techniques, for in-stance in environmental chemistry, food chemistry or in searches for new chemical compounds in biological materials. Despite the large amount of information contained in MS and IR data it is difficult to extract chemical structure information from the spectra. Reason for that is the complicated, and often not well known relationships between spectral data and chemical structures. For instance, many, less understood, competitive and consecutive fragmentation processes of ions cause the measured MS data; however, the ion abundances cannot be predicted sufficiently by known spectroscopic concepts. Also in IR it is often impossible to establish reliable rules for relationships between wavelengths of bands and the presence of functional groups.

The recognition of chemical structure information from spectral data challenged chemometricians since the very pioneering time of this discipline [4,5]. A lot of work has gone into the development of spectral classifiers based on a large variety of multivariate classification methods. Actually, in some cases such classifiers are capable to detect automatically, and with low error rates, the presence or absence of substructures from IR or MS data. However, the development of computer-assisted methods for structure elucidation of chemical compounds - and spectra simulation from structure data - are still research tasks in spectroscopy, computer chemistry and chemometrics [6]. Fig. 1 summarizes some aspects of the relationships between molecular spectral data and chemical structure information with respect to computer-assisted methods for spectra evaluation.

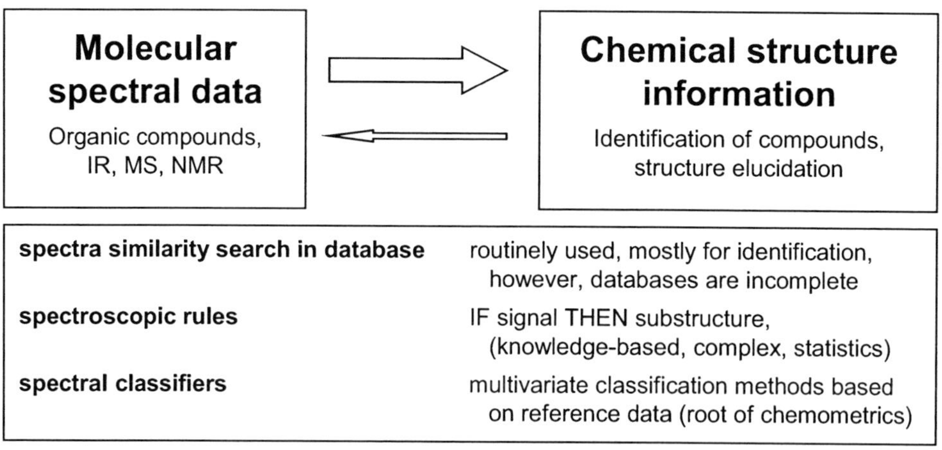

Fig. 1. From spectral data to chemical structure information by computer-assisted methods

Aim of spectra evaluation can be the *identification* of a compound (assuming the spectrum is already known and available) but also the *interpretation* of a spectrum if the chemical structure is unknown or the spectrum is not available in libraries. Identification of compounds is performed best by *library search* methods based on the computation of spectral similarities; numerous spectral databases and software products are offered for this purpose and are used routinely.

The more challenging task is the interpretation of a spectrum, for which no comprehensive solutions are available today. Different strategies have been suggested for the complex problem of substructure recognition or the recognition of more general structural properties from spectral data.

Knowledge-based methods try to implement spectroscopic knowledge about spectra-structure relationships into computer programs. Correlation tables containing characteristic spectral data (key ions, wavelengths) together with the corresponding substructures are widely used. However, a specific structural property does not always cause the same spectral signal, limiting the applicability of this approach.

Spectral classifiers are algorithms mostly based on multivariate classification methods or neural networks. Input of a classifier is spectral data and the result of classification indicates whether the unknown belongs to a certain class of compounds or not.

Interpretative library search techniques are able to provide structure information even if the unknown is not contained in the library. This technique assumes that reference compounds with spectra that are very similar to the spectrum of the unknown also have chemical structures similar to that of the unknown.

DATABASES AND SOFTWARE

The used IR database consists of 13,484 compounds and is part of the SpecInfo system [7]. Nominal molecular masses are between 18 and 962 with a mean of 242.2; the ranges of common elements in the compounds are C_{0-50}, H_{2-130}, N_{0-10}, O_{0-13}, F_{0-32}, Cl_{0-7}, Br_{0-4}, S_{0-5}, P_{0-3}. Each IR spectrum contains 801 absorbance values in the spectral range 500 to 3700 cm^{-1}, corresponding to a sampling interval of 4 cm^{-1}. Absorbances within each spectrum are scaled to the integer number range 0 to 255, corresponding to a resolution of 8 bit.

The used MS database consists of 106,955 compounds and is part of the NIST Mass Spectral Database [8]. Nominal molecular masses are between 2 and 1,218 with a mean of 271.6; the ranges of common elements in the compounds are C_{0-72}, H_{0-218}, N_{0-12}, O_{0-25}, F_{0-45}, Cl_{0-12}, Br_{0-12}, S_{0-14}, P_{0-10}. The mass spectra contain low resolution peak list data with integer mass numbers and intensities in the range 1 to 9,999.

Chemical structures in both spectral databases are encoded in the Molfile format [9,10]. Binary substructure descriptors for the characterization of chemical structures were calculated by software SubMat [11-13] running under operating systems Microsoft Windows 95/98/NT/2000/XP. SubMat uses two input files, one for the molecular structures, the other for the substructures; both have to be in Molfile format. The output file is in text format and contains a matrix with the descriptor vectors $\boldsymbol{y}$ for the molecular structures. Typical computing time with a Pentium IV 2.6 GHz is 1 second for 500 molecular structures and 200 substructures. SubMat has two operating modes. One is an interactive mode with a Windows style graphical user interface. The other is a remote mode by calling SubMat from another program and supplying necessary parameters via a command file in text format. Dedicated temporary files (so called semaphore files) are used for a communication between the calling program and SubMat, providing messages for progress, error and termination.

The isomer generator software used was Molgen [14] running under Microsoft Windows NT/2000. Molgen is capable to compute the complete set of connectivity isomers for a given molecular formula. Structural restrictions can be defined by a goodlist (separated into substructures that must not overlap and those that may overlap), and a badlist. Furthermore, lower and upper limits can be defined for bond multiplicity and ring size, as well as the number of H-atoms at C-, N-, and O-atoms. The construction of isomers is free from duplicates and fast; typically 4,000 - 8,000 isomers are generated per second.

Own software written in Matlab 6.0 [15] has been used for structure and spectra file handling, spectral similarity searches and statistics. For multivariate data analysis also the software products Scan [16] and Unscrambler [17] have been applied.

Global Models and Systematic Structure Elucidation

The most important systematic approach for structure elucidation of organic compounds dates back to the DENDRAL project and its successors [18-22]. It is based on a three-step strategy as summarized in Fig. 2. In the first step (*plan*) structural restrictions are derived from experimental data, usually from molecular spectra. Substructures, which are considered to be present in the unknown molecular structure, are collected in the *goodlist*, while forbidden substructures constitute the *badlist*. In the second step (*generate*) the molecular formula is necessary to generate all isomers that do not contradict with the given structural restrictions. If the number of generated candidate structures is too high a third step (*test/select*) is required in which the candidates are tested, mostly by spectra simulation, but also a cluster analysis of the candidate structures may be useful.

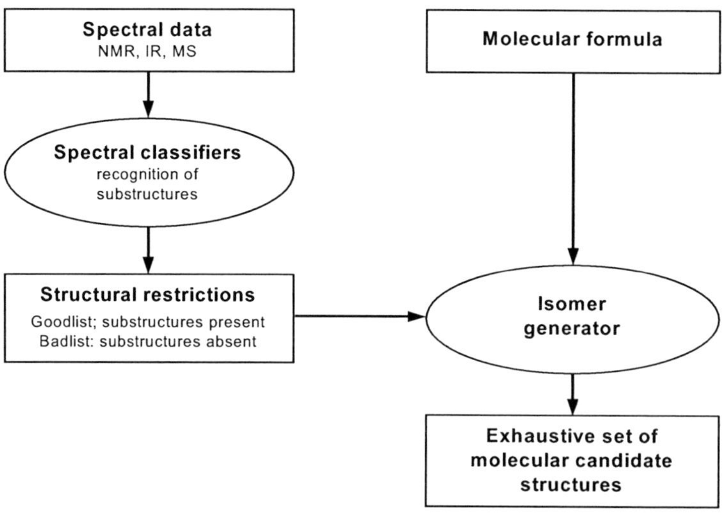

Fig. 2 Systematic structure elucidation using the molecular formula, substructure restrictions, and an isomer generator (*Dendral* approach).

Chemometrics is capable for contributing to this approach, for instance, with the development of spectral classifiers. A *spectral classifier* is an algorithm, with spectral data as input. Output is usually a discriminant variable that indicates presence or absence of a particular substructure. An essential step for mass spectra classification is an appropriate transformation of the peak data into a set of problem-relevant variables (spectral features). Each feature is a linear or nonlinear function of the peak intensities, with the definitions of features derived by spectroscopic and/or mathematical concepts. A summary for MS features is given in Table 1; details have been described elsewhere [23-25].

Table 1 Mass spectral features. Typically 400 - 1,000 features are used to represent a low resolution mass spectrum for the development of spectral classifiers or for spectra similarity searches.

Group	Feature description
1	Intensities (% base peak) at single masses
2	Intensities at single masses normalized to local ion current, window of 3 - 7 masses
3	Averaged intensities in mass ranges
4	Logarithmic intensity ratios with mass differences of 1 and 2
5	Modulo-14 summation for selected mass ranges
6	Autocorrelation with mass differences 1 - 60 in selected mass ranges
7	Spectra type features describing the distribution of peaks
8	Peak series features that characterize the joint presence of peaks

From the very beginning of using multivariate data analysis in chemistry, a lot of interest was focused on the recognition of substance classes or substructures from spectral data [26,27]. However, only much later, systematic structure elucidation of organic compounds has been tried by using spectral classifiers together with automatic isomer generation [18,28]. Several multivariate classification methods, such as LDA (linear discriminant analysis), classification by distance measurement to class prototypes, SIMCA, PLS (partial least-squares regression applied for classification), KNN-classification, ANN (artificial neural networks), and in the pioneering time of chemometrics also LLM (linear learning machine). Recently, feature selection by genetic algorithms [29] and boosting tree classification [30] have been tested for the development of spectral classifiers. Typical size of training and prediction sets is 100 - 500 spectra (objects) each.

Because simple yes/no-classifications cause unacceptable high error rates, classifiers with a rejection option have been recommended. The resulting value of a discriminant variable is compared with two thresholds, t_1 and t_2, to obtain one of three classification answers: „*yes*, substructure present", „*no*, substructure absent", or „answer not possible at a given maximum statistical risk" [28]. The classification thresholds can be determined by testing a set of mass spectra from known compounds (but not used for training); the thresholds refer to a certain maximum statistical risk α (typically 0.05 or 0.1) for giving a wrong result. Lowering the statistical risk increases the number of rejections and thereby decreases the number of *yes/no*-answers. Fig. 3 shows the probability density functions of a discriminant variable, obtained by LDA, for a recognition of the acetyl substructure from mass spectral data. In this example 300 acetyl and 300 non-acetyl compounds have been used. A set of 400 features has been reduced to 20 features by selecting the features with highest individual discriminating power (measured by the Fisher ratio [5,31]). A *y*-variable has been defined for encoding the class membership with value -1 for acetyl compounds, and +1 for non-acetyl compounds. PCR (principal component regression) has been used for the computation of a discriminant variable; the number of used principal components (eight) was found by cross validation with four segments. The large overlap of the distributions causes a relative high number of wrong classifications if only a single threshold would be applied. Definition of a rejection region reduces the risk of wrong classifications to a predefined limit; for instance a rejection region between -0.7 and +0.75 reduces the risk of wrong classifications to 0.05. One has to pay for this low error rate because for more than half of the spectra an assignment to classes is refused [32].

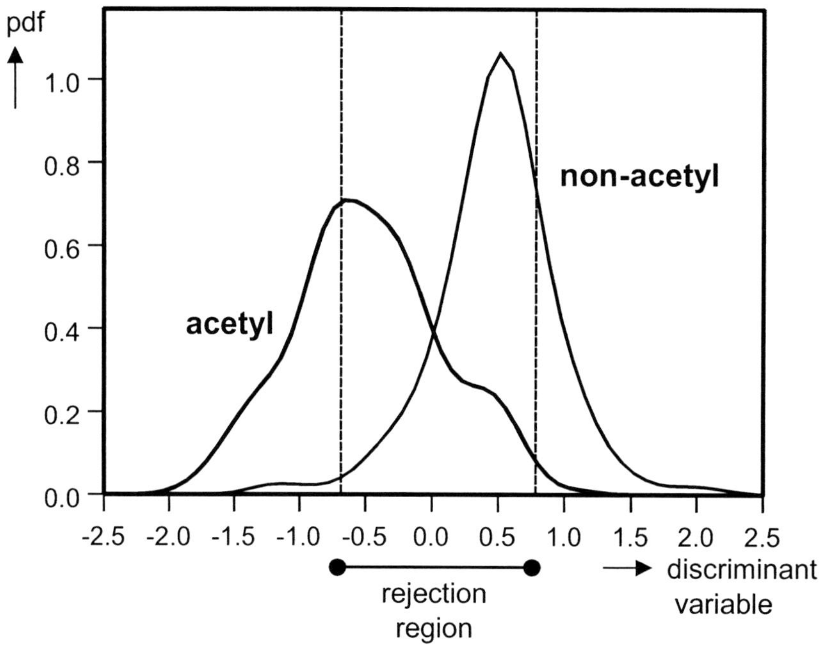

Fig. 3 Probability densities functions (pdf) for a discriminant variable, computed to recognize the acetyl substructure from mass spectral data.

An example for a successful systematic structure elucidation is shown in Fig. 4 [32,33]. The query compound, considered as unknown, is ethyl-o-methylbenzoate (CAS reg. no. 87-24-1) with molecular formula $C_{10}H_{12}O_2$. A very large number of more than 4.10^7 connectivity isomers exist for this formula, although not all isomers formally possible are stable chemical compounds. The used experimental data for the query is a low resolution mass spectrum and an infrared spectrum. Software MSclass [28], containing classifiers for 80 substructures, indicates the presence of two substructures, shown in the goodlist. With these restrictions only four molecular candidate structures are possible. Software IRclass [34]; containing classifiers for 20 substructures, yielded one substructure for the goodlist and one for the badlist. Considering these restrictions 512 molecular candidate structures are possible. The cross section of both candidate lists results in a single structure which is the correct solution.

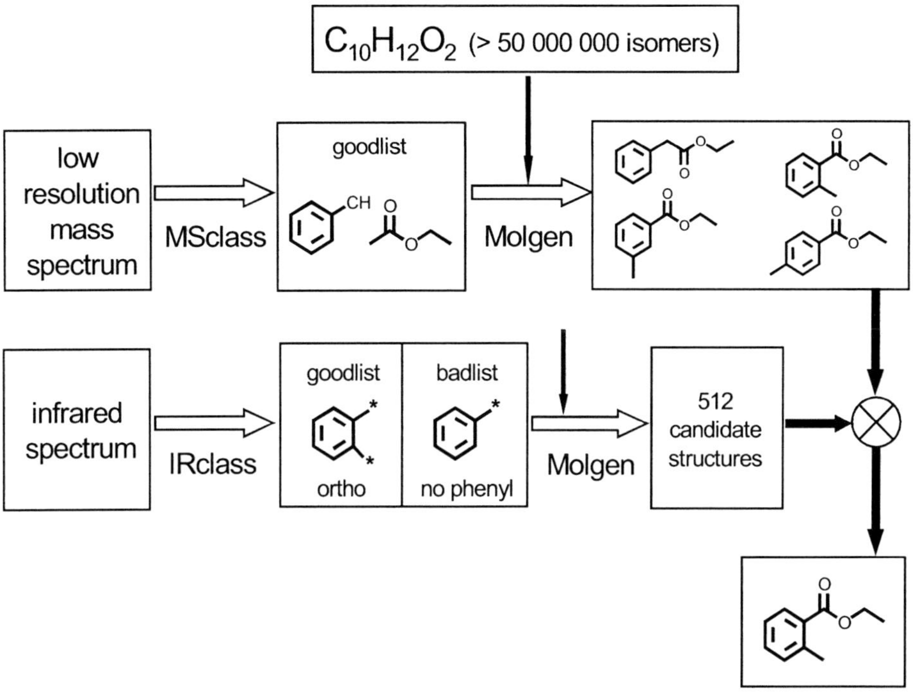

Fig. 4 Example for a systematic structure elucidation. For a query compound (considered as unknown) the molecular formula, the low resolution mass spectrum and the infrared spectrum are given. Application of multivariate spectral classifiers results in structural restrictions (goodlist, badlist) that finally give a single molecular structure candidate.

Automatic classification techniques based on multivariate statistics or neural networks are capable to some extent for recognizing the presence or absence of substructures from spectral data. The classification results can be used as structural restrictions in isomer generation. Examples demonstrate that for small molecules a systematic and sometimes almost complete structure elucidation is possible by this approach. Systematic means that each step of the structure elucidation process is clearly defined, and can be documented; furthermore, an exhaustive solution is obtained for defined structural restrictions. Systematic, however, does not mean that the result is always correct, or that an acceptable solution can be obtained in all cases. The crucial step is the chemical and spectroscopic part, namely the recognition of structural restrictions from spectral data.

LOCAL MODELS BASED ON SPECTRAL SIMILARITY HITLIST

Overview

Global models for the recognition of a structural property (presence/absence of a substructure, class of compounds) are usually derived from random samples of a database. A global model (for instance a spectral classifier) is considered to be applicable to "all unknowns" - at least as far as covered by the database. Fig. 5 shows two types of local models. One type is obtained from data belonging to a selected class of compounds; for example such a model has been built to predict the number of double bond equivalents in ketones [32].

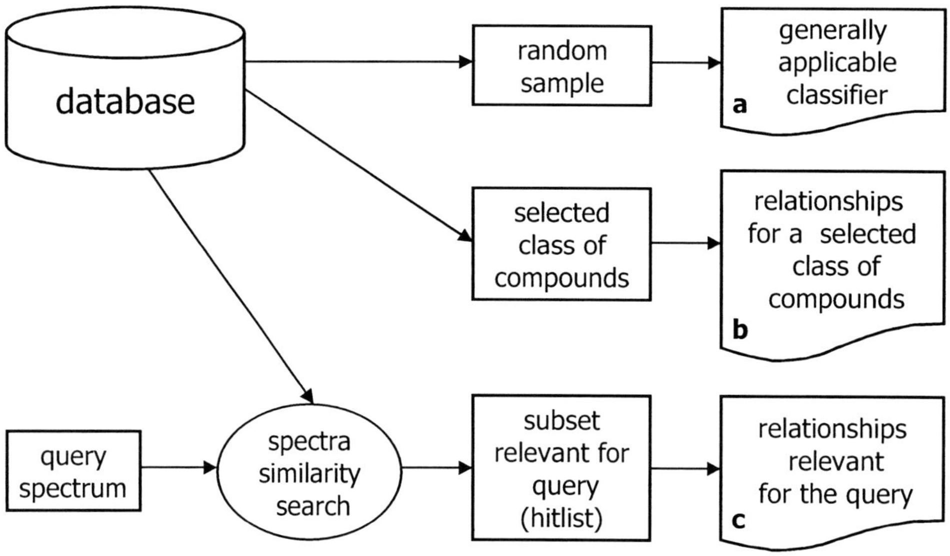

Fig. 5 Global models for spectra-structure relationships are derived from random samples of the database (a). Local models are derived from selected compounds belonging to a particular class (b) or from spectra similarity hitlists (c).

Another type of local models can be derived from spectra similarity hitlists. In this approach a query spectrum (spectrum from an unknown compound) is compared with the reference spectra of a database (library). In most library search methods a mathematically defined criterion for the similarity of two spectra is applied, and reference spectra are searched that are most similar to the query spectrum. The resulting subset from the library is called a hitlist; it is ordered by decreasing spectral similarity and typically contains five to 50 reference compounds. A number of powerful software products are available for this purpose and are routinely used. If the unknown is present in the library, the correct compound often shows a very high spectral similarity and appears as the first hit or among the first hits. However, if the unknown is not present in the library, spectra similarity hitlists may be less useful. An *interpretative power* is claimed by some spectral database systems, meaning that the hitlist structures are often similar to the structure of the query and therefore allow the extraction of structure information about the unknown. For a systematic evaluation of library search systems it is necessary to define similarity criteria for spectra as well as for chemical structures.

Similarity of Infrared Spectra and of Mass Spectra

Spectra are represented as vectors and a similarity measure for vectors is used as the spectral similarity. An IR spectrum has been characterized by a vector $\boldsymbol{x}$ with component x_j being the averaged absorbance in wave number interval j. A set of 801 wave number intervals of width 4 cm^{-1} between 500 and 3,700 cm^{-1} has been used, and the vector components scaled to the range 0 - 1. A low resolution mass spectrum has been characterized by a vector $\boldsymbol{x}$ with components x_j either being peak intensities, I_m, in % base peak intensity at selected masses m, or weighted peak intensities $I_m^{\,b}$ [25,35], with a typical value of 0.5 for exponent b. Alternatively, the vector components may be mass spectral features derived from the peak intensity data as described in Table 1.

A spectral similarity for IR spectra, that gives good hitlists [36], is defined as follows. Let $_{m}x_A$ be the arithmetic mean of the absorbances x_j in spectrum A, and $_{m}x_B$ the corresponding mean for spectrum B. Mean centered absorbance vectors, z, of spectrum A and B are given, respectively, by

$$\boldsymbol{z}_A = \boldsymbol{x}_A - \mathbf{1} \cdot {}_{m}x_A \tag{1}$$

$$\boldsymbol{z}_B = \boldsymbol{x}_B - \mathbf{1} \cdot {}_{m}x_B \tag{2}$$

with **1** being a vector of appropriate size with all elements set to 1. The used spectral similarity S_{IR} is based on the correlation coefficient, r_{IR}, calculated from vectors $\boldsymbol{z}_A$ and $\boldsymbol{z}_B$, and is scaled to the range 0 - 1,000.

$$r_{IR} = (\boldsymbol{z}_A^{\,T} \cdot \boldsymbol{z}_B) / (||\boldsymbol{z}_A|| \cdot ||\boldsymbol{z}_B||) \tag{3}$$

$$S_{IR} = 1000\,(r_{IR} + 1) / 2 \tag{4}$$

The spectral similarity used for mass spectra is also based on the correlation coefficient, r_{MS}, given by

$$r_{MS} = (\boldsymbol{x}_A^{\,T} \cdot \boldsymbol{x}_B) / (||\boldsymbol{x}_A|| \cdot ||\boldsymbol{x}_B||) \tag{5}$$

with $\boldsymbol{x}_A$ and $\boldsymbol{x}_B$ being the vectors representing the compared mass spectra A and B, respectively [37]. Because the vector components (peak intensities, weighted peak intensities, spectral features) used in this work are always positive, r_{MS} is in the range 0 to +1. For a better readability the used spectral similarity S_{MS} is scaled to the range 0 - 1,000.

$$S_{MS} = 1000\, r_{MS} \tag{6}$$

Similarity of Chemical Structures

Among the many approaches for representing chemical structures as vectors, the method using binary substructure descriptors plays an important role. Substructures are easily interpretable in terms of chemistry; they are capable to cover a great diversity of chemical structures, and substructure search is a standard operation in computer chemistry. Each chemical structure (two-dimensionally encoded) has been characterized by a vector $\boldsymbol{y}$ with components y_j being binary substructure descriptors. A set of 1,365 generally applicable substructures has been defined for this purpose [13]; the groups of substructures are listed in Table 2. The non-hydrogen elements contained in the substructures are C, N, O, S, P, F, Cl, Br, I, B, and Si; furthermore, pseudo elements A for hetero atoms and Q for non-hydrogen atoms have been used. The bond types are single, double, triple, aromatic, and not-defined (any bond type). A part of the substructures has been built systematically using the isomer generator Molgen [14]; others were defined on the basis of chemical and spectroscopic ideas.

Table 2 List of substructure groups for the representation of chemical structures by binary substructure descriptors

Group	Description	No. of substructures
1	Elements (single atoms)	46
2	Two-atom substructures	78
3	Single rings (not aromatic)	404
4	Condensed rings (not aromatic)	130
5	Aromatic rings	97
6	Other rings	39
7	Trees (chains and branches)	418
8	Functional groups	153
	Sum	1,365

Vector component y_j is 1 if substructure j is present in the molecule and 0 otherwise. The similarity of two chemical structures, represented by vectors $\boldsymbol{y_A}$ and $\boldsymbol{y_B}$, has been measured by the *Tanimoto index* [38], $t_{A,B}$, also called Jaccard similarity [39]

$$t_{A,B} = \Sigma\ \mathrm{AND}[y_{Aj}, y_{Bj}]\ /\ \Sigma\ \mathrm{OR}[y_{Aj}, y_{Bj}] \tag{7}$$

AND[.,.] is the result of the logical AND, and OR[.,.] the result of the logical OR of two binary variables. In contrary to the Euclidean distance, the Tanimoto index only considers substructures that are contained at least in one of the compared structures; it is in the range 0 - 1, with value 1 if all descriptors are pairwise equal. Fig. 6 shows the result of structure similarity searches in two databases using this approach. Software SubMat (see section 2) has been used for the generation of substructure descriptors.

Chemical Structure Information in Hitlists

For IR and MS has been shown that with appropriate spectra similarity measures, a high spectral similarity, in general also means a high structural similarity [25,40]. Consequently, the first hits in spectra similarity hitlists have highest structural similarities with the query. In particular cases, however, a high spectral similarity may also occur for very different chemical structures, and vice versa. A careful inspection of the hitlists is therefore advisable, eventually supported by chemometric methods as discussed in section 4.5.

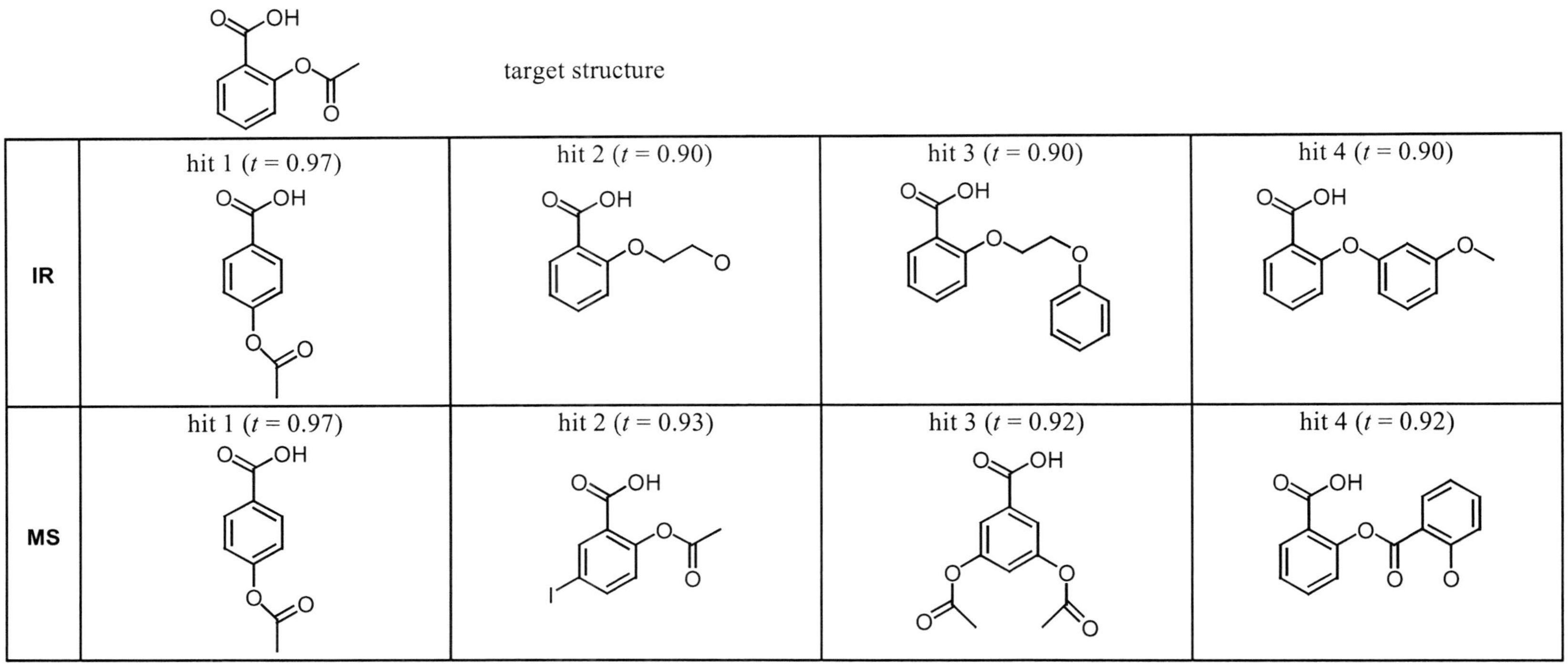

Fig. 6 Example for structure similarity searches. Target structure was salicylic acid acetate (aspirin, CAS reg. no. 50-78-2). Search has been performed in two databases, the IR database containing 13,484 compounds, and the MS database containing 106,955 compounds. The four database structures most similar to the target structure are shown; *t*, Tanimoto index between target structure and hit. The average structural similarity between target and hits 1 - 4 is 0.92 for the IR database, and 0.94 for the MS database

A quality measure for hitlists can been derived from the Tanimoto indices $t_{q,h}$ calculated for a query structure, q, and hitlist structures h ($h = 1$ to H). The averaged Tanimoto indices, $t_{q,1\text{-}h}$ for the first h hits (for instance $t_{q,1\text{-}5}$) are qualifiers for single hitlists. For an evaluation of spectra similarity search methods, random samples with n query spectra are used, and a search method can be characterized by the grand means, T_h,

$$T_h = (1/n)\ \Sigma\ t_{q,1\text{-}h} \tag{8}$$

With random samples containing 200 to 10,000 queries it was shown that T_h decreases with increasing h (Fig. 7). Although it is well known by practicing spectroscopists that compounds with similar chemical structures often have similar spectra (and hopefully vice versa), it is desirable to check and to quantify this hypothesis. The values of T_1 and T_5 are appropriate numbers for comparing search methods, because spectroscopists rarely consider more than the first five hits. In Table 3, T_5 for different similarity criteria of mass spectra are compared; best results have been obtained by using spectral features, next best by using the square root of peak intensities up to m/z 200 [25].

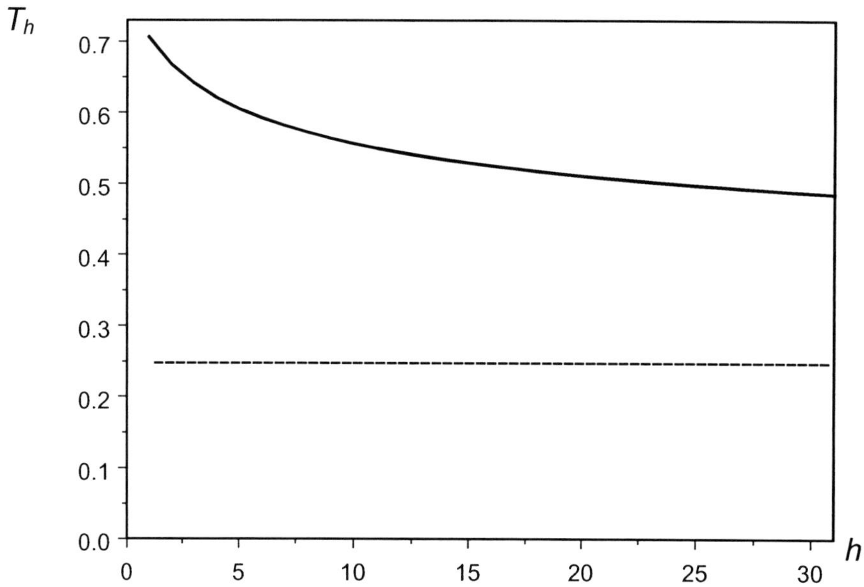

Fig. 7 Similarity T_h (grand mean of Tanimoto indices, Eq. 8) between query structures and the corresponding h hitlist structures. A set of 10,000 randomly selected mass spectra has been used as queries. Spectral similarity S_{MS} has been calculated from 862 mass spectral features (Table 1). The query compound itself has always been found as the first hit, and has been excluded from the hitlist in order to simulate a situation with the unknown not contained in the library. The averaged structural similarity between query and first hit is 0.71, while for random structure pairs it is only 0.25 (dotted line).

The same strategy has been applied to IR spectra for a selection of the best spectra similarity measure among four proposals [40].

Table 3 Comparison of similarity measures for mass spectra. T_5 is the averaged Tanimoto index (structural similarity) between query structure and the corresponding first five hitlist structures, calculated from 10,000 hitlists [25].

Spectral similarity calculated from	Mass range	T_5
Peak intensities I_m	1 - 1,000	0.48
Weighted peak intensities $(I_m)^{0.5}$	1 - 1,000	0.56
Peak intensities I_m	1 - 200	0.52
Weighted peak intensities $(I_m)^{0.5}$	1 - 200	0.58
862 spectral features	1 - 1,000	0.61

Evaluation of Hitlists

The most common utilization of a spectra similarity hitlist is the identification of an unknown compound. This task is easily applicable in routine work and is often successful. Requirements are a high spectral similarity between query spectrum and the first hit, together with much lower similarities for the other hits - and a reasonable chance for the unknown to be present in the database. If the spectral similarities and the contents of a hitlist do not evidently indicate an identification, the hitlist may still contain valuable chemical structure information about the unknown. Up to now only a few methods from chemometrics or computer chemistry have been suggested for supporting the evaluation of hitlists.

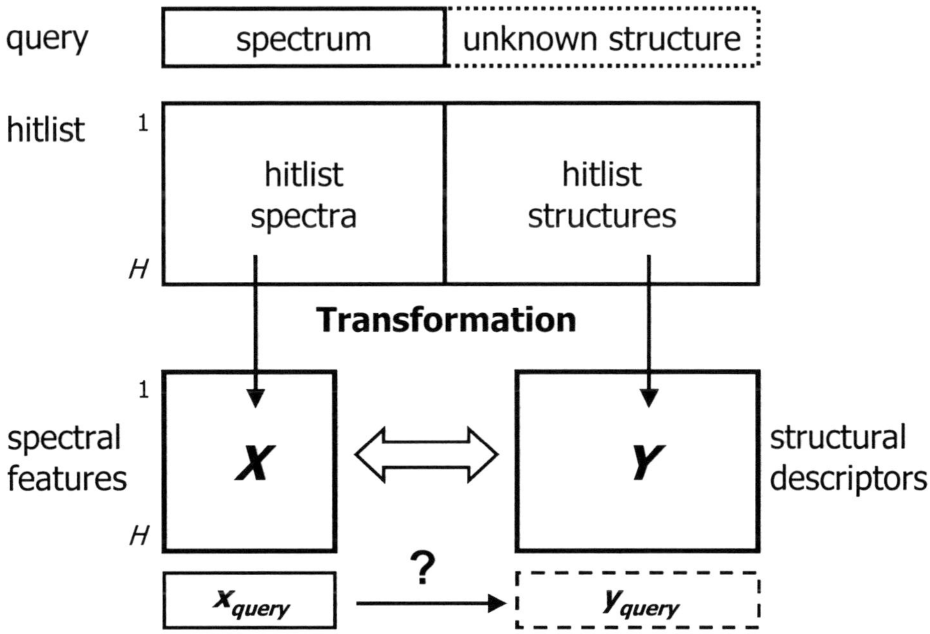

Fig. 8 Evaluation of spectra similarity hitlist data. Spectra and chemical structures of the reference compounds are transformed into vectors, giving the matrices X and Y. Relationships between X and Y can be used to predict some elements of the structure vector y_{query}

The situation for multivariate data analysis is summarized in Fig. 8. The hitlist consists of H reference compounds, for each are available the spectrum (represented by a vector $\boldsymbol{x}$), and the chemical structure (represented by a binary vector $\boldsymbol{y}$, if encoded by binary substructure descriptors). The number of objects (hits) is typically between 20 and 50; the lengths of the spectral and structural vectors can be several hundred elements. For the query only the spectrum vector, $\boldsymbol{x}_{query}$, is known, but not the structure vector, $\boldsymbol{y}_{query}$. Prediction of the structure vector would be desirable but seems to be hopeless because of the complicated and mostly unknown relationships between chemical structures and spectral data. More reasonable approaches for gathering structure information are as following.

(a) Exploratory data analysis of spectral data or structural data provides an insight into relationships within the spectra and within the structures. A cluster analysis of the spectral data (matrix $\boldsymbol{X}$) may indicate different groups of spectra and thereby facilitate the interpretation of the query spectrum; in this data analysis the query can be used as one of the objects. In a similar way a cluster analysis of the chemical structures (matrix $\boldsymbol{Y}$, of course without the query) may show for instance that two or more substance classes are contained in the hitlist because they have similar spectra. Knowing this, it may be easier to assign the unknown to one of these classes. Appropriate tools for such examinations are for instance score and loading plots from principal component analysis (PCA), Kohonen mapping and cluster analysis by dendrograms.

An example illustrates the application of PCA mapping for the evaluation of structural hitlist data. Salicylic acid acetate (aspirin, Fig. 6) has been used as query for a spectral similarity search in the IR database. The structures of the first four hits are shown in Fig. 9.

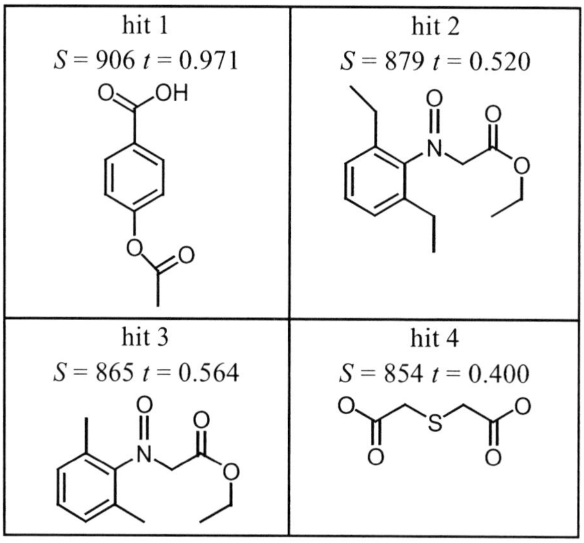

Fig. 9 Hitlist of an IR spectra similarity search. Query was salicylic acid acetate (see Fig. 6). S, spectral similarity between query and hit as given in Eq. 4. t, Tanimoto index between query structure and hit; average of the four hits is 0.61.

Note that the first hit is identical with the first hit of a structure similarity search (shown in Fig. 6). However, even this short hitlist contains very diverse structures and up to hit 30 does not contain another of the best structural hits. The chemical structures of the first 30 hits have been represented by 1,365 binary substructure descriptors, and for PCA the 50 descriptor with highest variances have been selected. The PCA score plot is shown in Fig. 10.

A separation of substance classes can be recognized along PC1 (first principal component) with aromatic compounds (containing a benzene ring) at the left hand side, alicyclic compounds in the middle part, and aliphatic carboxylic acids at the right hand side. Although a clustering of compounds with similar structures can be recognized, a direct hint for the substance class of the query cannot be obtained because no spectral data have been used in this type of hitlist evaluation.

(b) A powerful method for the evaluation of hitlist data is PLS mapping because spectral and structural data can be used together [24,41]. PLS algorithms extract latent variables which maximize the covariance between two multivariate data spaces, the *x*- and the *y*-space. Two latent variables from the *x*-space can be used to span a projection plane for the spectra, and the corresponding two latent variables from *y*-space span a projection plane for the structures. PLS directs compounds to lie in similar regions of both mappings such connecting structural and spectral similarity. In other words, the position of the spectral projection plane is influenced by the structures, and vice versa. For this - rarely performed application of PLS - the PLS components are preferably calculated as eigenvectors of the matrix products $\boldsymbol{X}^T\boldsymbol{Y}\boldsymbol{Y}^T\boldsymbol{X}$ and $\boldsymbol{Y}^T\boldsymbol{X}\boldsymbol{X}^T\boldsymbol{Y}$ [42] as summarized in Fig. 11.

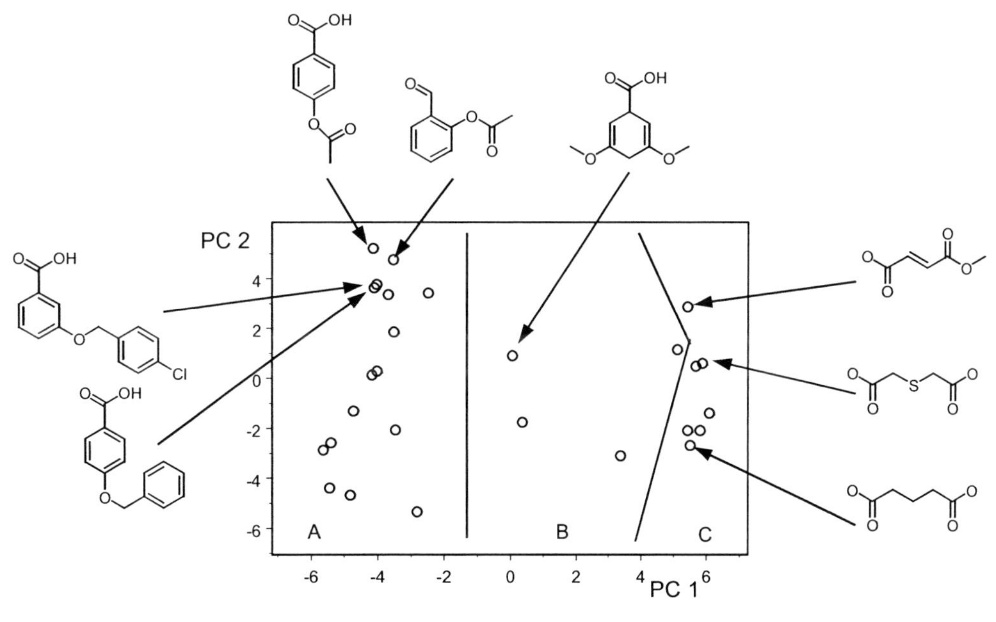

Fig. 10 PCA score plot of chemical structure data of the first 30 hits from a IR spectra similarity search (the first four hits are shown in Fig. 9). Query was salicylic acid acetate (see Fig. 6). Each chemical structure has been represented by 50 binary substructure descriptors (selected by maximum variances from 1,365 descriptors). PC1 and PC2 preserve 44.7% and 16.8% of the total variance, respectively. Examples for chemical structures are given; similar structures are often close. A, compounds containing a benzene ring; B, compounds with alicyclic ring; C, aliphatic compounds (six dicarboxylic acids, one oxo-carboxylic acid, and two ester).

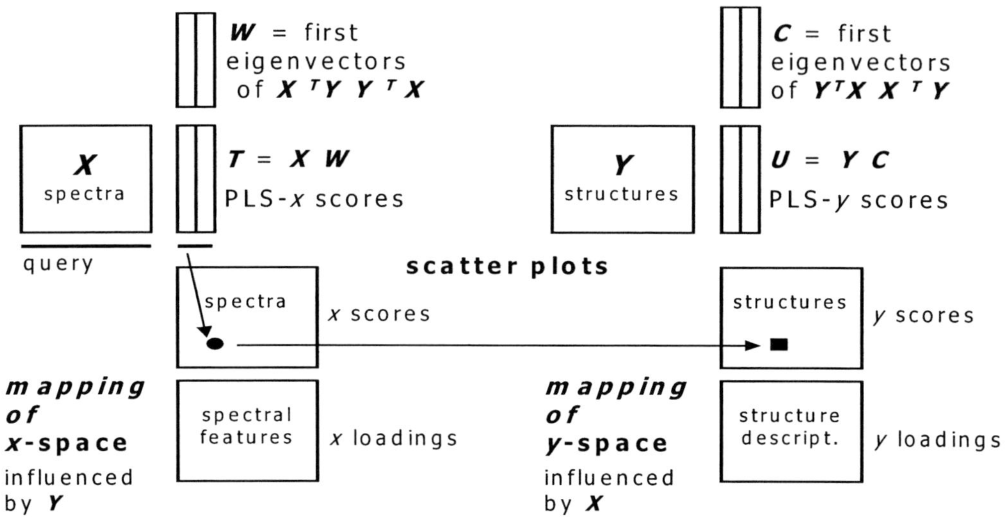

Fig. 11 PLS mapping of spectral and structural data from a set of compounds, for instance hitlist compounds. X, spectral data (peak intensities, absorbances, spectral features); Y, molecular descriptors. The position of the query in the y-score plot is calculated from the x-scores using the linear inner relationship between *x*-scores and *y*-scores.

The resulting directions of the components in *x*-space and in *y*-space are orthogonal. No deflation procedure is necessary; PLS-*x* scores can be calculated by ***X.W***, and PLS-*y* scores by ***Y.C***. The score plots for ***X*** and ***Y*** can be used for an exploratory data analysis of spectra and chemical structures, respectively. Because the two plots are related, the result may be spectroscopically more relevant than results obtained from separate PCA score plots. Furthermore, the query can be projected into the PLS-*x* score plot. Applying the (linear) inner relationship between *x*-scores and *y*-scores a position of the query in the PLS-*y* score plot can be calculated. Chemical structures in the neighborhood of the query may be the most informative ones. Furthermore, the loading plots provide useful information about relationships between spectral features and substructures.

(c) Other chemometric methods for the extraction of chemical structure information from hitlist data are for instance local classification models (derived from the hits) that recognize presence or absence of substructures in the query. An approach based on the application of graph theory is a search for substructures that are significantly more frequent in the hitlist structures than in random samples from database; such substructures can be found by searching maximum common substructures among the hitlist structures [36,43]. Furthermore, the topological hierarchy of the hitlist structures - presented as a graph - may be helpful in the evaluation of a hitlist [12].

(d) An example from a mass spectra similarity search demonstrates the combined use of hitlist evaluation methods. As before, the query was salicylic acid acetate (aspirin, Fig. 6), and the first 30 hits have been used for further processing. The structures of the first four hits are shown in Fig. 12.

Note that none of the reference compounds exhibiting highest structural similarity with the query (Fig. 6) are among the first four hits. However, the first four hits comprise salicylic acid and three alkyl esters of it, and thereby provide relevant structural information about the query. A PCA scores plot of the spectral features is shown in Fig. 13. From the 1,086 features used, a subset of 288 has been selected that have maximum variances and cover 90% of the total variance. PC1 and PC2 preserve only 19.9% and 12.7%, respectively, of the total variance of the used features; therefore exploitation of the plot has to be cautious. The hits in the neighborhood of the query have very similar structures to the query; from the six selected neighboring compounds five are salicylic acid alkyl esters. The averaged structural similarity between these hits and the query is 0.70. While the salicylic skeleton can be easily recognized from these hits, the ester function may be misleading, and the acetate substructure is not indicated at all.

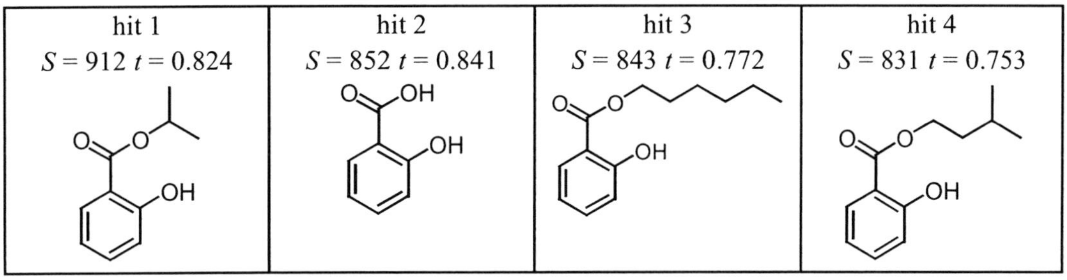

Fig. 12 Hitlist of a mass spectra similarity search. Query was salicylic acid acetate (see Fig. 6). *S*, spectral similarity between query and hit (Eq. 6), using 1,086 spectral features (Table 1). *t*, Tanimoto index between query structure and hit; average of the four hits is 0.80.

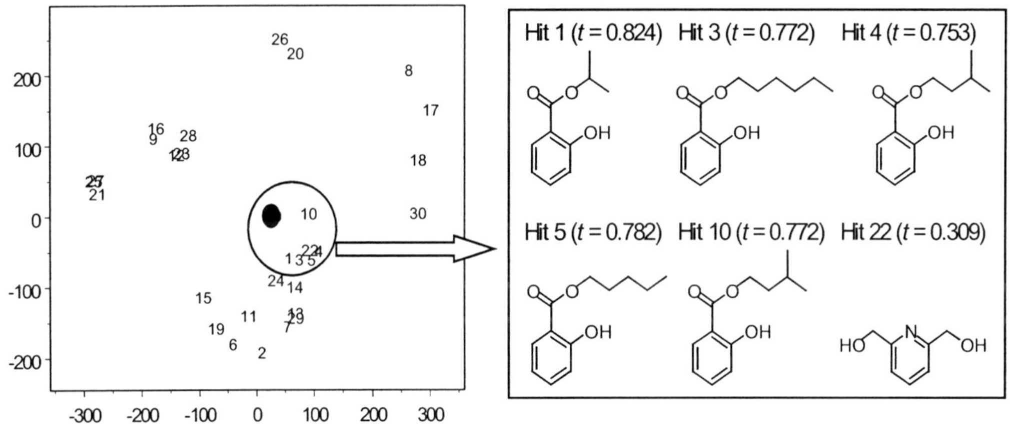

Fig. 13 PCA score plot of spectral features for the first 30 hits from a mass spectra similarity search (the first four hits are shown in Fig. 12). Query was salicylic acid acetate (see Fig. 6). Each reference spectrum has been represented by 288 spectral features (Table 1) covering 90% of the variance of the 1,086 features used in the spectra similarity search. PC1 and PC2 preserve 19.9% and 12.7%, respectively, of the total variance of the 288 features. Numbers in the plot are hit numbers (1 - 30). The filled circle is the query spectrum. The ellipse selects six hits that are close to the query in this plot; they give evidence about a salicylic substructure in the query. *t*, Tanimoto index between query structure and hit; average for these six hits is 0.70.

For a data evaluation by PLS an additional ***Y***-matrix containing 105 binary descriptors has been used covering 90% of the variance of all 1,365 descriptors. The variances preserved by the first two PLS components are for PLS-x-1 and PLS-x-2, 15.8% and 12.2%, respectively, of the total variance of the 288 features, and for PLS-y-1 and PLS-y-2, 33.4% and 9.6%, respectively, of the total variance of the 105 descriptors. Fig. 14 shows the score plots for the *x*- and *y*-space obtained by the method described in Fig. 11. Both plots show informative clustering of compounds with similar spectra and similar structures. In the *x*-score plot four compounds in the neighborhood of the query have been manually selected; their averaged structural similarity with the query is 0.85. The position of the query in the *y*-score plot has been calculated from the *x*-scores using the inner relationship. The selected five compounds in the neighborhood of the query provide additional structural information because two of them contain the acetyl substructure which was not found before among the first hits. The average structural similarity of these five compounds with the query is 0.87. Thus PLS was capable in this example to select hits with high structural similarity to the query. In summary, the PLS score plot for the descriptors yielded best results with an averaged Tanimoto index of 0.87. The best possible value is 0.94 for the four reference structures which are most similar to the query (Fig. 6) and the MS similarity search gave a value of 0.80 (Fig. 12). PCA of the mass spectral features gave a value of only 0.70 (Fig. 13), and the PLS score plot for the spectral features 0.85. For comparison, the best possible value for the IR database is 0.82 (Fig. 6) and for the IR spectra similarity search only 0.61 (Fig. 9).

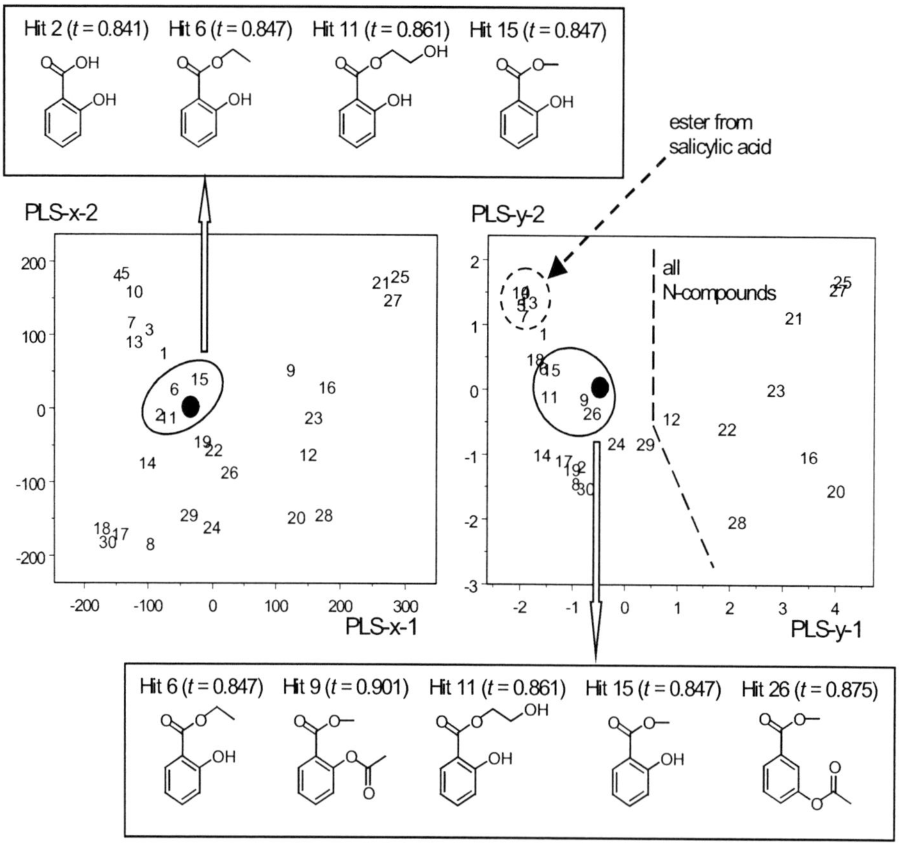

Fig. 14 PLS score plots of spectral features (***X***) and structural descriptors (***Y***) for the first 30 hits from a mass spectra similarity search (the first four hits are shown in Fig. 12). Query was salicylic acid acetate (see Fig. 6). Each reference spectrum has been represented by 288 spectral features (Table 1) covering 90% of the variance of the 1,086 features used in the spectra similarity search; each chemical structure has been represented by 105 binary descriptors (Table 2) covering 90% of the variance of the 1,365 descriptors used. Numbers in the plot are hit numbers (1 - 30). The filled circle is the query spectrum. The ellipses select hits that are close to the query; they give evidence about an aceto-salicylic structure in the query. *t*, Tanimoto index between query structure and hit.

SUMMARY AND CONCLUSIONS

Chemical structure information cannot be recognized from spectral data solely by applying fundamental relationships between spectral and structural data, because these relationships are very complex and not sufficiently known. Therefore, chemometric methods - based on learning from data - play an important role. Global models try to find relationships between spectral data and structure properties; such models can be applied without special restrictions. Examples for global models are spectral classifiers that are - to some extent - capable to recognize the presence of substructures. Results from such classifiers can be used as restrictions in automatic isomer generation leading to a complete set of possible structures for an unknown. However, this systematic structure elucidation is limited because of severe problems in the development of reliable classifiers, and the often occurring combinatorial explosion during isomer generation.

Another heuristic approach for spectra interpretation is based on spectral library search. This method is routinely used in IR and MS for the identification of compounds. If the query spectrum from an unknown compound is not present in the database, some structure information can still be extracted from the hits. A method has been developed for a quantitative evaluation of the structural similarities between query and hits. Structural similarity has been measured by the Tanimoto index, calculated from 1,365 binary substructure descriptors. An averaged Tanimoto index for the first five hits and for up to 10,000 queries has been used to select and optimize spectra similarity measures. The hits constitute a sample from the database which is often highly relevant for the query. Spectral and structural data of the hits can be used to develop local models. For instance, PCA maps and PLS maps give insights to relationships between spectral data and the substance classes of the hits, and help the spectroscopist in the evaluation of hitlists. Visualization of hitlist data by these chemometric methods is a promising tool for a better exploitation of spectra similarity search result in cases the unknown is not present in the database

ACKNOWLEDGMENTS

Thanks to W. Demuth, M. Karlovits, P.N. Penchev, H. Scsibrany and W. Werther for software and method developments; as well as A. Kerber and R. Laue for the isomer generator software Molgen, R. Neudert and E. Pretsch for the SpecInfo IR database, and S. Stein for the NIST MS database. The work was supported by the Austrian Science Fund, project P14792-CHE.

REFERENCES

[1] W. Robien, in: G. Gauglitz, T. Vo-Dinh (Eds.), *Handbook of spectroscopy*, Vol. 2, Wiley-VCH, Weinheim, 2003, p. 469-487.

[2] N. Davies, in: G. Gauglitz, T. Vo-Dinh (Eds.), *Handbook of spectroscopy*, Vol. 2, Wiley-VCH, Weinheim, 2003, p. 488-504.

[3] S. Thiele, R. Salzer, in: G. Gauglitz, T. Vo-Dinh (Eds.), *Handbook of spectroscopy*, Vol. 2, Wiley-VCH, Weinheim, 2003, p. 441-468.

[4] P. C. Jurs, T. L. Isenhour, *Chemical applications of pattern recognition*. Wiley, New York, 1975.
[5] K. Varmuza, *Pattern recognition in chemistry*. Springer, Berlin, 1980.
[6] J. Gasteiger, T. Engel, *Chemoinformatics - a textbook*. Wiley-VCH, Weinheim, 2003.
[7] Chemical Concepts, *Spectroscopic database system SpecInfo*. Chemical Concepts, http://www.chemicalconcepts.com/, Weinheim (Germany), 1996.
[8] NIST, *Mass spectral database 98*. National Institute of Standards and Technology, http://www.nist.gov/srd/nist1a.htm, Gaithersburg, MD (USA), 1998.
[9] Dalby, J. G. Nourse, W. D. Hounshell, A. K. I. Gushurst, D. L. Grier, B. A. Leland, J. Laufer, *J. Chem. Inf. Comput. Sci.* 32 (1992) 244-255.
[10] MDL Information Systems Inc., *CT file format*. MDL Information Systems Inc., http://www.mdli.com/downloads/literature/ctfile.pdf, San Leandro, CA (USA), 2002.
[11] H. Scsibrany, K. Varmuza, *Software SubMat*, generation of binary substructure descriptors. Laboratory for Chemometrics, Institute of Chemical Engineering, Vienna University of Technology, http://www.lcm.tuwien.ac.at, Vienna (Austria), 2004.
[12] K. Varmuza, H. Scsibrany, *J. Chem. Inf. Comput. Sci.* 40 (2000) 308-313.
[13] H. Scsibrany, M. Karlovits, W. Demuth, F. Müller, K. Varmuza, *Chemom. Intell. Lab. Syst.* 67 (2003) 95-108.
[14] Kerber, R. Laue, *Software Molgen*, isomer generator. Institute for Mathematics II, University of Bayreuth, http://www.mathe2.uni-bayreuth.de/molgen4/, Bayreuth (Germany), 2000.
[15] The Mathworks Inc., *Software Matlab*, Natick, MA (USA), 2000.
[16] Minitab Inc., *Software SCAN*, State College, PA (USA), 1995.
[17] Camo Process AS, *Software Unscrambler*. http: //www.camo.no, Oslo (Norway), 2002.
[18] N. A. B. Gray, *Computer-assisted structure elucidation*. Wiley, New York, 1986.
[19] R. E. Carhart, D. H. Smith, N. A. B. Gray, J. G. Nourse, C. Djerassi, *J. Org. Chem.* 46 (1981) 1708-1718.
[20] K. Funatsu, N. Miyabayashi, S. I. Sasaki, *J. Chem. Inf. Comput. Sci.* 28 (1988) 18-28.
[21] L. A. Gribov, M. E. Elyashberg, V. N. Koldashov, I. V. Pletnjov, *Anal. Chim. Acta* 148 (1983) 159-170.
[22] B. D. Christie, M. E. Munk, *J. Chem. Inf. Comput. Sci.* 28 (1988) 87-93.
[23] K. Varmuza, P. Penchev, F. Stancl, W. Werther, *J. Mol. Struct.* 408/409 (1997) 91-96.
[24] W. Werther, W. Demuth, F. R. Krueger, J. Kissel, E. R. Schmid, K. Varmuza, *J. Chemom-etrics* 16 (2002) 99-110.
[25] W. Demuth, M. Karlovits, K. Varmuza, *Anal. Chim. Acta* (2004), in print.
[26] L. R. Crawford, J. D. Morrison, *Anal. Chem.* 40 (1968) 1469-1474.
[27] P. C. Jurs, B. R. Kowalski, T. L. Isenhour, C. N. Reilley, *Anal. Chem.* 42 (1970) 1387-1394.
[28] K. Varmuza, W. Werther, *J. Chem. Inf. Comput. Sci.* 36 (1996) 323-333.
[29] H. Yoshida, R. Leardi, K. Funatsu, K. Varmuza, *Anal. Chim. Acta* 446 (2001) 483-492.
[30] K. Varmuza, P. He, K. T. Fang, *J. Data Science* 1 (2003) 391-404.
[31] R. O. Duda, P. E. Hart, *Pattern classification and scene analysis*. Wiley, New York, 1973.
[32] K. Varmuza, *Anal. Sci.* 17 (2001) i467-i470.

[33] K. Varmuza, W. Werther, in: E. J. Karjalainen, A. E. Hesso, J. E. Jalonen, U. P. Karjalainen (Eds.), *Advances in mass spectrometry*, Vol. 14, Elsevier, Amsterdam, 1998, p. 611-626.

[34] P. N. Penchev, G. N. Andreev, K. Varmuza, *Anal. Chim. Acta* 388 (1999) 145-159.

[35] S. E. Stein, D. R. Scott, *J. Am. Soc. Mass Spectrom.* 5 (1994) 859-866.

[36] K. Varmuza, P. N. Penchev, H. Scsibrany, *J. Chem. Inf. Comput. Sci.* 38 (1998) 420-427.

[37] K. Varmuza, in: J. C. Lindon, G. E. Tranter, J. L. Holmes (Eds.), *Encyclopedia of spectros-copy and spectrometry*, Academic Press, London, 2000, p. 232-243.

[38] P. Willett, *Similarity and clustering in chemical information systems*. Research Studies Press, Letchworth, 1987.

[39] B. G. M. Vandeginste, D. L. Massart, L. C. M. Buydens, S. De Jong, J. Smeyers-Verbeke, *Handbook of chemometrics and qualimetrics*: Part B. Elsevier, Amsterdam, 1998.

[40] K. Varmuza, M. Karlovits, W. Demuth, *Anal. Chim. Acta* 490 (2003) 313-324.

[41] W. Werther, K. Varmuza, Fresenius *J. Anal. Chem.* 344 (1992) 223-226.

[42] Höskuldsson, *J. Chemometrics* 2 (1988) 211-228.

[43] K. Varmuza, P. N. Penchev, H. Scsibrany, *Vib. Spectrosc.* 19 (1999) 407-412.

In: Progress in Chemometrics Research
Editor: Alexey L. Pomerantsev, pp. 89-102
ISBN 1-59454-257-0

Chapter 7

INVERSE PROBLEM IN POTENTIODYNAMIC ELECTROCHEMICAL IMPEDANCE

A. S. Bondarenko and G. A. Ragoisha
Belarusian State University, Minsk, Belarus, bondarenkoas@bsu.by

ABSTRACT

A computer program for solution of inverse problem of potentiodynamic electrochemical impedance spectroscopy (PDEIS) has been developed and integrated with virtual instruments of PDEIS spectrometer. The spectrometer gives, in a single potential scan, a dependence of real and imaginary impedance of electrochemical system on frequency and electrode potential. The built-in analyser processes the multivariate data and decomposes the total response into responses of constituents related with different elements of equivalent electric circuits. As a result of the decomposition, parameters of different processes are obtained and presented as functions of the potential, which enables separate monitoring of simultaneously proceeding interdependent processes on electrochemical interface.

Keywords: virtual instruments, electrochemical impedance, analysis of multivariate electrochemical response, complex nonlinear least squares.

INTRODUCTION

Potentiodynamic electrochemical impedance spectroscopy (PDEIS) [1-5] is a new digital technique for characterisation and analysis of electrochemical response of interfacial processes and structures. PDEIS uses virtual instruments both for the implementation of experiment and analysis of acquired data. A virtual PDEIS spectrometer probes the electrochemical system, by means of a common potentiostat, and gives in a potential scan a combination of a cyclic voltammogram, which characterises dc response, and 3D surfaces in the spaces of Nyquist and Bode plots, extended from 2D to 3D, due to a variable potential. The latter surfaces characterise the frequency response variation with the potential. Hence, PDEIS acquires, just in a single potential scan, the dc response, as a function of the potential, and the variation of ac response with frequency and potential. PDEIS spectra, like other kind of spectra, can be used either directly, as peculiar 3D signatures of electrochemical systems, or more profoundly, as multivariate data for further analysis guided by the theory of electrochemical impedance.

In this work we consider the analysis of PDEIS spectra based on the theory of equivalent electric circuits (EEC) of electrochemical systems. The concept of equivalent circuits [6,7] was originally developed for the analysis of 2D electrochemical impedance spectra, obtained in stationary states, usually at constant electrode potential. We have extended the two-dimensional analysis to the multivariate case for the analysis of the ac part of data obtained at the output of the PDEIS spectrometer.

Section 2 presents general information about PDEIS spectroscopy and shows examples of PDEIS spectra of different kind of electrochemical systems. Section 3 introduces the program, which analyses the 3D frequency response data by decomposing the constituent responses related to EEC elements. In Section 4 new opportunities provided by multivariate analysis of potentiodynamic impedance data is illustrated by the investigation of constituent ac responses related to simultaneously proceeding electrochemical processes. PDEIS has provided for the first time the separate monitoring of simultaneous cation and anion electrosorption in underpotential deposition of monoatomic layers, and has facilitated the analysis of multivariate ac response in various complex phenomena on electrochemical interface.

Potentiodynamic Electrochemical Impedance Spectroscopy

Fig. 1 shows a function chart of PDEIS spectrometer operation. During a potential scan the program of the spectrometer probes the electrochemical system with complex signals and analyses responses in real time. A typical view of the spectrometer screen is shown in Fig. 2. Three 2D graphs on the right side of the screen are Nyquist and Bode plots. Variable content of these plots shows frequency response as the dependence of imaginary impedance on real impedance (Nyquist plot) and frequency dependences of impedance amplitude and phase angle (Bode plots) at a current value of potential. The graph on the left side (Fig. 2) accumulates data on frequency response in the potential scan in the form of 3D surface. The surface is plotted as a variable potential extension of any of the three 2D plots. Fig. 2 shows the PDEIS spectrum as the extended Nyquist plot, together with a cyclic voltammogram (the voltammogram is plotted on the transparent front facet of the parallelepiped that embraces the spectrum).

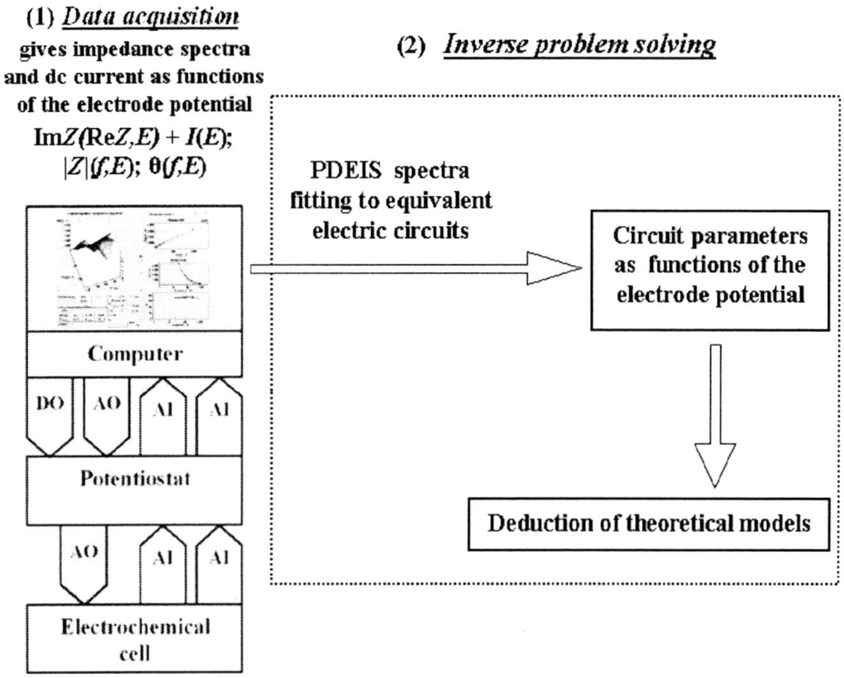

Fig. 1 Function chart of PDEIS spectrometer operation (DO – digital output, AO – analog output, AI – analog input).

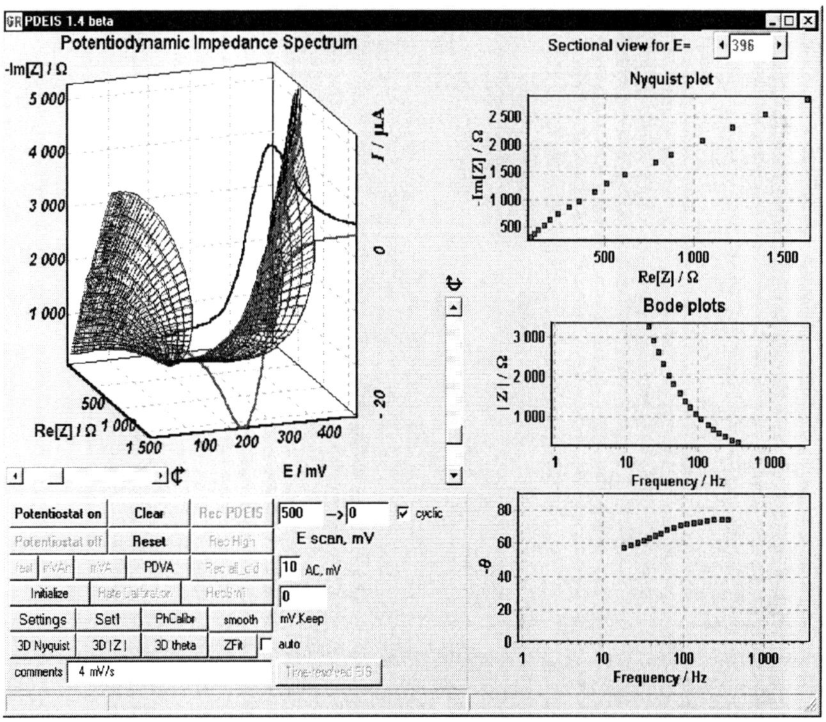

Fig. 2. Typical view of PDEIS spectrometer screen. The spectrum shows the potentiodynamic response of glassy carbon electrode in reversible redox transformations of $K_3[Fe(CN)_6]$.

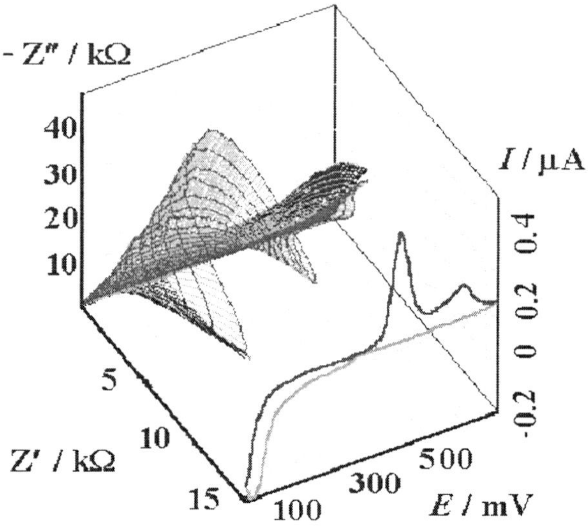

A

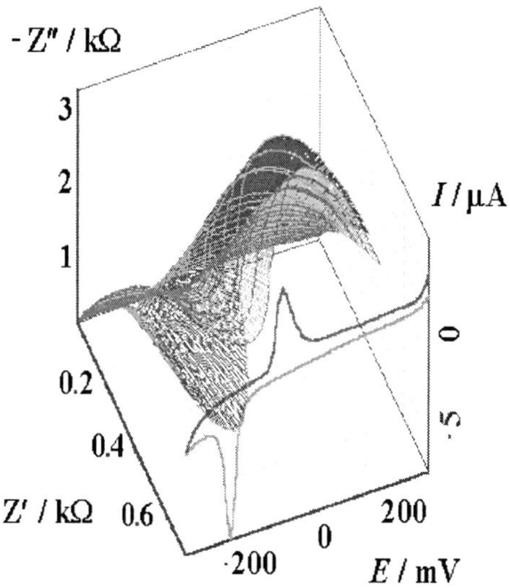

B

Fig. 3. PDEIS spectra of (a) Te monolayer and nanoparticles on Au, (b) Pb monolayer on Te.

Several examples of PDEIS spectra for different systems are shown in Fig. 3. The axes of coordinates in Fig. 3 are real impedance Z', imaginary impedance Z'', and electrode potential *E*. Actually both the real and imaginary parts of impedance are real physical quantities, complex notation is used in impedance spectroscopy just as a convenient form for depicting the two-dimensional nature of ac impedance. The real, or active, impedance characterises the in-phase component, while the imaginary, or reactive, impedance characterises the out-of-phase component in the ac response. The dependence of Z'' on Z' shows the frequency dependence of the impedance implicitly. The implicit form of presenting frequency dependence of impedance is convenient for solving inverse problem, as the various fragments of electrochemical circuits show characteristic signatures in Z''(Z') plots, e.g. a pair of resistance and capacitance, connected in parallel, gives a semicircle, and impedance of diffusion Z_w of reagents in electrochemical reactions often gives a straight diagonal line in Z''(Z') coordinates. In systems with simple spectra this gives the possibility for graphical determination of several important parameters.

PDEIS spectrum is characteristic to each system and can be used as a 3D signature for distinguishing from other systems, but the most exciting opportunities arise from analysis of PDEIS spectra as multivariate data. Frequency dependence of ac response enables the decomposition of the total response, which is presented by the spectrum, into constituents related to different processes (adsorption, charge transfer in electrochemical reaction, diffusion, etc.), and interfacial objects (double electric layer, space charge layer, etc.). The dependences of parameters of the constituents on the potential represent separately the variation of the responses of different actors on the interface and, thus, can disclose the details of electrochemical system behaviour, which are usually overlooked by other electrochemical techniques. In order to get this information, the data from PDEIS spectrum has to be analysed in terms of equivalent electric circuits. At each potential the analyser searches for the equivalent circuit that would have the same dependences of real and imaginary impedance on the frequency as the spectrum shows. In common impedance spectroscopy, the procedure of two-dimensional minimisation in finding equivalent circuit has been implemented by several programs [8,9], which used the technique of complex nonlinear least squares on a complex plane.

Unlike spectrum analysis in common EIS, which deals with single spectra and therefore can use manual fitting implemented e.g. in LEVM program [10], PDEIS analyser has to operate on thousands of 2D slices of a 3D spectrum. Though the fitting procedure in the analysis of 3D spectra also uses the minimisation method based on complex nonlinear least squares, it has to run in the automatic mode, and this imposes a restriction on manual interference of the operator. In order to comply with the requirements, the fitting routine had to overcome common traps of the minimisation problem – sensitivity of the solution to initial approximations and trapping by local minima in the search of a global minimum. Also several specific conditions had to be fulfilled by the spectrum analyser to work on 3D data.

SPECTRUM ANALYSER

Electric equivalent circuits of electrochemical systems contain resistors, capacitors and inductance coils, like the circuits used in electrical engineering. In addition to these common elements electrochemists use several specific elements in order to represent specific aspects of electrochemical system response with minimal number of parameters, e.g. Warburg element W represents impedance of the diffusion of electroactive particles. In the case of semi-infinite diffusion to flat electrode surface the impedance of diffusion Z_w has a very simple representation in complex impedance notation:

$$Z_w = A/(j\omega)^{0.5} \quad (1)$$

where j – imaginary unit, and ω – circular frequency, A is Warburg constant, which is the function of the electrode potential. Strong dependence of electrochemical parameters on the potential is one of the reasons why very low amplitude of ac probing (≤10 mV) is used in electrochemistry. This makes difficulties for probing procedures but, as a trade-off, the frequency response can give much more useful information, when it is analysed as a function of the potential. Warburg element contributes in equal parts to real and imaginary impedance, while impedance of charge transfer in electrochemical reaction is equivalent to a common resistor (the latter is also dependent on the potential) and the impedance of the electric double layer on the interface is usually equivalent to a capacitor with the value of capacitance C, which contributes only to the imaginary component of impedance:

$$Z_c=C^{-1}(j\omega)^{-1} \tag{2}$$

In many cases the behaviour of the double layer deviates from a simple capacitor model. Then a constant phase element (CPE) with fractional exponent n is used to represent it in the equivalent circuit. The impedance of CPE Z_{cpe} has the following frequency dependence in complex impedance notation:

$$Z_{cpe}=Q^{-1}(j\omega)^{-n} \tag{3}$$

Besides the above-mentioned elements the EEC of an electrochemical system may comprise more complex blocks, which are usually represented by combinations of resistors and capacitors, e.g. electrosorption of cations or anions on the electrode surface may manifest itself in the pair of sequentially connected resistor and capacitor. Also impedance of the diffusion may deviate from Eq. (1), resulting in more complex frequency dependences of diffusion impedance. As the parameters of all the elements are functions of the potential, it appears convenient, first, to decompose the total response into the constituents related to EEC elements on constant potential sections of PDEIS spectrum and, afterwards, analyse the parameters of the constituent responses on the potential scale.

The analysis of multivariate PDEIS data was decomposed into a series of 2D tasks. First, the 3D spectrum was smoothed along the potential axis using a smoother width comparable with the ac probing amplitude. The latter was typically from 5 to 10 mV, while the steps in the potential scan along *E* axis were several times smaller. Due to the smoothing, each of the 2D sections of the 3D spectrum attained information from neighbouring sections, and this improved the accuracy of the subsequent EEC fitting on 2D slices of the 3D spectrum.

The fitting procedures, implemented in the analyser routine, operate in automatic or semi-automatic mode and process each of the 2D sections of the PDEIS spectrum, moving ahead and backward (for analysis of a cyclic scan) along *E* axis. Fig. 4 shows the analyser built-in routine window of the virtual PDEIS spectrometer in this work.

In semi-automatic mode the equivalent circuit to be tested is assembled in the EEC input window manually with a mouse from available elements and the routine tries to find the combination of parameters of the EEC which fits the data with a specified accuracy. If the analyser succeeds, then it shows the spectrum of the equivalent circuit overlaid on the experimental points and tries to apply the same kind of equivalent circuit to the next 2D section of the spectrum, so that the fitting runs afterwards automatically along the *E* axis, showing the correspondence of the experimental data and simulated spectrum for current slice of the plot. As soon as the misfit becomes alarming, the automatic process stops and the operator has to make a decision on changing the type of the circuit. Due to preliminary smoothing of the 3D spectrum, the fitting procedure in each step takes into account the response recorded not only for that particular value of potential but for a certain layer of the spectrum. This adds stability to the fitting, because of noise reduction, and provides sufficient accuracy of fit. Fig. 5 shows some examples of the fitting on the slices of the PDEIS spectrum of Cu monoatomic layer electrochemical formation on gold [2], accompanied by coadsorption of sulphate anions (PDEIS analyses dynamic systems, so it characterises processes, not just stationary objects).

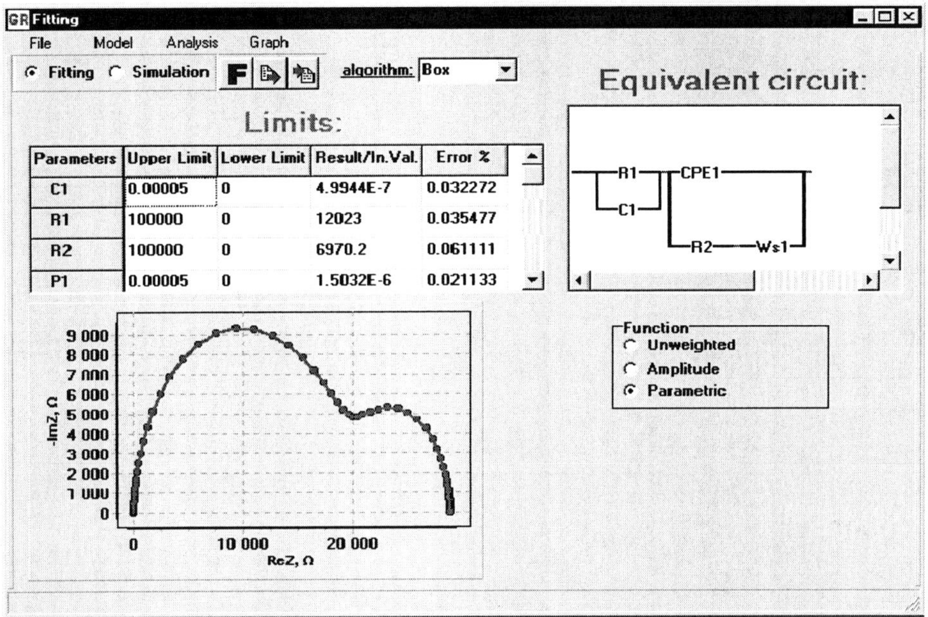

Fig. 4. The spectrum analyser window of the PDEIS virtual spectrometer. Solid circles in the plot show the data for a current slice of a 3D spectrum, and the line shows the simulated spectrum using the parameters obtained by the fitting routine. The automatic run is controlled by parameters of discrepancy.

In the automatic mode, which has been also implemented in the PDEIS analyser, about 20 predefined most often found circuits are tried automatically at predefined potential and, after rating the fits, the analysis goes along the *E* axes with the circuit, which got the highest rating.

The analysis of data in constant potential 2D sections of the PDEIS plot is implemented with a complex nonlinear least squares fitting routing, which uses three different minimisation algorithms (Box, Newton and Powel). Due to a great number of sections, the minimisation problem has to be solved from several hundred to several thousand times in a series. Therefore the stability of data processing appears to be very important. Moreover, the function for minimization has to be given implicitly, as the equivalent circuit cannot be foreknown. From the three implemented algorithms Box algorithm has shown the best performance in solving the minimisation problem with minimal information on initial approximations. Box algorithm does not require calculation of matrices of derivatives. It uses just objective functions and constraint function values in each iteration, also it is less sensitive to initial approximation than the other two algorithms.

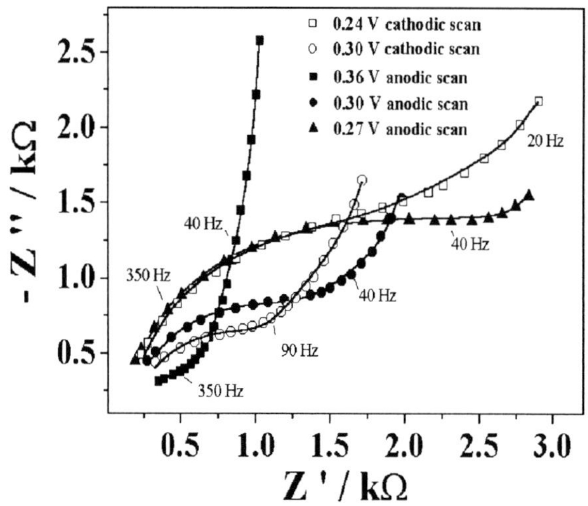

A

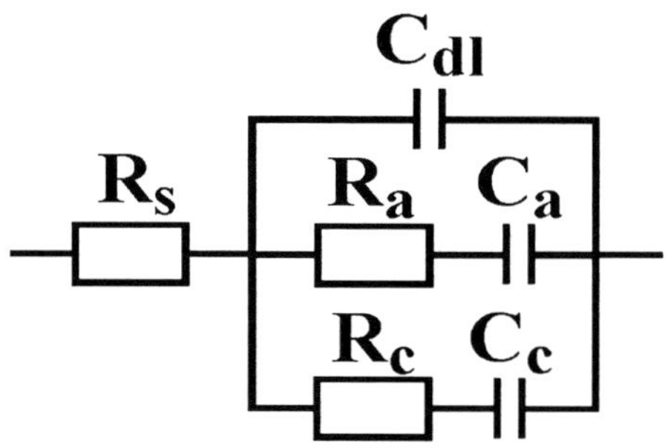

B

Fig. 5. (a) Experimental data and simulated plots for several values of potential in the processing of the PDEIS spectrum of Cu monolayer underpotential deposition in presence of sulphate; (b) the EEC for the data in Fig. 5a; R_s –resistance of solution, C_{dl} – double layer capacitance, R_c and C_c – resistance and capacitance of copper cation electrosorption, R_a and C_a – resistance and capacitance of anion coadsorption.

To secure reliability of fitting we supplied each of the algorithms of complex nonlinear least squares fitting with the three minimisation functions – unweighted (Eq. 4), amplitude (Eq. 5) and parametric (Eq. 6):

$$\sum_{i=1}^{N}(Z_i^{'} - Z_{i,calc}^{'})^2 + (Z_i^{"} - Z_{i,calc}^{"})^2 \tag{4}$$

$$\sum_{i=1}^{N} \frac{(Z_i^{'} - Z_{i,calc}^{'})^2 + (Z_i^{"} - Z_{i,calc}^{"})^2}{Z_i^{'2} + Z_i^{"2}} \tag{5}$$

$$\sum_{i=1}^{N} \frac{(Z_i^{'} - Z_{i,calc}^{'})^2}{Z_i^{'2}} + \frac{(Z_i^{"} - Z_{i,calc}^{"})^2}{Z_i^{"2}} \tag{6}$$

where *i* is the index in the row of N data points of a 2D slice of the PDEIS spectrum, and subscript *calc* indicates the calculated value of impedance.

APPLICATIONS OF THE SPECTRUM ANALYSIS

PDEIS spectrum analysis provides a unique opportunity of separate monitoring of simultaneous nonstationary processes. The opportunity comes from the decomposition of the total response into the constituents related to different elements of equivalent circuits. The processes on the interface that show different frequency responses, thus, can be distinguished also on the potential scale. Fig. 6 illustrates this new opportunity by the separate presentation of the responses of Bi monoatomic layer formation on gold in underpotential deposition (the cathodic reduction of Bi^{3+} above Nernst potential) and the accompanying adsorption of anions.

Bi monolayer forms on gold above Nernst potential due to a so-called underpotential deposition (upd), a process characteristic to a great number of systems [11,12]. The equivalent circuit in the case of Bi appeared to be the same as for Cu monolayer formation (Fig.5b). The capacitance C_c results from electrosorption of cation – the process that produces a monolayer of metal atoms, while C_a represents the capacitance of anion coadsorption (perchlorate coadsorption in the case shown in Fig. 6a). Fig. 6a shows that these two processes are very interrelated in the beginning of the monolayer formation in the cathodic scan around 300 mV, but in later stages the interrelation becomes less obvious. This happens because perchlorate does not tend to adsorb on bare gold, but its adsorption on gold is favoured by bismuth adatoms, produced as a result of upd. Therefore, the two processes are perfectly synchronised in the initial stage of the cathodic scan. Later in the scan, Bi forces out the anions from the surface of gold. However, with the formation of Bi monolayer anions start to adsorb on Bi, which results in complex dependences shown in Fig. 6a. Fig. 6b shows how the double layer capacitance C_{dl} varies in the potential scan during Bi monolayer formation and destruction in presence of perchlorate and nitrate. Because of the above-mentioned coadsorption of anions in Bi upd, the anions affect also strongly the C_{dl}.

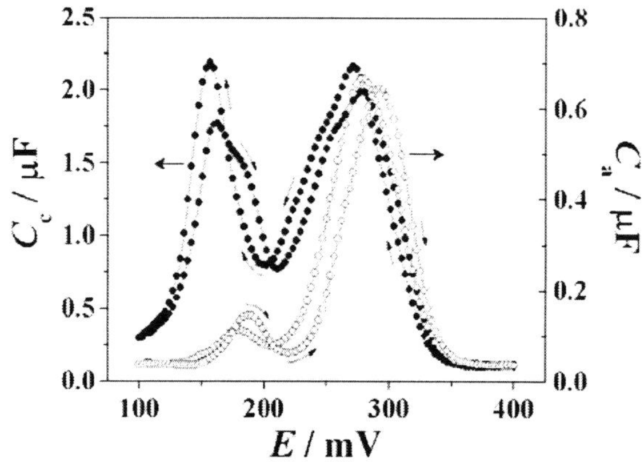

a

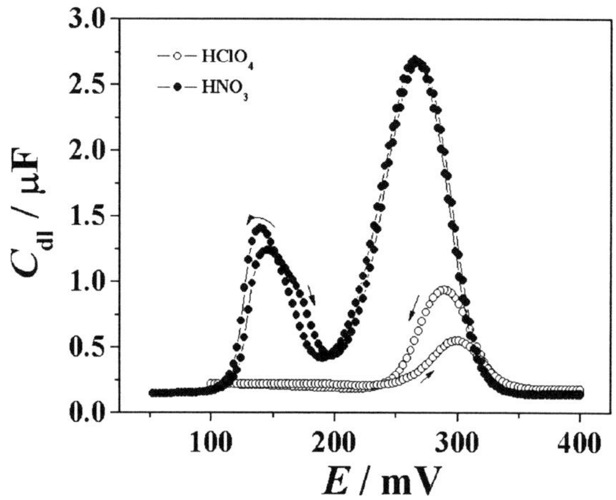

b

Fig. 6. (a) Electrosorption capacitances of Bi^{3+} and perchlorate anions on Au as functions of electrode potential in the cyclic potential scan; (b) double layer capacitance in Bi upd on Au in perchloric and nitric media.

Fig. 7 shows the result of the analysis of the PDEIS spectrum shown in Fig. 2 – the equivalent circuit (Fig. 7a) and the variation of inverse Warburg constant A^{-1} (Fig. 7b) - the parameter of Warburg impedance Z_W, which has been introduced by Eq. (1). Fig. 7b shows $A^{-1}(E)$ dependences for cyclic scans at different scan rates. It appears that A^{-1} (E) is independent on the scan rate and scan direction, and this is different from what was shown in the previous example, where the direction of the scan was important. The effect of the direction of the scan in Fig. 6 originated in the irreversibility of the underlying processes, while Fig. 7 represents a typically reversible system.

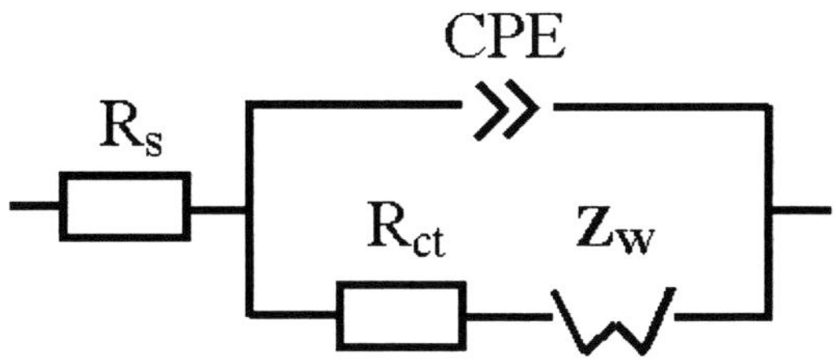

a

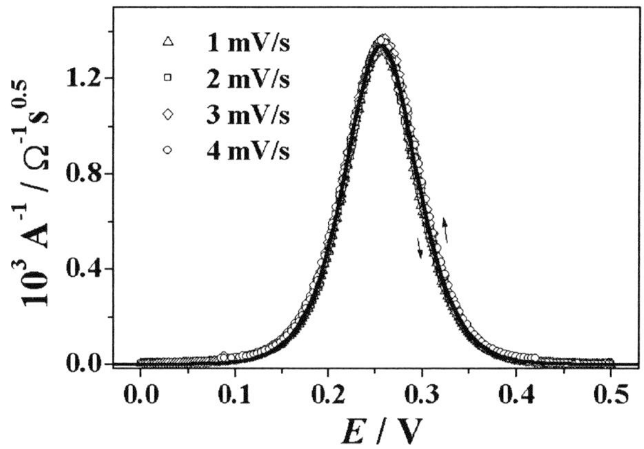

b

Fig. 7. (a) EEC of $K_3[Fe(CN)_6]$ reversible redox transformations on glassy carbon and (b) the dependences of inverse Warburg constant on electrode potential in cyclic scans at different scan rates.

Warburg constant for diffusion impedance of a reversible reaction is characterised by Eq. (7) [13]. Eq. (8) gives the dependence on the potential for the other component of Faraday impedance – charge transfer resistance (R_{ct} in Fig. 7a).

$$\mathrm{A}(E) = \frac{4\mathrm{RT}}{n^2F^2A\sqrt{2D_{Ox}}c_{Ox}}\cosh^2\left(\frac{nF}{2RT}\left(E - E_{1/2}\right)\right) \quad (7)$$

$$\mathrm{R_{ct}}(E) = \frac{\mathrm{RT}}{n^2F^2A\cdot k_0c_{Ox}}\cdot\frac{1+\exp\left[\frac{nF}{RT}\left(E - E_{1/2}\right)\right]}{\exp\left[\left(1-\alpha\right)\frac{nF}{RT}\left(E - E_{1/2}\right)\right]} \quad (8)$$

The important point in these dependences is that the second one contains the rate constant k_0, while the first one is unaffected by the rate constant. Fig. 8 shows how this difference becomes apparent in the constituent responses of ferrocyanide reduction on bare Pt surface and Pt pretreated with thioglycerol. The adsorbed thioglycerol on the surface of the electrode affects charge transfer in ferrocyanide reduction, and this results in the depression of the curve that characterises the part of the Faraday impedance related with the charge transfer (Fig. 8a). However, the diffusional component of the Faraday impedance remains almost unaffected by the passivation of Pt (Fig. 8b).

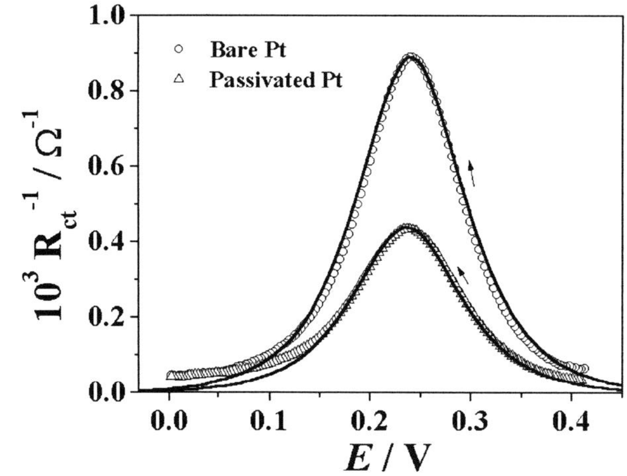

a

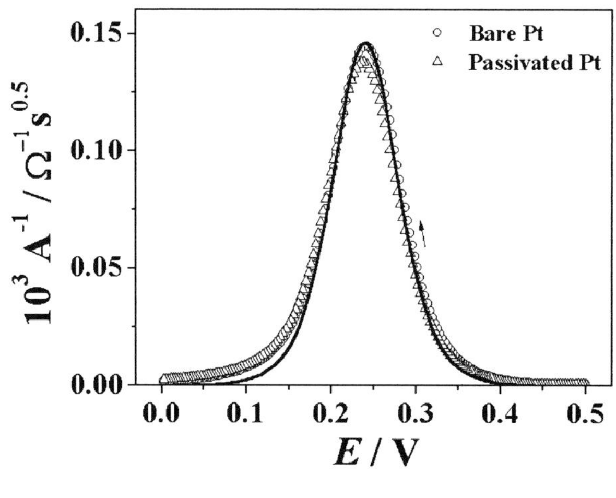

b

Fig. 8. $K_3[Fe(CN)_6]$ redox transformations on Pt (bare surface and passivated with thioglycerol); inverses of (a) charge transfer resistance and (b) Warburg constant variation in the potential scan. Solid lines were simulated using Eqs. (7) and (8).

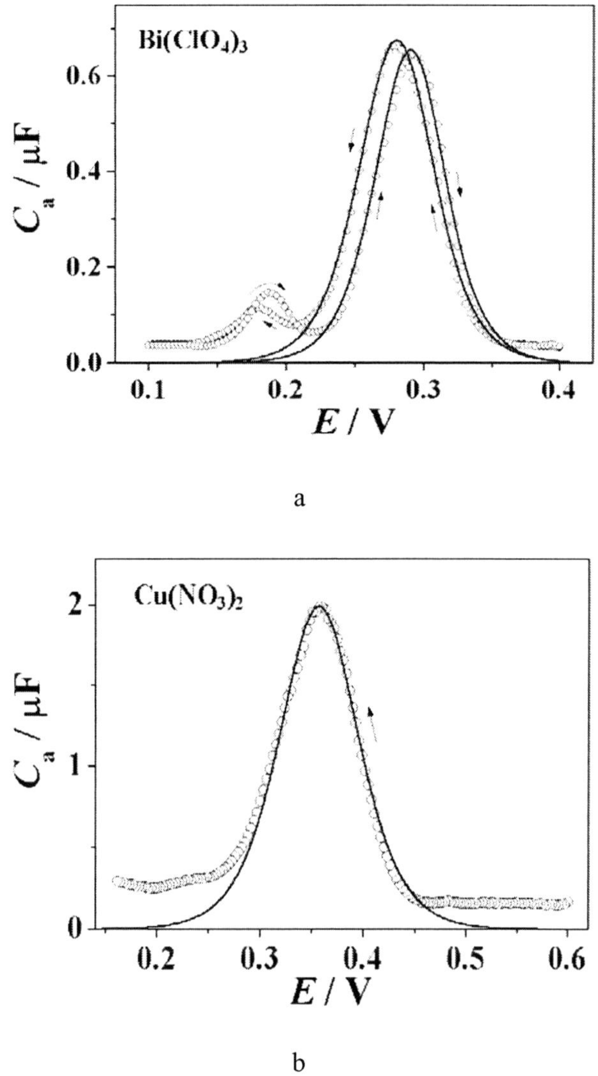

Fig. 9. Capacitance of anion coadsorption in (a) bismuth and (b) copper upd on gold. Experimental data is shown with open symbols, solid lines show the simulated dependences based on Langmuir adsorption model.

The second round in the inverse problem solving, hence, comes after the decomposition of the PDEIS spectrum into constituents. The objects of the analysis in that stage are the dependences of the parameters of constituent responses on the potential. Fig. 9 shows the second stage of the inverse problem solution for the adsorption capacitance of anion coadsorption in Bi and Cu upd. The solid lines show how the corresponding parameter should depend on the potential in the case of Langmuir type adsorption. Good fit of the experimental data to the model gives evidence to Langmuir type adsorption in the initial stages of perchlorate coadsorption in bismuth upd and of nitrate coadsorption in copper upd.

Thus, the PDEIS spectra analysis enables separate monitoring of simultaneous nonstationary processes, moreover, the transformations of extremely low quantities of reacting materials can be not only detected but also compared with theoretical models.

CONCLUSIONS

The analysis of PDEIS spectra is a multi-stage process. First, the 3D spectrum has to be pre-processed and cut into a great number of 2D slices. Then the fitting to equivalent circuits with complex nonlinear least squares fitting routine is implemented on each of the 2D slices, which gives the dependences of the equivalent circuit parameters on the potential. The dependences obtained from the decomposition of the frequency response characterise the constituent responses of different processes and individual responses of different objects in the electrochemical system. Finally, the dependences of the parameters obtained are compared with expected model dependences. In total, this gives the possibility for detailed characterisation of the constituents, which may be initially hidden in chorus of many simultaneous responses of various interfacial structures and processes.

ACKNOWLEDGEMENT

A.S.B. acknowledges the support of Scientific Production Association "Centre" (Minsk) for payment of WSC3 conference fee, and support of Fundamental Research Foundation of Belarus (grant X03M-013).

REFERENCES

[1] G.A. Ragoisha, A.S. Bondarenko, *Solid State Phenom.* 90-91, 103-108 (2003).

[2] G.A. Ragoisha, A.S. Bondarenko, *Electrochem. Comm.* 5, 392-395 (2003).

[3] G. A. Ragoisha, A. S. Bondarenko, N. P. Osipovich, E. A. Streltsov, *J. Electroanal. Chem.*, doi:10.1016/j.jelechem.2003.10.014.

[4] G.A. Ragoisha, A.S. Bondarenko, *Physics, Chemistry and Application of Nanostructures*, World Scientific, 373-376 (2003).

[5] G.A. Ragoisha, A.S. Bondarenko, *Chem. Problems of the Development of New Materials and Technol.*, Minsk 1, 138-150 (2003).

[6] J.R. Macdonald, *Impedance Spectroscopy*, John Wiley & Sons, N.Y., 1987.

[7] Z.B. Stoynov, B.M. Grafov, B. Savova-Stoynova, V.V. Elkin, *Electrochemical Impedance* (in Russian), Nauka, Moscow 1991.

[8] B.A. Boukamp, *Solid State Ionics*. 20, 31-44 (1986).

[9] J.R. Macdonald, L.D. Potter, *Solid State Ionics*. 24, 61-79 (1987).

[10] R. Macdonald, LEVM Manual, *J. Ross Macdonald and Solartron Group Limited*, 1999.

[11] D. M. Kolb, M. Przasnyski, H. Gerischer, *J. Electroanal. Chem.* 54, 25 (1974).

[12] E. Herrero, L. J. Buller, H. D. Abruna, *Chem. Rev.* 101, 1897 (2001).

[13] Lasia, Electrochemical Impedance Spectroscopy and its Applications, *http:// alfa.chem.uw.edu.pl/LasiaLecture/EIS1.pdf*

In: Progress in Chemometrics Research
Editor: Alexey L. Pomerantsev, pp. 103-109
ISBN 1-59454-257-0

Chapter 8

APPLICATION OF PATTERN RECOGNITION METHODS IN ANALYSIS OF COMPLEX HYDROCARBON MIXTURES FOR IMPROVEMENT OF KINETIC DATA QUALITY

***D. Kubička*[1]*, S.-P. Reinikainen*[2] *and D. Yu. Murzin*[1]**

[1] Åbo Akademi University, Finland; *dkubicka@abo.fi*

[2] Lappeenranta University of Technology, Finland

ABSTRACT

A SIMCA model was developed for classification of C_{10} alkyl-naphthenes to assist evaluation of kinetic experiments involving complex mixture of C_{10} hydrocarbons. The influence of different pre-processing methods, including logarithmic and square root scaling and autocorrelation transformation, on classification of C_{10} alkyl-naphthenes was investigated.

Keywords: SIMCA, pattern recognition, pre-processing, mass spectra, C_{10} alkyl-naphthenes

INTRODUCTION

Detailed analysis and identification of reaction products are essential for interpreting the kinetic data in order to disclose relevant reaction mechanisms. Although detailed analysis of products could be accomplished owing to availability of advanced analytical methods, such as mass spectrometry and infrared spectroscopy, the extraction of relevant information for products identification is challenging especially as the complexity of the mixture to be analyzed increases. Despite application of separation techniques, such as gas chromatography, in connection with the above-mentioned spectroscopic methods, the identification of all individual products can be achieved only rarely. Therefore, classification into product groups, based on the similarity of the reaction components, is then the only possibility to extract relevant information, which can be subsequently used in derivation of reaction mechanisms. Moreover, the obtained classification data may be used for prediction of physico-chemical properties of the product mixtures, which would otherwise require additional analysis, often too laborious and expensive.

Mass spectrometry provides useful information about functional groups in organic molecules. Additionally, information about molecular conformation, allowing identification of isomers, can be extracted from mass spectra as well. However, extraction of such information from complex mass spectra requires using pattern recognition methods to find similarities in the collected data. The similarities usually represent characteristic features common for the analyzed organic compounds. Soft Independent Modeling of Class Analogy (SIMCA) is a pattern recognition method relying on independent modeling of all (defined) classes by means of principal component analysis (PCA) [1-2]. For new (unknown) object the distance to all defined classes can be calculated. If the distance of an object lies within limits defined for a class then it is considered to belong to the class. Furthermore, objects that are members of new classes and outliers can be identified by SIMCA. Previously, SIMCA of mass spectral data has been applied in classification and identification of hazardous organic molecules in ambient air [3-4] and in classification according to chemical structure [5].

The need for identification and classification of C_{10} alkyl-naphthenes has arisen due to current research aimed at improvement of diesel fuel quality [6]. A successful classification may shed more light on the mechanism of transformation of C_{10} hydrocarbons. Moreover, the identification of all products could be used in prediction of important diesel fuel properties, such as cetane number. Accordingly, this study focuses on the applicability of SIMCA in complex chemical structure classification, e.g. classification of C_{10} alkyl-naphthenes. Based on the available mass spectral data, nine classes have been defined, reflecting the number of carbon atoms in the naphthenic ring and the length of the longest alkyl-substituent. In addition, the influence of scaling and transformation of the raw spectral data on the classification is discussed.

EXPERIMENTAL

Data

The mass spectral data of C_{10} alkyl-naphthenes were collected from two main sources: mass spectral (MS) library [7] and literature [8-9]. The data published in scientific articles were essential for development of SIMCA models since the number of mass spectra of C_{10} alkyl-naphthenes in the MS library was limited. Totally 136 spectra were collected, from which 81 originated from literature and 42 from MS library. The remaining 13 spectra belong to C_{10} alkyl-naphthenes that were synthesized and analyzed in course of this study. The 14 most relevant mass spectral fragments of C_{10} alkyl-naphthenes were kept in the datasets to be used in modeling. This had to be done because the data published in literature did not contain full-range spectra and only the main 14 fragments were given.

The examination of collected data has shown that nine classes could be defined for C_{10} alkyl-naphthenes based on two criteria: size of the naphthene ring and number of carbon atoms in the longest alkyl-substituent. Two classes of alkyl-naphthenes can be defined according to the size of the naphthene ring – alkyl-cyclopentanes (alkyl-cC5) and alkyl-cyclohexanes (alkyl-cC6). Each of these classes can be further divided into 5 and 4 classes, respectively, based on the alkyl-substituent length. Thus, following nine classes of C_{10} alkyl-naphthenes were suggested: methyl-cC5 (4), ethyl-cC5 (14), propyl-cC5 (3), butyl-cC5 (2), pentyl-cC5 (5), methyl-cC6 (32), ethyl-cC6 (33), propyl-cC6 (29) and butyl-cC6 (14). The number in parentheses determines the number of spectra collected for a particular class. Typical structures representing each class are given in Fig. 1. While the four cyclohexane classes consist of sufficient amount of data to describe the suggested classes, the amount for data for cyclopentane classes is not sufficient. Accordingly, SIMCA models were developed for all alkyl-cC6 classes and just one cyclopentane class, namely ethyl-cC5.

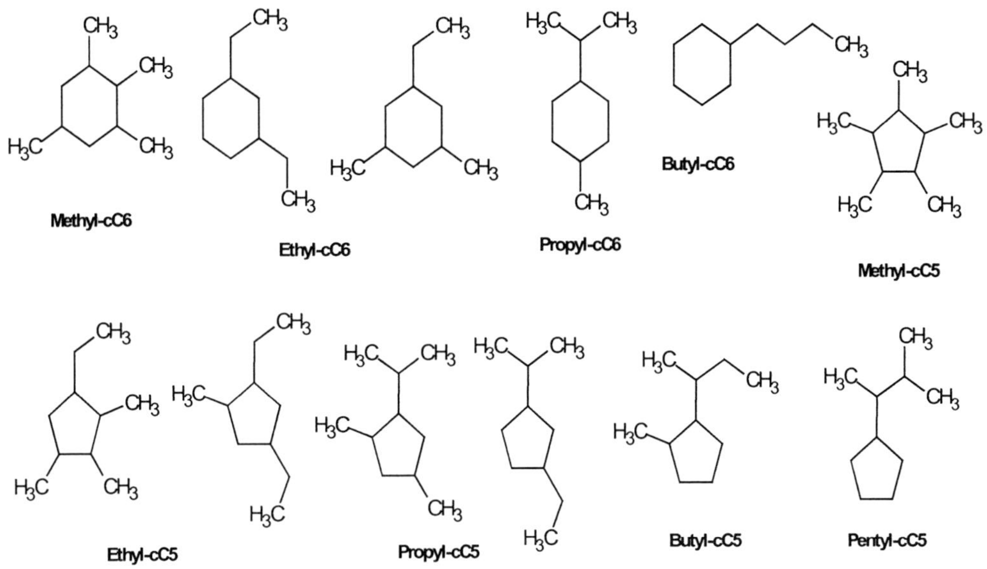

Fig. 1. Typical structures of possible classes of C_{10} alkyl-naphthenes

Methods

All spectra were normalized to a base peak. The 14 most relevant fragments for C_{10} alkyl-naphthenes in the collected dataset were selected to model the individual classes. Prior to modeling by SIMCA [10], data were pre-processed. The pre-processing consisted of scaling of mass spectral intensities and data transformation. Since the ion mass spectral intensities were normalized to a base peak, the ions of high intensity might have gained too much weight as compared to ions with low intensities. Therefore, two scaling procedures –square root and logarithmic scaling of spectral intensities – were applied to investigate the effect of intensities scaling. For comparison purposes, the data were modeled without intensities scaling as well.

Autocorrelation transform (Eq. 1) was used for transformation of intensities scaled and non-scaled data. The autocorrelation transform of a mass spectrum has been proposed to be more suitable for pattern recognition applications than the original mass spectrum since it reveals the periodicities in the spectrum [3-5]. The new set of variables, resulting from autocorrelation transform of a mass spectrum, describes losses of particular fragments, such as CH_2^+, from the parent molecule. Thus, the information in the autocorrelated spectrum can be easily interpreted from the chemical point of view [5]. The used autocorrelation transform is given in Eq. 1:

$$x_{ij} = \sum_{k=1}^{K}(I_{ik} I_{i(k+j)}) / \sum_{k=1}^{K}(I_{ik})^2 \tag{1}$$

where I_{ik} and $I_{i(k+j)}$ are the intensities of ions k and k+j, respectively, in the i-th spectrum and x_{ij} is the autocorrelation coefficient for ions separated by j mass units. Once more, the data were modeled without autocorrelation transformation to see the impact of the transformation on class separation.

Each class was autoscaled separately after pre-processing; i.e., first the column means were subtracted from all values in the corresponding columns and then the mean centered data were divided by standard deviations of the values in the corresponding columns. Independent SIMCA models using the pre-processed and autoscaled data were calculated according Eq. 2:

$$x_{ij}^{q} = x_{mj}^{q} + \sum_{a=1}^{Aq} t_{ia}^{q} p_{aj}^{q} + e_{ij}^{q} \tag{2}$$

where q, i and j are the class, object and variable index, respectively, x_{mj} is the mean value of variable j and A is the number of principal components used, t, p and e are the model scores, loadings and residuals, correspondingly. The score plots of all models were examined visually to reveal possible outliers in the training datasets.

Validation of the class models had to be carried out by a cross validation procedure since a completely independent dataset for testing could not be obtained. Therefore, data for each class were divided into three sub-datasets. Two of the sub-datasets were merged forming a training dataset for each class, and independent sub-models of each class were built up. The third sub-dataset was used as a testing dataset to validate the sub-models. Each of the sub-datasets was used once and only once as a testing dataset; i.e. three sub-models of each class were calculated and validated. The validation results of the three sub-models were combined together providing thus a basis for comparison of the applied pre-processing methods. The sub-models were validated, in addition to cross validation, using a dataset created from spectra of compounds known to belong to classes, which were not modeled; i.e., methyl-cC5, propyl-cC5, butyl-cC5 and pentyl-cC5.

Classification of a compound (object) was based on comparison of its residual variance (Eq. 3) with residual variances of the class models (Eq. 4):

$$(s_z^q)^2 = \sum_{j=1}^{N_j}(e_{zj}^q)^2 / (N_j - A^q) \tag{3}$$

$$(s_0^q)^2 = \sum_{i=1}^{N_i}\sum_{j=1}^{N_j}(e_{ij}^q)^2 / ((N_i - A^q - 1)(N_j - A^q)) \tag{4}$$

where N_j and N_i are the numbers of variables and objects, respectively, of class q. The subscript z is the index of objects in the testing dataset and s^2 is the residual variance. The meaning of other symbols is the same as in Eq. 2.

An object from the testing dataset was assigned to a class when its residual variance was smaller than the 99% confidence limit of the class residual variance. The residual variance of an object represents the distance of the object to the class it is being projected into. The objects can be either within the space occupied by a class, i.e., their distance to the class is smaller than the limit distance, or outside the class space, i.e. they do not belong to the class.

RESULTS AND DISCUSSION

SIMCA models were developed for each of the five defined classes. Three principal components explaining between 60 – 90% of variation in data, regardless the pre-processing method applied, were used for modeling of all classes but ethyl-cC5, which could be described by only two principal components. Visual inspection of the score plots for each class revealed just two clear outliers, one in methyl-cC6 and another in ethyl-cC6 class, and these were excluded from further development of individual class models. Both of them were, however, used for evaluation of class models. While all models suggest that the outlier from methyl-cC6 class does not belong to any of the known classes, the outlier from ethyl-cC6 class is unambiguously classified as methyl-cC6 by all models regardless the pre-processing method used.

The classification results obtained by the cross validation procedure are reported in Table 1. All models, irrespective of the pre-processing method, were found to describe the spectral data well with the number of correctly classified objects being between 87 – 93%. More importantly, the incorrectly classified objects were assigned in vast majority to the unknown class; i.e., they were not misclassified into any of the known classes. In fact, these so-called false positives were observed only when logarithmic scaling with autocorrelation transformation and autocorrelation transformation alone was used. While in the case of logarithmic scaling with autocorrelation transformation eight false positives were discovered, just one was found in the case of pre-processing the data with autocorrelation transform.

Square root scaling of the original spectra resulted in the most successful classification followed by square root scaling in combination with autocorrelation transformation and autocorrelation transformation. Based on these results, it could be concluded that the transformation of original mass spectra using autocorrelation transform is not necessary for a successful classification of the mass spectra of C_{10} alkyl-naphthenes. Moreover, even the models relying on autoscaled original mass spectra can classify the alkyl-naphthenes reasonably well. On the other hand, logarithmic scaling in combination with autocorrelation transformation provides less successful classification with false positives being the main problem.

Table 1 Number of correctly classified objects for different pre-processing methods: logarithmic (log) and square root (sqr) scaling, autocorrelation transform (at) and no pre-processing (raw).

Class name	Number of objects	Pre-processing method raw	at	sqr	sqr+at	log	log+at
Methyl-cC6	31	*27*	*27*	*29*	*29*	*26*	*29*
Ethyl-cC6	32	*30*	*30*	*30*	*30*	*30*	*29*
Propyl-cC6	29	*24*	*27*	*25*	*25*	*25*	*25*
Butyl-cC6	14	*11*	*11*	*11*	*11*	*11*	*11*
Ethyl-cC5	14	*12*	*11*	*13*	*11*	*12*	*12*
Unknown	42	*42*	*42*	*42*	*42*	*42*	*36*
Total	162	*146*	*148*	*150*	*148*	*146*	*142*
Correct classification, %		*90.1*	*91.4*	*92.6*	*91.4*	*90.1*	*87.7*

The class of butyl-cC6 has the least successful model regardless the pre-processing method. The number of correctly assigned objects is just 79%. Closer examination of the class data reveals that the problem is due to tert-butylcyclohexane since its spectrum is close to being an outlier. As the dataset of mass spectra for butyl-cC6 is small, the spectrum of tert-butylcyclohexane gains high importance in the models. This indicates that it would be beneficial to keep tert-butylcyclohexane out of the butyl-cC6 class model and identify/classify it by different means. On the other hand, no problems were observed for the other groups; e.g. the classification into ethyl-cC6 class was not hindered by the fact the class is formed by diethyl-cyclohexanes and dimethyl-ethyl-cyclohexanes.

The mass spectra of all other C_{10} alkyl-cyclopentanes than ethyl-cC5 were correctly classified as members of yet unknown class/classes in course of validation. The correct classification was obtained for all pre-processing methods except for logarithmic scaling combined with autocorrelation transformation, which yielded six false positives, five of them into the butyl-cC6 class. The discrimination of these C_{10} alkyl-cyclopentanes suggests that with more spectral data available new C_{10} alkyl-cyclopentanes classes could be defined and modeled.

CONCLUSIONS

SIMCA has been successfully applied for modeling of original and pre-processed mass spectra of C_{10} alkyl-naphthenes. The class modeling has proved that the two parameters suggested for classification of C_{10} alkyl-naphthenes, namely the size of the ring and length of the longest alkyl-substituent, could be used for classification purposes. Unfortunately, the lack of data for C_{10} alkyl-cyclopentanes has prevented the development of SIMCA models for any C_{10} alkyl-cyclopentanes class, but ethyl-cC5. Nevertheless, the spectra of methyl-, propyl-, butyl- and pentyl-cyclopentanes remained unassigned to any of the defined classes – methyl-, ethyl-, propyl- and butyl-cC6 and ethyl-cC5 – confirming their discrimination from the five modeled classes.

Just small differences in classification have been found as a result of modification of data pre-processing. All models have classified correctly between 87 – 93% of samples (objects).

Simple square root scaling of the spectral data has yielded the best model. Logarithmic scaling combined with autocorrelation transformation has shown the highest misclassification of objects. The results suggest that the autocorrelation transformation of spectral data is not necessary for successful classification of the mass spectra of C_{10} alkyl-naphthenes.

ACKNOWLEDGMENTS

This work is part of the activities at Åbo Akademi Process Chemistry Centre within the Finnish Centre of Excellence Programme (2000–2005) by the Academy of Finland. The authors are thankful to Michael Sjöström (University of Umeå) for encouragement and helpful discussions.

REFERENCES

[1] S. Wold, *Pattern Recognition*, 8, 127-139, (1976).

[2] S. Wold, M. Sjöström, in B.R. Kowalski (Ed.), Chemometric: Theory and Application, ACS Symposium Series, No. 52, *Am. Chem. Soc.*, 243-282, (1977).

[3] D.R. Scott, W.J. Dunn III, S.L. Emery, *Environ. Sci. Technol.*, 21, 891-897, (1987).

[4] W.J. Dunn III, S.L. Emery, G. Glen, *Environ. Sci. Technol.*, 23, 1499-1505, (1989).

[5] S. Wold, O.H.J. Christie, *Anal. Chim. Acta*, 165, 51-59, (1984).

[6] D. Kubička, N. Kumar, P. Mäki-Arvela, M. Tiitta, V. Niemi, T. Salmi, D.Yu. Murzin, *J. Catal.*, 222, 65-79, (2004).

[7] *NIST98 mass spectral database*.

[8] L.S. Golovkina, G.V. Rusinova, A.A. Petrov, *Izvestiya Akademii Nauk SSSR, Seriya Khimicheskaya,* 1979 (1), 68-73 and 73-80.

[9] G.V. Rusinova, L.S. Golovkina, O.A. Arefev, A.A. Petrov, *Izvestiya Akademii Nauk SSSR, Seriya Khimicheskaya,* 1981 (8), 1809-1814 and 1981 (9), 2044-2054.

[10] *SIMCA tool* (ver. 0.5), S.-P. Reinikainen, Lappeenranta University of Technology, (2004)

In: Progress in Chemometrics Research
Editor: Alexey L. Pomerantsev, pp. 111-118
ISBN 1-59454-257-0

Chapter 9

BAND AND POINT STATISTICAL ESTIMATION OF CHANNEL-FORMING POLYPEPTIDES POTENTIAL

***A. Dmitriev*[1], *V. Baryshnikov*[1], *I. Markov*[1] *and V. Tverdislov*[2]**
[1]Lipetsk State Pedagogical University, Russia, a_v_dmitriev@mail.ru
[2]Lomonosov Moscow State University, Russia, tverdislov@mail.ru

ABSTRACT

The algorithm of electrostatic potential adequate statistical estimation in channel-forming polypeptides of biological and model membranes was created. We have obtained the formula, defining potential's area of true values, which is explained by X-ray diffraction and NMR experimental errors of polypeptides atomic coordinates.
The offered approach allows:

1. to conduct demonstrative band matching of polypeptides potential distributions;
2. to obtain an adequate cumulative distribution function of polypeptides potential;
3. to conduct demonstrative anti-aliasing of local maximums and minimums of polypeptides potential.

Keywords: gramicidin channels, electrostatic potential, estimation error

INTRODUCTION

The goal of this work is to define electrostatic field characteristics of transmembrane channels on the basis of experimental data analysis and estimated electronic density distribution data of the appropriate polypeptides.

Transmembrane channels are integral membrane proteins, capable of selective transition of Na^+, K^+, Ca^{2+}, Cl^- ions, and small polar molecules inside and outside the cell [1,2]. Biological significance of these processes, first of all, is explained by cell's vital activity support. For example, ion channels take part in spawning ionic asymmetry inside and outside the cell, which is a necessary condition of cell's stability [3]. Porines [1], one of the transmembrane channel classes, allow nutritious substances to penetrate into the bacteria cell's body, and for the wastes to come out into the exterior environment.

Modern experimental methods give the opportunity to obtain integral characteristics of transmembrane transition, to settle channel's atomic structure but don't give the information about ionic motion inside the channel. For example, the problem of the structure and the character of partially dehydretated ion's motion inside protein's pore still remains unsolved.

Theoretical research method of ion motion inside protein's pore is modeling. There are two approaches, considered to be traditional ones [4,5]: kinetic, based on Eyring's theory, and dynamic, based on one of the dynamic equation's solution. The common idea of these approaches is the necessity of knowing electrostatic potential distribution in protein's pore, which is evaluated according to coordinate values and charges of molecular atoms of transmembrane channel.

The role of charges in plain calculations is played by the parameters of electrostatic interactions of one of the force fields: AMBER [6,7,8,9], CHARMM [10,11,12], OPLS [13,14,15]. More accurate computation requires using quantum-chemical methods [16,17]. Atomic structure of the channel's molecule can be determined using different experimental methods. These methods include [2]: biochemical protein isolation, cloning, definition of amino acid sequence, point mutation for changing its structure in particular areas of protein's molecule, channels' expression in alien cells. Final information about atomic coordinates of the molecule is given by X-ray diffraction or NMR [18].

Despite the well-known accuracy of these methods, atomic coordinates of the transmembrane channel's molecule can be evaluated only to a certain degree error. That error leads to appearance of a reliability interval for each atom's coordinate, which evidently, reflects on the evaluation accuracy of electrostatic potential distribution. Therefore, it becomes necessary to solve the problem of defining the interval potential estimate $U \pm \Delta U$, where U – a point potential estimate, ΔU – the absolute error of potential estimate.

The obtained interval potential estimate allows demonstratively solve the problems of approximation, potential smoothing and also comparison of channelforming polypeptides potentials.

EXPERIMENTAL

Materials

All the computations were conducted for the gramicidin channel, which is formed by building in gramicidin A peptide antibiotic in a lipid bilayer. Gramicidin A is a hydrophobic peptide, consisting of 15 L- and D-amino acids. It has the following amino acid sequence [19]:

HCO-L-Val-Gly-L-Ala-D-Leu-L-Ala-D-Val-L-Val-D-Val-L-Trp-D-Leu-L-Trp-D-Leu-L-Trp-D-Leu-L-Trp-$NHCH_2CH_2OH$.

Numerous research using flat membranes and liposome showed that the channel is a peptide dimer. Since gramicidin consists of unusual D-amino acids, it's impossible to be used as a structure model, which is found in membrane proteins. However, this peptide still allows to get a general idea of one of the ways of forming a selective pore in bilayer. This is an easily accessible polypeptide, that's why it's frequently used as "membrane protein model" while studying the perturbation effect of membrane proteins on lipids.

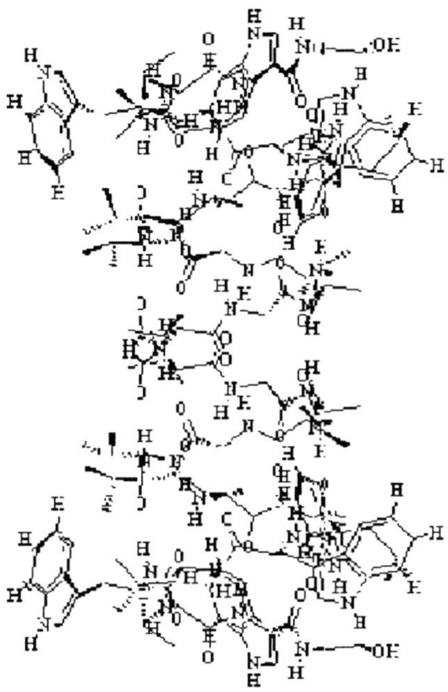

Fig. 1. Structure of HHSH-configuration gramicidin channel

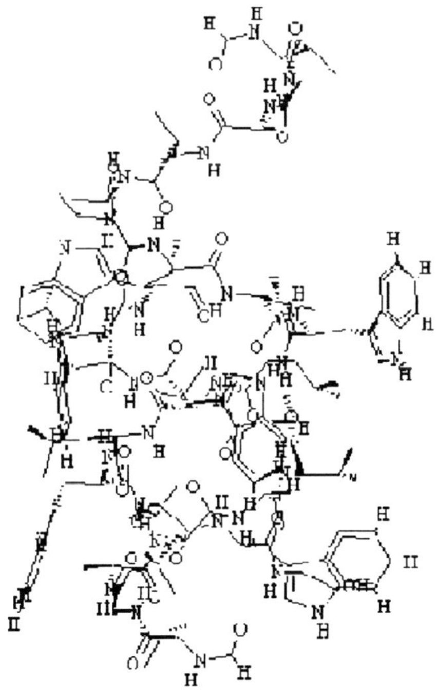

Fig. 2. Structure of DSDH-configuration gramicidin channel

We have studied two principally different configurations of gramicidin channel. In the first case, transmembrane channel is formed in the manner of head-to-head single-stranded helix (HHSH-configuration) head-to-head single-stranded helix [20,21,22]. Such channel (fig. 1) is formed by building gramicidin A in hydrated DMPC-bilayer. In the second case – in the manner of double-stranded double helix (DSDH-configuration) [23]. This channel (Fig. 2) is formed in the complex of gramicidin A with cesium chloride. Our choice can be explained by the fact that these channelforming polypeptides had the most exactly determined atomic coordinates.

Experiments

Atomic coordinates in HHSH-configuration of gramicidin channel were defined by Solid State NMR method with 0.86Å resolution [20,21,22], DSDH-configuration – by X-Ray diffraction method with the 1.4Å resolution [23]. The following coordinates and their absolute errors were taken from the Protein Data Bank (Brookhaven National Laboratory, USA) [24].

Distribution of polypeptides' electronic density in their fixed configuration was computed using semiempirical CNDO/2 method of quantum chemistry in standard parameterization [25]. This is explained by the fact, that this method is specially parameterized according to the distribution of electronic density and gives good atomic charge calculation results for different molecules. CNDO/2 method had it's most successful implementation in the study of C-, H-, N-, O-, F-, Cl- containing organic molecules' electronic distribution and in the study of the induction effect [25,26,27,28,29]. We suppose that, CNDO/2 method allows to obtain an accurate (with zero error) electronic density distribution. The obtained polypeptide electronic density distributions were used to estimate partial (point) atomic charges.

Polypeptide electrostatic potential distribution was calculated along Z-axis of axial symmetry in gramicidin channel using 0.5Å step. Considering that ionic motion in gramicidin channels is single scale along Z-axis, researching potential distribution along that axis is the biggest interest. In our computations we have used the formula of Coulomb's potential.

$$U(Z) = 14\sum_{i=1}^{N} \frac{q_i}{\sqrt{x_i^2 + y_i^2 + (z_i - Z)^2}} \quad (1)$$

where q_i – a point charge on i atom of polypeptide, x_i, y_i, z_i - atom's coordinates. In this case atomic point charge approximation is substantial, because we have researched potential distribution over the limits of Van-der-Waals radii of atoms which cover the channel's cavity [30].

Data Analysis

ΔU value was calculated using:

$$\Delta U = 14\Delta R \sum_{i=1}^{N} \frac{q_i^2}{x_i^2 + y_i^2 + (z_i - Z)^2} \tag{2}$$

where ΔR – maximum absolute error of polypeptide atomic coordinates estimation (ΔR=0.03Å for HHSH-configuration and ΔR=0.035Å for DSDH-configuration of gramicidin channel). We have got the following formula as a formula for computing indirect U estimation absolute error.

Results And Discussion

Polypeptide electrical potential distribution along the symmetry axis of gramicidin HHSH- (■) and DSDH-configuration (▲) is shown in figure 3. The pattern of potential to gramicidin HHSH-configuration axis coordinates relationship is within the limits of absolute error and corresponds with symmetrical two-barrier model of ion channel. This model reflects a typical existence of two potential pits of 0.5V depth, with minimums located at a distance of 8Å from center in two directions, and also two potential barriers of 0.5V and 0.8V height. Moreover, the central minimum of potential function can be "smoothed" within the limits of potential estimate absolute error, which is impossible to do using point estimation. In case of gramicidin channel's DSDH-configuration considering absolute estimation error it's possible to indicate the existence of three barriers and three potential pits. It's typical for a model like that to have potential barriers of 0.1, 0.6, 0.025V height with ion moving from left to right, barriers of 0.025, 0.1, 0.65V height with ion moving from right to left, and also potential pits of 0.1, 0.1, 0.025V depth. An important feature of gramicidin channel's DSDH-configuration potential function is in its asymmetry towards the channel's center, which may be the cause of conduction differences of that channel in two directions.

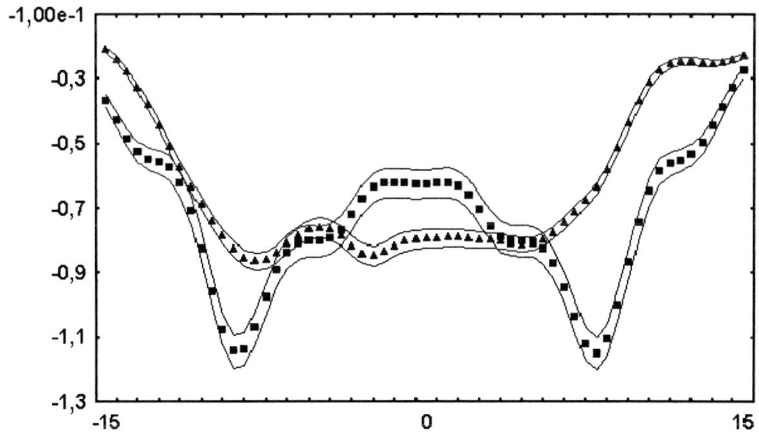

Fig. 3. Electrostatic potential distribution of gramicidin channel

The obtained interval potential estimates of both observed configurations of gramicidin channel allow to conduct an interval comparison of polypeptide's potential distributions.

Fig.2 shows the overlapping of intervals in two 4Å size areas and located at a distance of 3Å from channel's center in both directions. Considering that the form of potential function determines ion's dynamic in the channel, it's possible to suppose the equivalence of ionic motion in these two areas for different configurations of gramicidin channel.

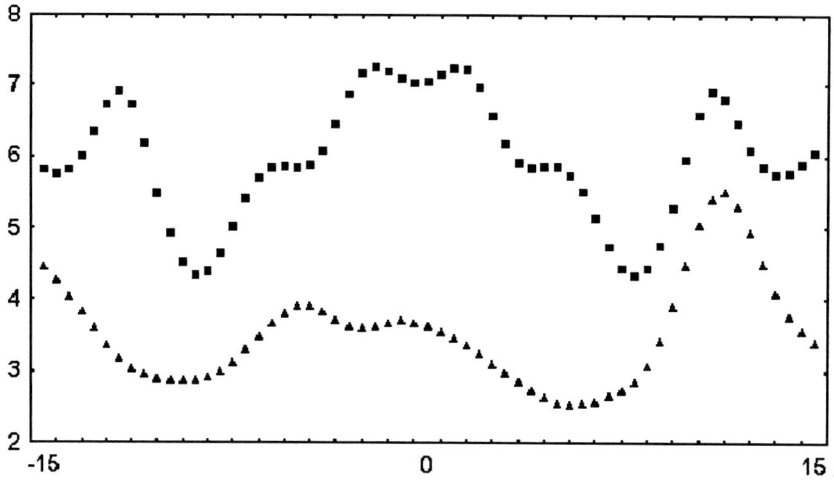

Fig . 4. Gramicidin channel relative error distribution of potential

Figure 4 presents the relative error distribution of potential estimates for HHSH- (■) and DSDH-configurations (▲) of gramicidin channel. Maximum values (about 7%) of potential estimates relative errors for HHSH-configuration is observed at distance of 2Å from center in two directions and at a distance of 4Å from both sides of the channel, minimum values (about 4.5%) at a distance of 7Å from both sides of the channel. Maximum values (about 5.5%) of potential estimate relative errors for DSDH-configuration is observed at a distance of 3Å from right side, minimum values (about 2.5%) – at 9Å from right and 5Å from left side distances.

Conclusions

Maximum relative electrostatic potential estimation error for HHSH-configuration of gramicidin channel equals to not more than 7%, for DSDH-configuration – not more than 5.5%.

The obtained electrostatic potential interval estimate for HHSH- configuration of gramicidin channel allows to eliminate ("smooth") central minimum of potential function.

The areas of 4Å size, located at a distance of 3Å from channel's center in both directions are characterized by equal polypeptide potential distribution.

For DSDH-configuration of gramicidin channel, there is a relationship between conduction and ion motion direction.

ACKNOWLEDGMENTS

This work has been partially supported by grants 00-15-97852 from the Russian Foundation for Basic Research.

REFERENCES

[1] D. Aidley, P. Stanfield, *Ion Channels. Molecule in Action*, Cambridge University Press, Great Britain, 1996
[2] J. Nicholls, A. Martin, B. Wallace, P. Fuchs, *From Neuron to Brain. Fourth Edition*, Sinauer Associares, USA, 2001
[3] L. Yakovenko, V. Tverdislov, *Biophysics*, 48, 1137-1145, (2003)
[4] D. Levitt, *J. Gen. Physiol.*, 113, 789-784, (1999)
[5] W. Nonner, D. Chen, B. Eisenberg, *J. Gen. Physiol.*, 113, 773-782, (1999)
[6] W. Cornell, P. Cieplak, C. Bayly, I. Gould, K. Merz, D. Ferguson, D. Spellmeyer, T. Fox, J. Caldwell, P. Kollman, *J. Am. Chem. Soc.*, 117, 5179-5197, (1995)
[7] P. Kollman, R. Dixon, W. Cornell, T. Fox, C. Chipot, A. Pohorille, In *Computer Simulation of Biomolecular Systems*, W. F. van Gunsteren, P. K. Weiner, A. J. Wilkinson, eds., 3, 83-96, (1997)
[8] G. Moyna, H. Williams, R. Nachman, A. Scott, *Biopolymers*, 49, 403-413, (1999)
[9] W. Ross, C. Hardin, *J. Am. Chem. Soc.*, 116, 4363-4366, (1994)
[10] MacKerell, J. Wiorkeiwicz-Kuczera, M. Karplus, *J. Am. Chem. Soc.*, 117, 11946-11975, (1995)
[11] MacKerrell, *J. Phys. Chem. B*, 102, 3586-3616, (1998)
[12] S. Feller, D. Yin, R. Pastor, A. MacKerell, *Biophys. J.*, 73, 2269-2279, (1997)
[13] W. Jorgensen, J. Tirado-Rives, *J. Am. Chem. Soc.*, 110, 1657-1666, (1988)
[14] J. Aqvist, *J. Phys. Chem.*, 94, 8021-8024, (1990)
[15] J. Chandrasekhar, D. Spellmeyer, W. Jorgensen, *J. Am. Chem. Soc.*, 106, 903-910, (1984)
[16] *Semiempirical Methods of Electronic Structure Calculation*, Edited by G. Segal, Part A: Techniques, Plenum Press, 1977
[17] *Semiempirical Methods of Electronic Structure Calculation*, Edited by G. Segal, Part B: Applications, Plenum Press, 1977
[18] *Accurate Molecular Structures. Their Determination and Importance*, Edited by A. Domenicano, I. Hargittai, Oxford University Press, 1992
[19] R. Sarges B.Witkop, *J. Am. Chem. Soc.*, 87, 2011-2020, (1965)
[20] K. Lee, S. Huo, T. Cross, *Biochemistry*, 34, 857-867, (1995)
[21] W. Hu, N. Lazo, T. Cross, *Biochemistry*, 34, 14138-14146, (1995).
[22] R. Ketchem, W. Hu, T. Cross, *Science*, 261, 1457-1460, (1993).
[23] Burkhart, N. Li, D. Langs, W. Pangborn, W. Duax, Proc. Natl. Acad. Sci. USA, 95, 12950-12955, (1998) *http://www.ncbi.nlm.nih.gov/entrez*
[24] J. Pople, G. Segal, *J. Chem. Phys.*, 44, 3289-3296, (1966)
[25] Pankratov, *Ind. J. Chem.*, 34B, 689-694, (1995)
[26] Pankratov, A. Shchavlev, *Monatsh. Chem.*, 129, 1007-1017, (1998)
[27] Clark, D. Kilcast, *J. Chem. Soc. A.*, 21, 3286-3290, (1971)

[28] P. Politzer, S. Whittenburg, T. Warnheim, *J. Phys. Chem.*, 86, 2609-2613, (1982)
[29] J. Park, K. No, M. Jhon, H. Scheraga, *J. Comput. Chem.*, 14, 1482-1490, (1993)

In: Progress in Chemometrics Research
Editor: Alexey L. Pomerantsev, pp. 119-135
ISBN 1-59454-257-0

Chapter 10

SOFTWARE FOR INTERACTIVE CURVE RESOLUTION USING SIMPLISMA

***Andrey Bogomolov*[1]*, Michel Hachey*[2] *and Antony Williams*[2]**
Advanced Chemistry Development, Inc., [1]Russia, [2]Canada, bogomol@acdlabs.ru

ABSTRACT

Self-modeling curve resolution is one of the most sophisticated methods of multivariate data analysis. The approach can be applied to a wide range of practical problems facing the modern analytical laboratory and has proven to be a powerful tool for qualitative mixture analysis and characterization. However, in spite of the widespread applicability of the approach, curve resolution algorithms mostly remain an advanced theory rather than a practical tool in the hands of chemists. One reason for this situation is the lack of a convenient interface for routine use that can be applied by chemists and spectroscopists without requiring an extensive statistical expertise. The creation of user-friendly and intuitive software tools provides a bridge between statistics and chemistry, theory and practice.

Software for interactive curve resolution based on our implementation of the SIMPLISMA algorithm (SIMPLe-to-use Interactive Self-modeling Mixture Analysis) has been developed. Two practical mixture samples taken from the available literature were used to demonstrate how curve resolution can be easily and responsibly employed by a non-expert operator through an interactive interface. The advantages of SIMPLISMA compared to other curve resolution techniques are discussed. Method limitations are considered and possible workarounds suggested for complex analytical cases.

Keywords: curve resolution, SIMPLISMA, software

INTRODUCTION

In the modern analytical laboratory hyphenated techniques and multi-dimensional experiments are now common. Indeed, for maximum data content, they dominate the classical forms of analysis and the resulting data matrices are now common outputs. The necessity to extract hidden information about a mixture sample from large experimental data sets, such as matrices of spectral data, has given birth to a number of powerful methods and algorithms.

In some cases, the solution is sought from incomplete data where the problem is mathematically ambiguous. In other cases, no explicit information about a mixture sample is available *a priori* other than the measured data matrix itself. Nonetheless, a wide number of applications frequently require the ability to overcome such obstacles in a way that resolves the data matrix in question into its underlying set of physical response curves. This problem, known as curve resolution, has particular practical importance. For example, it can be applied to extract concentration profiles and the related spectra of pure components from a series of spectra which have been generated as a function of time. Suitable data matrices for study can be obtained by monitoring a chemical reaction spectroscopically, or through the process of chromatographic separation.

Mathematically, curve resolution is expressed by Eq. (1).

$$D = S \times P + E \qquad (1)$$

where D is the (*rxc*) raw data matrix with spectra in columns, S the (*rxn*) matrix of pure component spectra (*n* is the number of components), P the (*nxc*) matrix of the component concentration profiles, and E the (*rxc*) matrix of errors. The general curve resolution problem is schematically shown in Fig. 1.

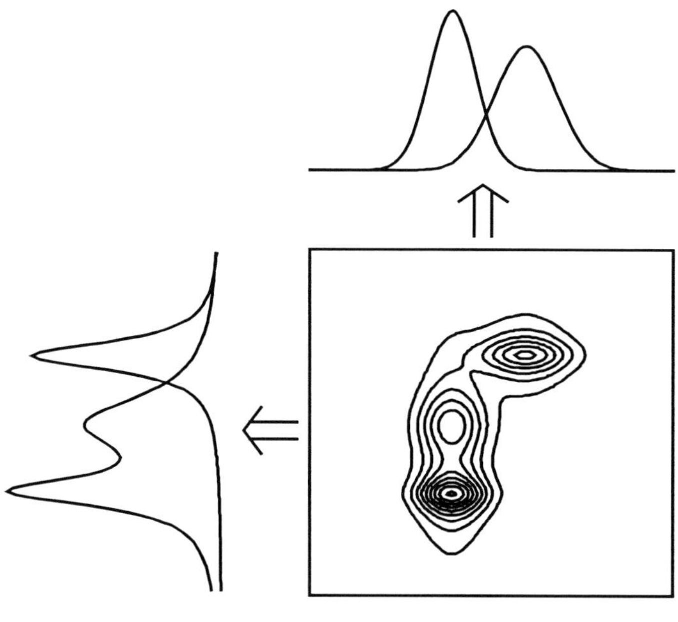

Fig.1. Curve resolution schematic

During the last two decades, a number of excellent works have been published on the curve resolution problem. Comprehensive reviews were made by Hamilton and Gemperline [1] and in the book by Malinowski [2]. Those techniques that do not use any reference information, such as component spectra or known concentrations, to perform the resolution, belong to the family of self-modeling (also called soft-modeling) methods. Capable of deriving full qualitative characterization of an unknown mixture where extreme data deficiency exists, self-modeling curve resolution remains one of the greatest challenges of chemometrics.

Most of the self-modeling methods are based on the apparatus of factor analysis and consist of the following main blocks varying mostly in how this or that step is performed. Firstly, the number of components is determined, usually, by PCA. Then, initial estimates of the component spectra or concentration profiles are obtained. Finally, an iterative improvement process is run that imposes specific constraints on the spectra and/or profiles followed by least-squares regression to obtain new estimates. The optimization process is repeated until a convergence criterion is met. The need to apply limiting constraints is caused by the necessity to reduce the mathematical ambiguity of the system in the absence of reference information. This is characteristic of self-modeling curve resolution. Indeed, there are an infinite number of ways to decompose a matrix in accordance with Eq. (1). To find a unique solution (or an acceptably narrow set) that belongs to the physical data space, some assumptions about the nature of the data and results are typically made. The appropriate constraints are chosen individually in each case, depending on the system properties. For example, the requirement of non-negativity can be applied to the spectra and the concentrations, unimodality (a curve shape with a single maximum) can be required of chromatographic concentration profiles of individual components. The more system-specific constraints are applied, the lesser the ambiguity and the more reliable the solution.

The main weakness of this approach is that it is extremely sensitive to the number of components preliminarily determined by PCA. Moreover, the assumptions about the spectra and concentration profiles often are not strict enough to provide a unique solution. This effect, known as "rotational ambiguity", may lead to distortions in the result.

The present paper is devoted to a curve resolution algorithm that does not follow the above commonly-used strategy. SIMPLISMA (SIMPLe-to-use Interactive Self-Modeling Mixture Analysis) algorithm was invented by Willem Windig and implemented in the MATLAB program. The code which was published [4] and made available to the scientific community allowed wide recognition for the approach. SIMPLISMA possesses a number of advantages compared to factor-based methods. One of the most important features of the approach is interactivity, which allows the operator to guide the process through every step.

A number of practical problems associated with mixture analysis can be addressed by curve resolution approaches and, in particular, SIMPLISMA. It provides a powerful means for analysis of samples that undergo changes with respect to time without requiring prior information of either the mixture composition or component spectra and concentrations. One application is to extract component spectra and interpret them directly, or search for likely matches in a reference library. This provides some qualitative information on the mixture constituents and allows the chemical composition of an unknown mixture to be suggested. In a second application, the concentration profiles of the components may serve for interactive process monitoring. Since it is a fast resolution technique, the SIMPLISMA algorithm can be used in real-time monitoring systems. A third application is to use SIMPLISMA to study chemical reactions. Kinetics and mechanism or equilibrium constants can be deduced using the resolved concentrations. In yet another application, the method can be used to resolve co-eluting signals in hyphenated techniques such as high-performance liquid chromatography with diode array detection (HPLC/DAD). Using curve resolution results in quantitative analysis is also possible. This does, however, require an additional stage of calibration for each individual component.

Even with the multitude of possible applications for SIMPLISMA, it still remains an instrument for experts to apply rather than as a general laboratory tool. A number of papers have been published on various extensions to the basic algorithm capable of improving its performance [5-7]. Nevertheless, the area of practical usage of the approach is still limited to a small group of enthusiastic individuals. This situation is due, in part, to the mistaken perception that a high learning curve exists for applying such an algorithm. Unfortunately, implementations using expert generic software tools such as MATLAB functions or Excel macros are not necessarily suited to change this perception for the mainstream users. As the complexity of the mathematical apparatus for data mining continues to grow, the availability of software designed to provide greater ease of use can become the key factor in putting advanced theory into practice. Modern methods for multivariate data analysis offer an excellent illustration of this trend, and SIMPLISMA is not an exception.

In this paper, we will present the first commercial implementation of SIMPLISMA in a ready-to-use intuitive interface designed to make the algorithm more readily accessible to chemists as a routine analysis tool. Two experimental datasets available from the literature have been used to demonstrate the software. Some practical aspects of the application of this method are considered. Possible limitations of the approach are analyzed, and workarounds suggested in complex cases.

EXPERIMENTAL

Basic Algorithm Definitions

The basic idea behind the algorithm is fairly simple. SIMPLISMA assumes that there are variables in the data (such as spectral wavelengths or wavenumbers) that experience contributions from only one of the components in the mixture. Such variables are called pure variables. The spectral intensity of this pure variable represents a concentration profile (provided that Beer's Law is obeyed). If every mixture component has at least one pure variable in the data, then the corresponding component spectra can be resolved in the least-squares sense.

To find the pure variables, SIMPLISMA calculates a purity function as given by Eq. (2):

$$p_{j1} = \frac{\sigma_j}{\mu_j + \alpha} \tag{2}$$

where p_{j1} is the first purity value of the j-th variable; σ_j the standard deviation of the j-th variable calculated by Eq. (3); μ_j the mean of the j-th variable calculated by Eq. (4); and α a constant offset. The meaning of the α parameter in Eq. (2) will be explained shortly. The standard deviation, σ_j, is given by

$$\sigma_j = \sqrt{\tfrac{1}{c}\sum_{i=1}^{c}(d_{ij} - \mu_j)^2} \tag{3}$$

where c is the number of spectra, and d_{ij} an element of the data matrix (spectra in rows) that belongs to the i-th row and the j-th column. The mean, μ_j, is given by

$$\mu_j = \tfrac{1}{c}\sum_{i=1}^{c} d_{ij} \tag{4}$$

Windig discovered that the ratio of the standard deviation to the mean for the same variable j correlates to the purity of the variable. Because the purity, the standard deviation, and the mean have the same set of variables as spectra, they can be represented as spectral-like curves along the horizontal spectral axis, which therefore facilitates their interpretation. The first pure variable is taken as a maximum of the first purity spectrum.

SIMPLISMA resolves the components one by one. To find the next pure variable, the effect of the chosen component is mathematically eliminated from the purity spectrum. Simultaneously, it is also eliminated from the standard deviation spectrum, and thus, a so-called purity-corrected standard deviation spectrum is obtained. This curve is a reflection of the residual variance in the data not modeled by the previously chosen components.

To illustrate the meaning of the α constant in Eq. (2), consider a non-informative spectral region that does not contain any noticeable signal for a given mixture, only random noise around zero. In accordance with Eq. (4), the mean for such variables tends to zero. Since the purity spectrum is the ratio of the standard deviation to the mean, this would then result in division by a value close to zero if α was also zero. The result in this case would be an abnormal overestimation of the apparent purity value. Adding a constant to the denominator of the ratio fixes the problem, suppressing the effect of overestimated purity. Usually, the α value is expressed as a percent of the mean value. To adequately treat the apparently high purity of "silent" spectral regions, the α value is set to 3% for "regular" spectral data using the conventional SIMPLISMA approach, and α=20% for the inverted 2nd derivative method (see below). Alternatively, non-informative regions can be manually marked as "inactive" and thus, excluded from the variable search.

One of the most important parameters when analyzing the mixtures by multivariate methods is the number of components n. Both an insufficient number or excessive number of factors may lead to incorrect results. SIMPLISMA offers a number of criteria to help the user to decide when to stop modeling. The decision is made by comparing the results at the current step with those of the previous one and returning back if necessary. The main principles of deducing n will be shown below using experimental data. Note, however, that because SIMPLISMA factors are non-orthogonal, the number of components does not necessarily coincide with the number found using PCA.

One of the main assumptions of the SIMPLISMA algorithm is that there is a pure variable for each mixture component. In real world mixtures, with many components and a high degree of signal overlap, this requirement often cannot be met. Fortunately, in most cases the limitation may be overcome by a modified SIMPLISMA approach based on the 2nd derivative. The 2nd derivative has wonderful properties: it improves the resolution of overlapped signals while keeping the linearity of the response. The only inconvenience in terms of analysis is that the minima in the 2nd derivative correspond to positive peaks in the data. Since SIMPLISMA works on positive data only, prior to running the algorithm the derivatives are inverted and the negative values are set to zero. The inverted 2nd derivative is then used by SIMPLISMA instead of the raw data to detect pure variables and corresponding component concentration profiles.

Data

Two datasets taken from the literature are used in this paper to demonstrate curve resolution using SIMPLISMA implemented in the software being described.

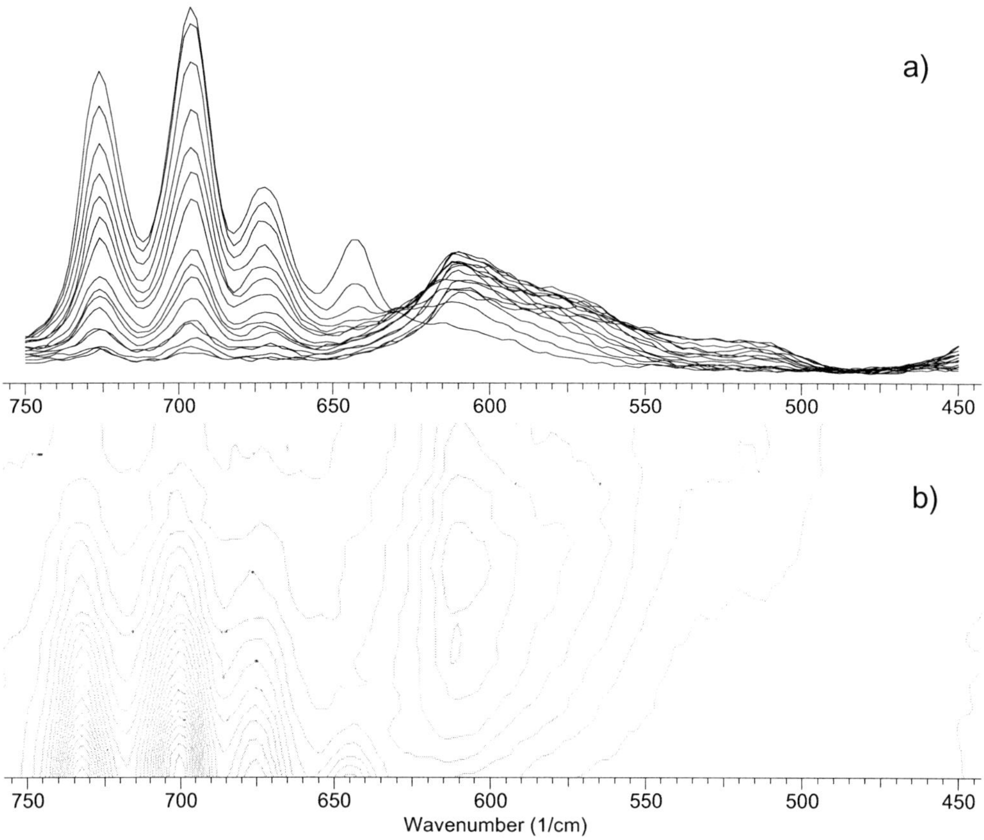

Fig. 2a,b. TMOS data: a) series of Raman spectra; b) contour map.

One of these data sets (hereafter called TMOS data) consists of sixteen Raman spectra following a hydrolysis reaction of TMOS (tetra-methyl orthosilicate) over time (Fig. 2a,b). The peak at wavenumber 644 cm^{-1} represents the original TMOS. Under the proper reaction conditions, hydrolysis products are formed, with peaks at 673, 696, 726 cm^{-1}. The next reaction results in a condensation product, with a peak at 608 (a shoulder at 586 cm^{-1}), followed by a further condensation product associated with a peak at 525 cm^{-1} [1]. These spectra result from a mixture of TMOS, hydrolysis products, and condensation products.

Another dataset (hereafter called the NIR data) includes 70 mixture samples from the paper [5]. An orthogonal mixture design was used to create the samples including the following five components: 2-butanol, methylene chloride, methanol, dichloropropane, and acetone. The samples were analyzed by NIR spectroscopy. The series of NIR spectra is shown in Fig. 3

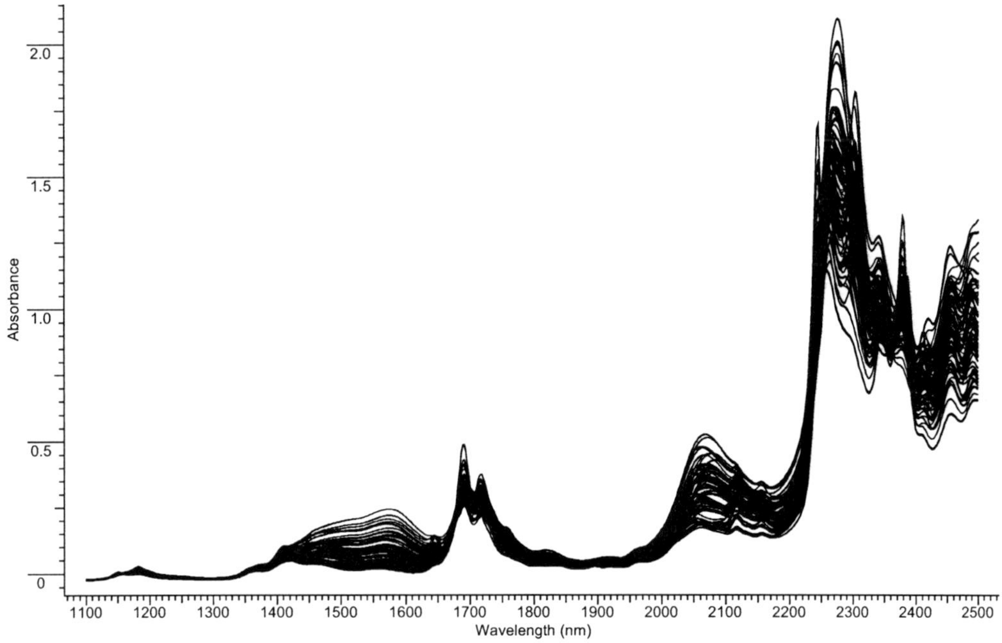

Fig. 3. NIR data.

Both datasets were taken from an open source [4] and imported into the software using built-in converters.

Software

The SIMPLISMA algorithm was integrated into the interface of ACD/UV-IR Processor software, version 8, a product of Advanced Chemistry Development, Inc [8]. Being a module of ACD/SpecManager, the global spectroscopy software, ACD/UV-IR Processor has been designed for working with spectra in the optical range (UV-Vis, NIR, IR, Raman, and related techniques). The product includes a comprehensive set of tools for collecting data (importing from a large number of instrumental formats), visualization, processing, analysis, reporting, and storage. The optionally supplied Spectral Data Management System (this combination is called ACD/UV-IR Manager) adds databasing capabilities including spectral searching of reference libraries and user-built databases of spectra.

Results and Discussion

Resolving Curves

Fig. 4 shows the standard deviation (b), the mean (c), and the first purity spectra (d) of the TMOS reaction data along with the raw series (a). The offset value is set to 3% as recommended earlier.

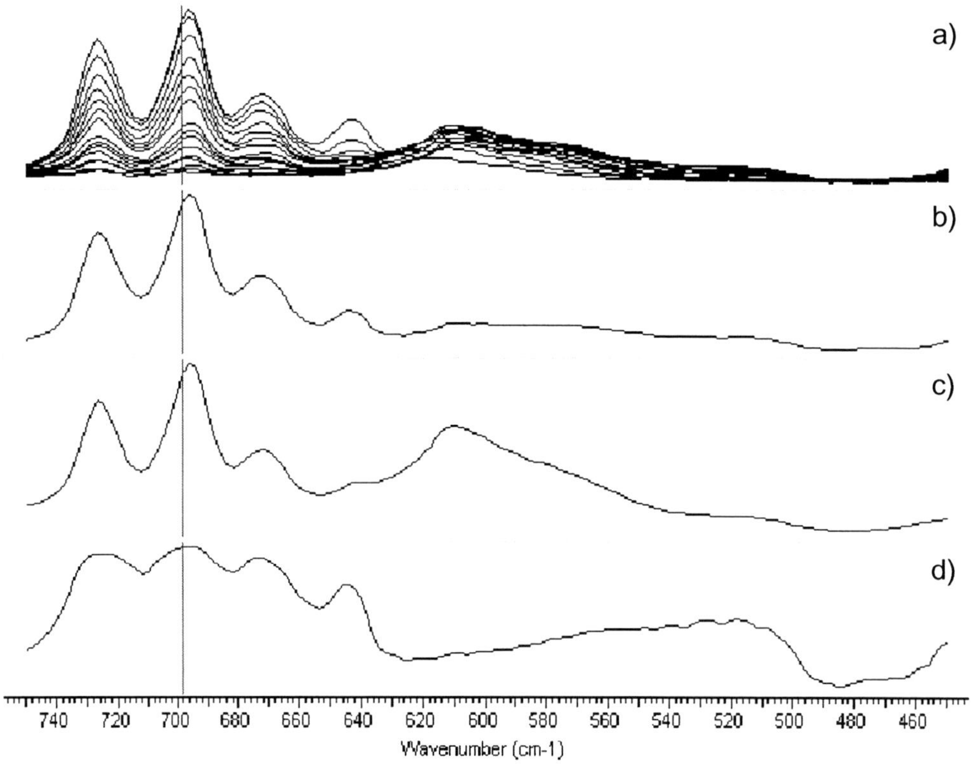

Fig. 4a-d. Resolving TMOS data, step 1: a) raw data series, b) standard deviation spectrum, c) mean spectrum, and d) first purity spectrum (α=3%).

The first purity variable is detected as a maximum at 698 cm^{-1} on the purity spectrum. Pressing the Next button in the software interface calculates the second purity spectrum and finds the second pure variable at 516 cm^{-1} (Fig. 5). Since the effect of the previous variable was removed, the second purity spectrum and purity-corrected standard deviation have zero intensity at the first pure variable (Fig. 5c,d). This is a common rule for all previously selected pure variables. The purity-corrected standard deviation spectrum (Fig. 5d) can be thought of as residual information not explained by all previously chosen variables.

The next figure illustrates the effect of the offset α. As we take the next step, the third purity spectrum is calculated (Fig. 6b, solid line) giving a third pure variable maximum at 644 cm^{-1}. However, setting α=0% results in significant changes in the purity spectrum shape (Fig. 6b, dashed line) so that its maximum moves to 478 cm^{-1} (indicated by the dashed vertical line in Fig. 6). It is evident that this pure variable is abnormal and highly suspect because it belongs to a data region that looks like it should contain only noise. This can be seen in the raw data (Fig. 6a) and confirmed by the purity-corrected standard deviation (Fig. 6c). It offers a vivid demonstration of the power the offset has to suppress misleading purity findings in "silent" spectral regions.

The fourth pure variable is found at 578 cm^{-1} (Fig. 7). Resolving four components results in the component spectra shown in Fig. 7c. The process of component extraction can be continued. For the final solution, however, it is necessary to decide how many significant constituents are present in the mixture and, therefore, how many components should be retained in the model. The next section considers this problem in detail.

It is important to understand that every incremental step of the algorithm results in the recalculation of the whole set of resolved spectra and concentrations (not only adding a subsequent pair). Thus, until all of the significant components are included into the model, the resolved curves are only an intermediate solution. These should not be thought of as spectra and concentrations of the mixture components.

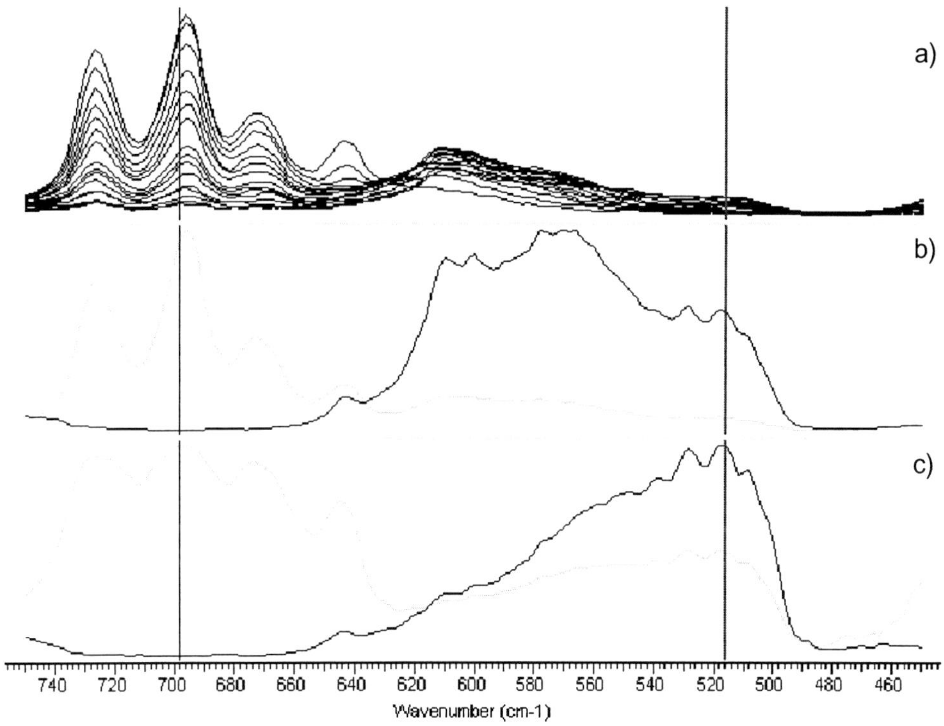

Fig. 5a-c. Resolving TMOS data, step 2: a) raw data series, b) second purity-corrected standard deviation spectrum, and c) second purity spectrum (α=3%). Previous standard deviation and purity spectra are shown in the corresponding plots in gray (equally scaled).

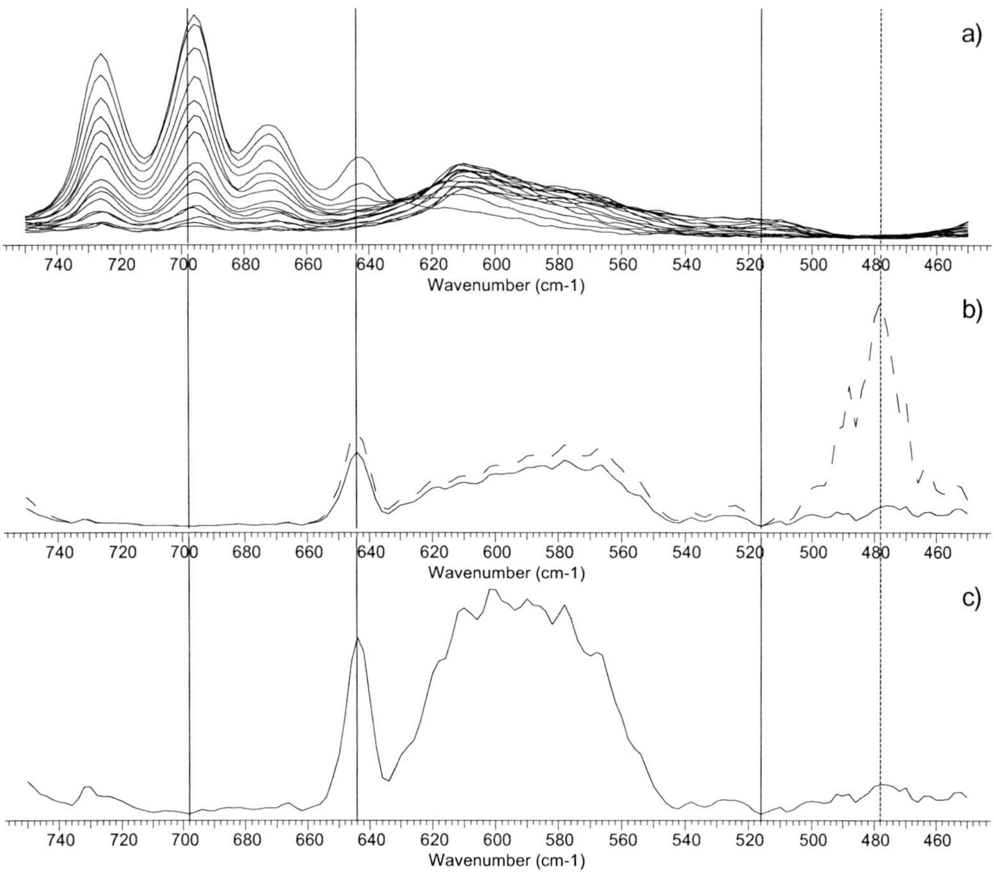

Fig. 6a-c. Resolving TMOS data, step 3: a) raw data series, b) third purity spectrum at α=3% (solid line) and α=0% (dashed line), and c) third purity-corrected standard deviation spectrum.

Deducing the Number of Components

One of the main conditions for obtaining satisfactory results with SIMPLISMA is that the number of components *n* is detected correctly. An underdetermined model (insufficient *n*) results in incorrect shapes for the resolved curves. In the case of overfitting, excessive components start modeling the error (noise) in the data. Either case leads to a distortion of the solution; thus, it is always recommended to pay particular attention to deducing the number of mixture components.

Determination of the number of mixture components in the SIMPLISMA algorithm is carried out by analyzing the results and diagnostics of the current algorithm step: purity and purity-corrected standard deviation spectra, resolved curves, residual data, and statistics. Often, an extra step should be performed to attempt to detect an excessive component in the model. If the $n+1$ component is excessive but *n* is not, the number of mixture components is equal to *n* and the results of the *n*-th step are therefore considered to be the final solution. Therefore, correct recognition of an excessive component is a key feature of the entire algorithm.

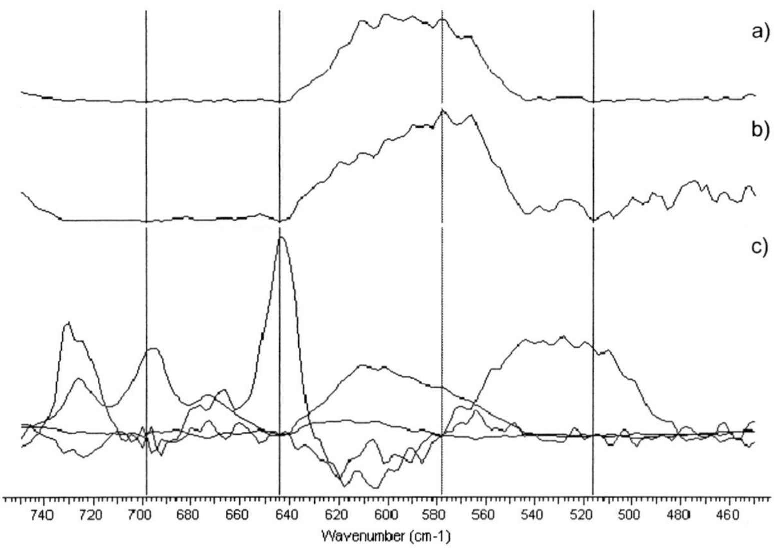

Fig. 7a-c. Resolving TMOS data, step 4: a) fourth purity-corrected standard deviation spectrum, b) fourth purity spectrum (α=3%), and c) resolved pure component spectra.

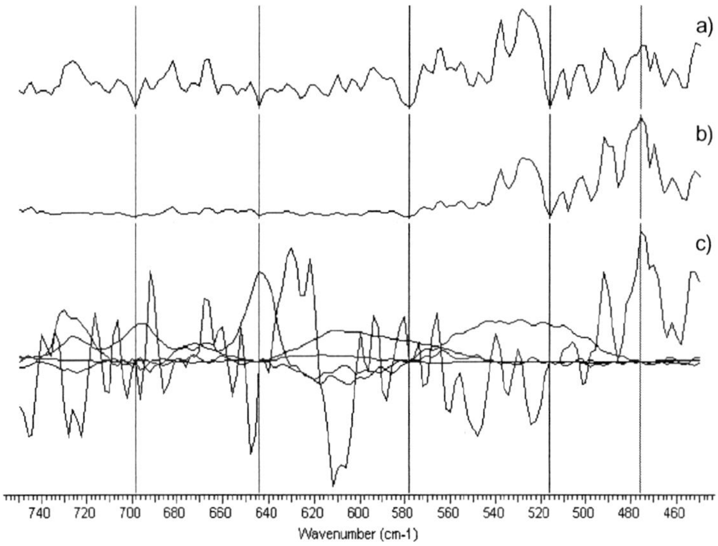

Fig. 8a-c. Resolving TMOS data, step 5: a) fifth purity-corrected standard deviation spectrum, b) fifth purity spectrum (α=3%), and c) resolved pure component spectra.

SIMPLISMA offers a number of tools capable of reflecting the information about the number of components to be used in the model. Some of them are generally accepted standards for multivariate modeling while others are genuine attributes of the approach. Here, we are going to represent only some of them having, in our opinion, the most practical value.

The residual standard deviation (RSD) is a straightforward criterion for deducing n when the experimental error is known. This method is based on an evident statement: modeling accuracy should not exceed that of the data. In this approach, the known error is used as a cut-off threshold for the n components. While the RSD is above the error value, the components are considered significant for the model and are retained. In contrast, those lying below the threshold are not used because they are considered redundant. Note that the data should have an error distribution close to normal to use this method. For the real data, however, the experimental error often cannot be calculated with acceptable accuracy or its distribution function departs from normality. For this reason, various indirect statistical or empirical estimates are widely used in practice.

Some preliminary conclusions can be made on the basis of visual inspection of the residuals, resolution results, and subsidiary plots. It is always recommended to view the residuals (difference between the raw and reproduced data) at each step. When the modeling is incomplete, the residuals have both systematic positive and negative features, which often look like a mirror reflection. With the correct model residuals contain only extracted errors and should therefore look like random noise. Additionally, in a successful model, the absolute values of the residuals should be comparable to the known or expected experimental error. Also, negative peaks or other unexpected features in resolved concentrations and spectra can often be evidence of an underdetermined model.

The characteristics of an excessive component are as follows. A purity-corrected standard deviation spectrum lacks explicit features (peaks) and looks noisy. The purity spectrum shape may be diagnostic in a similar manner. Thus, comparing Fig. 7a,b and 8a,b one may conclude that the fifth component is excessive for TMOS data modeling. Non-monotone or erratic behavior of the resolved concentration profiles (when systematic changes are expected) can also be characteristic of overfitting. Similarly, if the last resolved spectrum demonstrates the effect of random changes, the variable should be called into question as redundant. Based on this, Fig. 8c serves an additional confirmation to the above conclusion that the number of components necessary for TMOS data does not exceed four.

The total spectral intensity (TSI) plot is one of the specific diagnostic tools of SIMPLISMA. The TSI represents each spectrum as a point on a curve by adding its intensities. This plot is a simplified view of the total variance in the data with respect to samples. Similar plots can be calculated from regenerated data at any application step of the algorithm and plotted over TSI. Visual comparison of the curves in Fig. 9 suggests that three components are not enough for modeling TMOS data; however, the model with four components adequately describes the data. A quantitative measure of the difference between the curves *vs.* component number can be used instead of visual comparison. For this purpose, Windig suggested using the relative residual sum of squares (RRSSQ), the square root of the squared differences between the original and reproduced TSI divided by the square root of the sum of squares of the TSI. Because RRSSQ is relative, it is varied between 0 and 1, coming to a visual leveling off at the appropriate n. TSI plots may also serve for the detection of unusual or outlying samples.

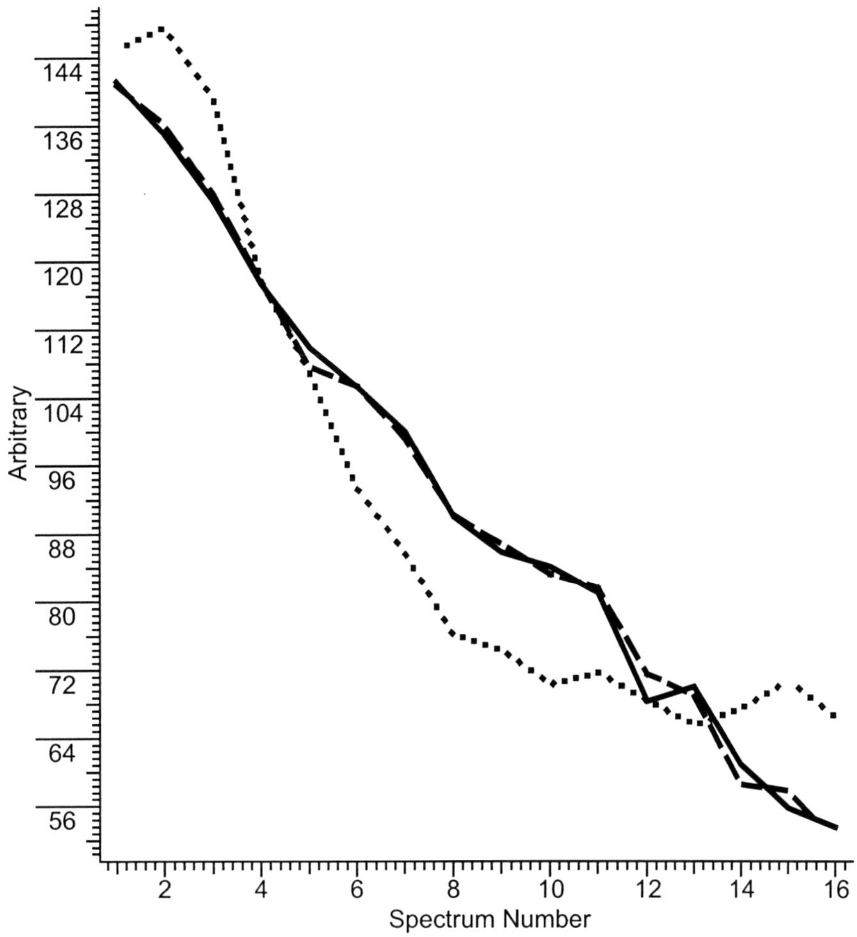

Fig. 9. Total spectral intensity (TSI) plots for TMOS dataset: raw data (solid line), and regenerated data for three components (dotted line) and four components (dashed line).

A new diagnostic tool has been suggested by the authors of the present paper. We used an empirical indicator function (IND) by Malinowski [2]. Originally applied to principle component analysis (PCA), the IND function has shown good potential for automatic detection of the number of components, which is taken as a first minimum of the function. We used the same formula but replaced the PCA factors by SIMPLISMA components. The resulting IND function for TMOS data has a minimum at n=4 (Fig. 10).

The analysis performed in this section leads to the conclusion that the TMOS data describe a four-component system. The resolved concentration profiles are shown in Fig. 11. One may also conclude that the studied reaction includes more than one stage. While two components are consumed, one product is formed. Besides, there is an intermediate product corresponding to a curve with a maximum.

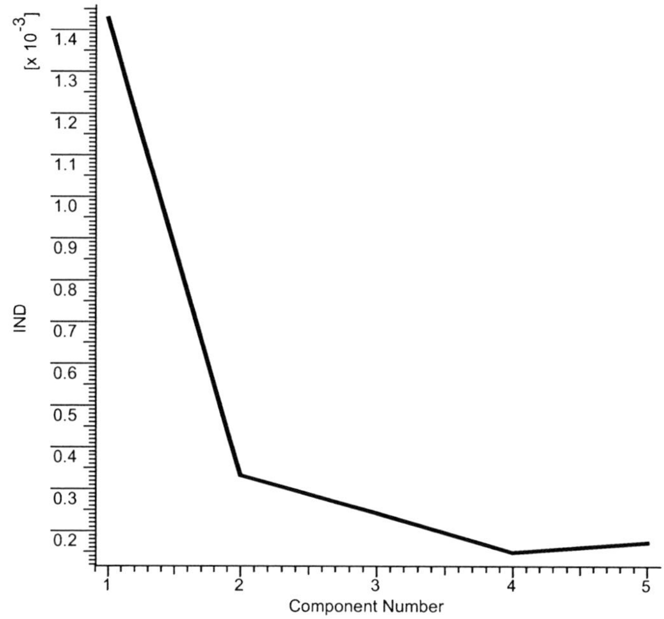

Fig. 10. Indicator function (IND) for TMOS data

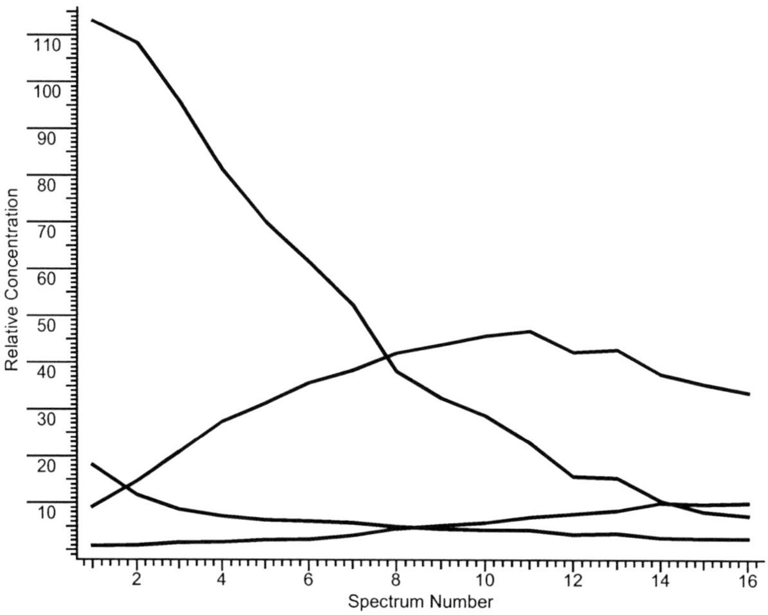

Fig. 11. Resolved concentration profiles for TMOS data.

SIMPLISMA with Inverted 2nd Derivative

One of the main assumptions of the SIMPLISMA algorithm is that there is a pure variable for each mixture component. In real world mixtures, with many components and a high degree of signal overlap, this requirement often cannot be met. Fortunately, in most cases, the limitation may be overcome by a modified SIMPLISMA approach based on the 2nd derivative [5]. This is possible due to a wonderful property of the derivative to improve resolution of overlapped signals while keeping the linearity of the response. As an example, we used NIR data for a five-component mixture shown in Fig. 3.

The second derivative was calculated using the method of Savitsky-Golay [9] applying necessary smoothing to suppress the noise. In the 2nd derivative the minima correspond to positive signals in the data. Since SIMPLISMA works on positive data only, prior to running the algorithm the derivatives are inverted and the negative values are set to zero as shown in Fig. 12a. The inverted derivative is then used by SIMPLISMA instead of the raw data. Fig. 12b shows the pure component spectra resolved using the inverted 2nd derivative approach.

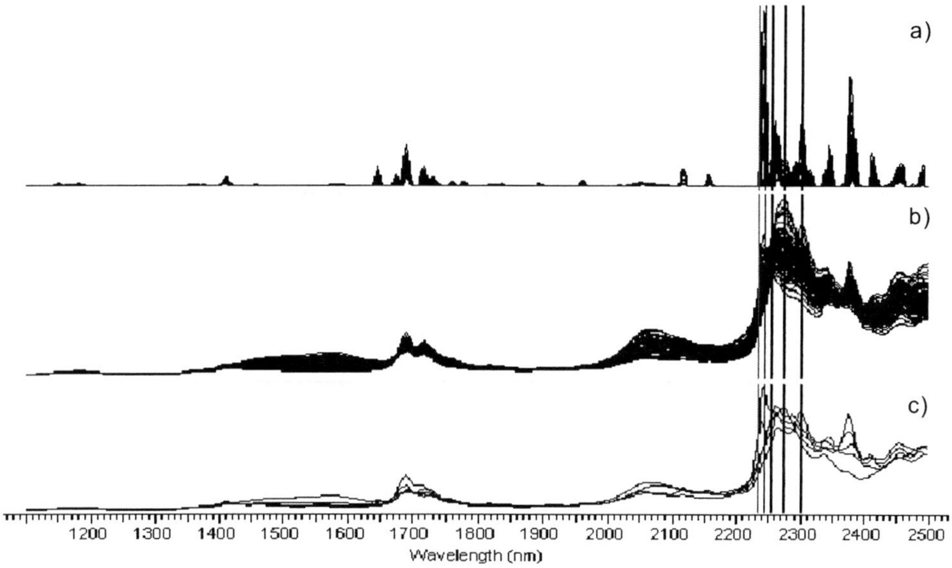

Fig. 12a-c. Resolving NIR data (α=10%): a) inverted second derivative, b) raw data, and c) resolved pure component spectra.

Advantages and Limitations of SIMPLISMA

What are the advantages of SIMPLISMA compared to other curve resolution techniques and what makes it unique? First of all, its interactivity—the ability to control the resolution process at every step and interact with the process if necessary by changing options and returning to the previous step when appropriate. The diagnostic tools are intuitively understood by spectroscopists and chromatographers alike. SIMPLISMA is free of time-intensive cycles which make it much faster than most other techniques. Not using the curve shape constraints provides wide applicability. Finally, all of the above advantages in combination with the high level of intelligence of the algorithm provide a good basis to create automated systems.

SIMPLISMA is a powerful tool capable of solving a wide range of practical problems. Nevertheless, it has its own weaknesses and it would be unfair to exclude these from discussion. With an awareness of these limitations it may be possible to overcome them in certain cases. Since SIMPLISMA applies relative purity rather than absolute, there is no guarantee that all the selected variables are actually pure. To determine the validity it is possible to use one of the known algorithms for purity assessment. Since it is a variance-based method, it is insensitive to components whose concentration does not change noticeably from sample to sample compared to other components in the series (for example, a solvent). One suggestion to solve this problem is to add more measurements to make the component concentration vary.

In the end, some datasets may be too complex for even the 2nd derivative method to work. In such cases, one can try to split the problem into several subsets, e.g., resolve each group of overlapped peaks in an HPLC/DAD chromatogram individually.

Conclusions

Laboratory workers using SIMPLISMA report a time savings ranging from 15 to 50 hours per quarter for each initiated worker. With such an advantage in time available to the initiated why do many scientists who could benefit from this technology not utilize it? The answer appears to be that people remain intimidated by multivariate mixture analysis and the perception is that applications are necessarily complex. While SIMPLISMA was developed in a manner emphasizing ease of use (self-modeling, spectrally meaningful displays and mathematically simple) it has until now only been implemented within expert software systems which are not themselves user friendly.

The SIMPLISMA algorithm has been implemented into a commercial software package in order to facilitate its use. In this paper, we have reviewed some of the basic underlying principles and gave two illustrative examples that hopefully help demystify the application of SIMPLISMA. Due to the highly intuitive and interactive interface, which has been delivered, both spectroscopists and chemists now have access to the technology, not only statistical experts. This curve resolution method emphasizes spectral-like data displays which are physically meaningful without requiring any overt mathematical knowledge. By applying several rules of thumb semi-automatically, SIMPLISMA lets users produce results even in the absence of special knowledge.

The use of the IND function has been suggested as a new diagnostic tool for deducing the number of components to be used in a SIMPLISMA model. This tool may be used to identify the correct number of components simply by finding the minimum on the graph.

The advantages and limitations of the algorithm have also been analyzed and possible workarounds suggested that may be applied to complex analytical cases.

Acknowledgments

We are grateful to Willem Windig for his help with implementation of the SIMPLISMA algorithm. Elsevier Science V.B. is acknowledged for making the data available online. The SIMPLISMA algorithm has been licensed to Advanced Chemistry Development Inc. by the Eastman Kodak Company.

REFERENCES

[1] J. C. Hamilton, P. J. Gemperline, *J. Chemom.*, 4, 1-13, (1990).

[2] E. R. Malinowski, *Factor Analysis in Chemistry, Third Edition*, Wiley-Interscience, New York, 2002.

[3] W. Windig, J. Guilment, *Anal. Chem.*, 63, 1425-1432, (1991).

[4] W. Windig, *Chemom. Intell. Lab. Syst.*, 36, 3-16, (1997).

[5] W. Windig, D. A. Stephenson, *Anal. Chem.*, 64, 2736-2742, (1992).

[6] W. Windig, B. Antalek, J. L. Lippert, Y. Batonneau, C. Bremard, *Anal. Chem.*, 74, 1371-1379, (2002).

[7] S. Gourvénec, D. L. Massart, D. N. Rutledge, *Chemom. Intell. Lab. Syst.*, 61, 51-61, (2002).

[8] ACD/SpecManager software, version 8.0, ACD/Labs, Toronto, Canada, 2004.

[9] Savitsky, M. J. E. Golay, *Anal. Chem.*, 36, 1627-1639, (1964).

In: Progress in Chemometrics Research
Editor: Alexey L. Pomerantsev, pp. 137-149
ISBN 1-59454-257-0

Chapter 11

MODELING OF DATA STRUCTURE FOR MULTIVARIATE CALIBRATION IN ATOMIC EMISSION SPECTROMETRY

E. Shabanova and I. Vasilyeva
Institute of Geochemistry SB RAS, Irkutsk, Russia, shev@igc.irk.ru

ABSTRACT

The conditions of application of PCR-calibration were studied to purpose of the taking into account of spectral and nonspectral influences in atomic emission spectrometry. This paper presents different structure of experimental data used for multivariate calibration; and their effecting on the quality of a training and prediction. The "range of relative bias" assess was offered to select the best calibration at issue versions of *n*-dimensional calibrations.

Keywords: atomic emission spectrometry; modeling of structure of experimental data; multivariate calibration; range of relative bias

INTRODUCTION

The clear guiding principles to apply a multivariate statistical analysis of data (MSA) for concrete tasks in many knowledge domains are not available yet. The atomic emission spectrometry (AES) belongs to one of knowledge domains. However, the MSA methods seem to be promising in AES calibration, because they can ensure consideration of latent relationships, which essentially influence on physical processes of obtaining the arc atomic spectra.

THEORY

The atomic emission spectrometry is based on the simple link between entry chemical and output analytical signals. The feature of this link are described by the empirical formula Lomakin-Scheibe [1]

$$I = aC^b, \tag{1}$$

where I is an intensity of spectral line, C is contents of element, a and b are some constants depending on the properties of radiating atoms and excitation manner of spectrum. Eq. (1) is a linear calibration in logarithm coordinates within the range of one or two orders of the analyte content.

The array of analytical signals and element contents, which specify calibration samples, is named as "training" sampling. The analyst finds the calibration dependence as some regression function. Its coefficients are commonly assessed and calculated using the ordinary least square method (OLS). Naturally, that the calculated function of responses includes the random noise of the measuring system (sample preparation + spectrometer) and systematic bias caused by varying parameters, i.e. variations of total chemical composition, of training set of certified reference materials (CRM). The calibration coefficients obtained allow the analyte contents to be calculated in samples of training set and to assess the contribution of the systematic bias to the analytical results. For this purpose, the relative bias of result of analyte determination in the each j-th sample of training sampling is provided by the following formula

$$R_j = \frac{C_{cal.} - C_{cert.}}{C_{cert.}} \times 100\%, \tag{2}$$

where $C_{cal.}$ and $C_{cert.}$ are calculated and certificated contents, accordingly. For the entire set of CRMs within the interval $\Delta C = C_{\max} - C_{\min}$ using the R_j-values calculated, the range of relative bias (RRB) is computed as

$$\Delta R = R_{\max} - R_{\min}. \tag{3}$$

The lower ΔR values, the better the model function fits the experimental data obtained with training set, i.e.

$$\Delta R_{train} \to \min. \tag{4}$$

This interpretation considers the RRB estimation as the characteristic of the whole range ΔC rather than isolated point circumscribing analyte concentration C_j [1], [2]. Therefore the estimation (ΔR) allows the influence of data structure to be compared of training sampling to the calibration quality within interval ΔC_{train}.

It is known that spectral line intensity of analyte depends on: (*i*) its content in a sample; (*ii*) total chemical composition, i.e. major elements concentrations; (*iii*) contents of interferents; (*iv*) presence of molecular bands in spectra. However, the degree of influence of these factors on intensity of lines in spectra of new samples could not be theoretically calculated. The prediction ability of calibration is additionally estimated via the data of test sampling, which consists of the samples with known contents of analyte and displays ΔR_{test} of RRB (Eq. (3)) in interval ΔC_{test}. Starting from the analytical sense, the high quality of calibration model for training and test sets could be described by the next common equation:

$$\begin{cases} \Delta R_{train} \to \min, \\ \Delta R_{test} \to \min. \end{cases} \quad (5)$$

In practice, it is unreal to fulfill Eq. (5) for $\Delta C_{train} \sim \Delta C_{test}$, i.e. to make training and test sets of CRMs that are identical in number and major compositions. Therefore, the test sampling is usually smaller, than training sampling. It can include new spectra of samples from the training set. It does not contradict the strategy of segmentation of a data set, although the citation suggests, that "... if we make this model in order to understand the relationships in our systems, for screening, or to investigate the possibility of using multivariate modeling for future indirect measurements, perhaps we can live with insufficient validation of the screening model" [2], P. 168]. We consider that computing of RRB for X-variable of training and test samplings followed by their comparison (Eq. (2)-(5)) are useful evaluating both the calibration qualities and prediction abilities of model.

The modern spectrometers cause minimal random errors when measuring analytical signals (Y-variables). Due to a good repeatability, the change of intensity with the same content analyte mainly depends on matrix (nonspectral) and interference (spectral) effects. They cause an increase of a systematic error in analytical results. In this case the RRB needs to be regarded as the characteristic of distinctions of samples in chemical compositions. Its value points out a calibration ability to consider the variety of major composition of CRMs in training and test sets.

EXPERIMENTAL

The spectra of training and test sets were acquired with a complete evaporation of a10 mg substance from the channel of graphite electrode in a vertical arc of direct current. The spectral installation includes a diffraction spectrograph DFS-458, generator Vesuvius and multichannel analyzer of emission spectra MAES-10 [3]. The instrument is managed with the commercial software ATOM for recording and processing of the spectral information. In this software a calibration dependence of the 1-st (OLS-1) or 2-nd degrees (OLS-2) are fitted and their statistical characteristics are evaluated via the experimental data.

NUMERICAL EXPERIMENT AND DISCUSSION

Structure of spectral data and construction of *n*-dimensional calibrations were modeled for two analytical techniques: the determination of gold and platinum in insoluble carbonaceous substance; and definition of impurities in crystalline silicon and quartz, exemplified by boron.

Determination of Gold and Platinum in Insoluble Carbonaceous Substance

The insoluble carbonaceous substance (ICS) contains over 50 % of carbon and concentrates noble metals in black shales [4]-[6]. To reduce nonspectral (matrix) influence the calibration samples were prepared using a graphite powder. The calibration samples, marked by S, are prepared by a sequential dilution CRM hosting 0.5 % gold and platinum. Samples, marked as GP, were prepared by dropping standard solutions on a graphite powder. Samples SOP-Au, CHR-Pt^+, VP-2 are CRMs of ores.

Table 1 presents the independent variables of training and test samplings. The table rows present titles of samples. The table columns (X-variables) provide the gold and platinum concentrations. The most sensitive lines of analytes with near resemblance detection limits are selected as the dependent variables. These lines for gold are lines Au 267.595 nm and Au 242.795 nm denoted as 1Au and 2Au, accordingly (Table 2). Table 3 illustrates the results of simulation using platinum lines: Pt 265.945 nm (1Pt), Pt 306.471 nm (2Pt), and Pt 299.797 nm (3Pt). The columns (Y-variables) corresponding with influence of major compound and interferents were not included in the tables of the experimental data for two reasons. The analytical lines of gold and platinum are free from overlapping interferents and molecular bands (Fig 1). The non-spectral influences are not considered, as carbon content in calibration and ICS-samples investigated is a constant value.

Table 1 Training and test sets for the determination of gold and platinum in insoluble carbonaceous substance

Set	Sample	Au, %	Sample	Pt, %
Training	S/2	0.25	S	0.5
	S-1	0.25	S/2	0.25
	S-3	0.05	S-1	0.25
	S-4	0.025	S-3	0.05
	GP-5	0.001	S-4	0.025
	SOP Au	0.00418	CHR-Pt^+	0.0058
	SOP Au /2	0.00209	GP-6	0.003
	GP-3	0.0002	GP-4	0.0005
			GP-3	0.0002
Test	S-2	0.125	S-2	0.125
	GP-4	0.0005	CHR-Pt^+	0.0058
	CHR-Pt^+	0.00043	GP-5	0.001
			VP-2	0.000264

The line intensities of analytes in spectra of calibration samples vary from detection limit to maximal intensities of the same lines recorded in the spectra of ICS under study. So the ranges of contents of training sets were 0.0002-0.25 % for gold and 0.0002-0.5 % for platinum. Thus, these ranges go beyond the area of the linear relationship between variables (Eq. (1)), but ΔC_{test} belong to ΔC_{train}.

Table 2 Statistical characteristics of calibrations using Au analytical lines

Calibration model	Set Element line	Training *SDev*	Training *RMSEC*	Training *RRB*	Testing *RRB*
OLS-1	1Au	0.0176	0.1014	74	5
	2Au	0.0492	0.1603	108	44
OLS-2	1Au	0.0091	0.0076	52	3
	2Au	0.0284	0.1200	68	40
PCR	1Au & 2Au			129	97

Table 3 Statistical characteristics of calibrations using Pt analytical lines

Calibration model	Set Element line	Training *SDev*	Training *RMSEC*	Training *RRB*	Test *RRB*	Test *RRB* without CHR-Pt^{+}
OLS-1	1Pt	0.0046	0.0788	40	23	14
	2Pt	0.0098	0.0932	49	1206	40
	3Pt	0.0116	0.1075	71	327	6
OLS-2	1Pt	0.0046	0.0794	35	21	1
	2Pt	0.0096	0.0925	29	1256	46
	3Pt	0.0100	0.1043	63	378	3
PCR	1Pt & 2Pt			36	610	610
	1Pt & 3Pt			45	23	13
	2Pt & 3Pt			40	699	699
	1Pt & 2Pt & 3Pt			39	317	317

The numerical experiment was executed separately for lines of gold and platinum. The OLS-1 and OLS-2 calibration were constructed for each line of analytes. For PCR-calibrations the data structure was modified by a sequential addition of dependent variables, i.e. intensity of the following line of analytes. Tables 2 and 3 show the statistical characteristics of calibrations obtained by adapting the data structure. The standard deviation (SDev) and root-mean square error of calibration (RMSEC) were calculated for Y-variable and range of relative bias (RRB) was estimated for X-variable.

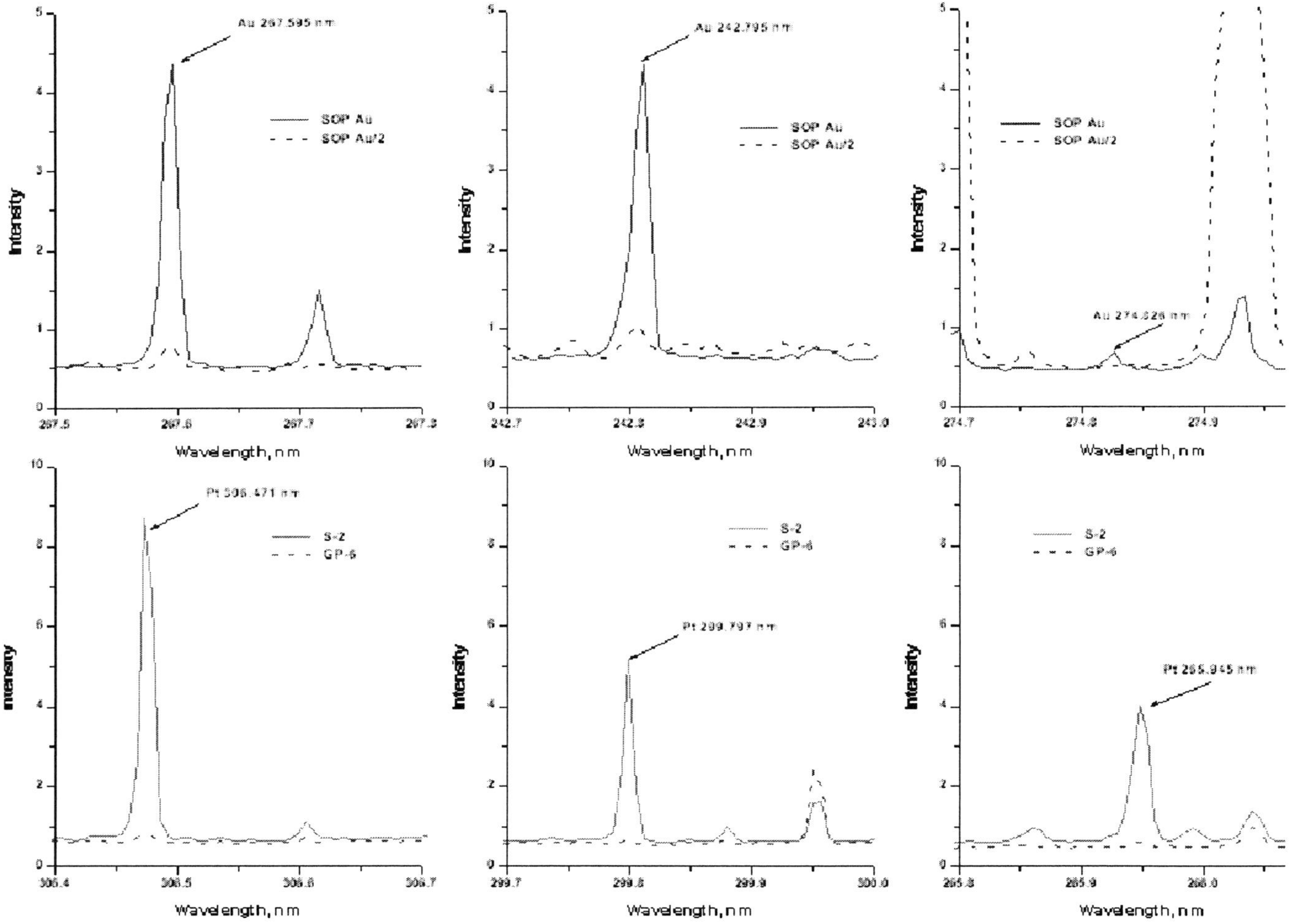

Fig 1 CRMs spectra for different lines of gold and platinum

Now consider OLS-1 and OLS-2 calibrations, constructed with training sampling for each line of gold (Table 2). The OLS-2 calibration for 1Au has the least values of SDev, RMSEC and RRB. Validation of test set displays that a minimal RRB is also observed for this model. The nonlinear calibration seems to be the best for gold determination. The Eq. (5) is realized only for OLS-2 (1Au) and distinguishes it as the best calibration. The model describes well the process of obtaining analytical signal of gold having low melting point and high speed of vapour transport to plasma of arc electric discharge. Consequently, the PCR-model, which is satisfactorily circumscribing linear relationship, gives maximal ΔR values for both training and test sets.

When training, the comparison of statistical characteristics of OLS-1 and OLS-2 models for platinum lines marks close values of SDev, RMSEC. Therefore, the process of obtaining analytical signal is satisfactorily described by the linear function of response. The application of PCR-calibrations offers the good results for training. The CHR-Pt^{+} sample of chromite does not correspond to ICSs in composition, but it was included in training and test samplings. Its presence in test sampling increases ΔR for data of 2Pt line, and also points out high sensitivity of this feature to compositionally heterogeneous samples. Eq. (5) is approximately implemented equally for models OLS-2 (1Pt), OLS-1 (1Pt), and PCR (1Pt&3Pt); and it specifies their good prediction ability.

Thus, numerical experiment shows that:

- the estimation "RRB" (Eq. (3)) is sensitive to the change (*i*) structure of experimental data and (*ii*) total chemical composition of samples;
- when processing spectral information through the RRB values, the variation of the number of spectral lines and a range of their intensities in structure of model set reflects the degree of consideration of latent relations;
- the execution of Eq. (5) provides choice of a good calibration model, which fully characterizes the analytical process.

Definition of Boron Impurity in Crystalline Silicon and Quartz

The data structure is simulated to study the possibility of take into account of both non-spectral and spectral influences by multivariate calibration.

In the training and test sets (Table 4) analyte is represented by boron lines of 249.772 and 249.677 nm, accordingly 1B and 2B.

The area of contents defined ranges from 0.00015 to 0.0059 % (less than two orders), and also it is well made by linear calibration (Eq.(1)). Both boron lines undergo the overlaps of SiO molecular band; besides, line Fe 249.653 nm is also imposed on line B 249.677 nm. It is well visible on spectra of BS-17 quartz and crystalline silicon samples (Fig.2). Subtracting the blank (bl) from spectra of training and test samplings, the contributions of molecular band was accounted. The blank (spectrum of IOTA) is the enriched high-pure quartz (99.99 % SiO_2) with the boron content (0.000004 %) that is much lower than the detection limit of the direct atomic emission analysis (0.00007 % B). The interferences on boron line were estimated by calibration constructed with and without subtracting the blank (Table 5). This table indicates SDev, RMSEC and RRB of training and test samplings. The minimal values of these statistical characteristics were obtained when training for OLS-1 (1B) and OLS-1 (2B) calibrations with subtracting the blank.

Table 4 Training and test sets for determination of impurities in crystalline silicon and quartz

Sample			Elemental content, %				
Set	Name	Substance	B	Al	Ca	Fe	Si
Training	Kv -1	quartz	0.00024	0.0079	0.0021	0.322	46.4
	BS-17	quartzite	0.00015*	0.00055	0.00055	0.00055	46.5
	Quartz "Lisma"	enriched high-purity quartz	0.0002	0.0054	0.004	0.03	46.5
	Silicon Standard (PV)	crystalline silicone	0.0019	0.141	0.0104	0.319	99.4
	SOP-1 (Si № 1)		0.0059	1.14	0.68	1.02	97.2
	SOP-3 (Si № 3)		0.0047	0.73	0.34	0.68	98.1
	SOP-4 (Si № 4)		0.0011	0.07	0.017	0.45	99.2
	SOP-12 (Si №12)		0.0015	0.056	0.0082	0.185	99.8
	Control probe 1	Mixture Kv-1+ BS-17 (1:1)	0.000165	0.0042	0.00132	0.161	46.5
	Control probe 3	Mixture SOP-4 + BS-17 (1:1)	0.000745	0.0353	0.00878	0.225	72.8
	SDPS-1	soil	0.0030	1.78	0.19	0.69	42.6
Test	SKv-1	quartzite	0.0004	0.063	0.071	0.44	46.2
	BS-17		0.00015	0.00055	0.00055	0.00055	46.5
	Kv-1	quartz	0.00024	0.0079	0.0021	0.322	46.4
	SOP-1 (Si № 1)	crystalline silicone	0.0059	1.14	0.68	1.02	97.2
	SOP-4 (Si № 4)		0.0011	0.07	0.017	0.45	99.2
	SOP-12 (Si №12)		0.0015	0.056	0.0082	0.185	99.8
	SDPS-1	soil	0.0030	1.78	0.19	0.69	42.6

* Recommended content is indicated as italic type

Because of number calibration samples of quartz and pure crystalline silicon is extremely restricted, the test set has some samples duplicated from the training set. However, we have included only new spectra of such samples in test sampling. In this case, based on the theoretical suppositions, the evaluation of the prediction ability for the test set can be identical or better than for the training set. In our experiment (Table 5) the RRB, obtained at testing, only for OLS-1 (1B without subtracting the blank) and OLS-1 (2B-bl), OLS-2 (2B-bl) are better.

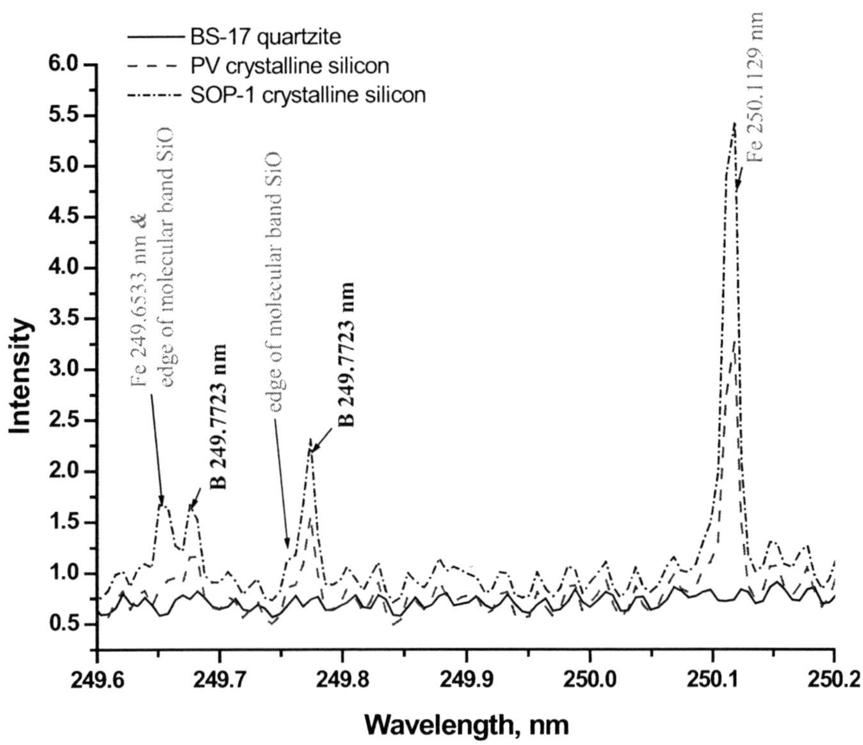

Fig.2 Intensity distribution of B, Fe and molecular band SiO on the wavelengths

Table 5 Statistical characteristics of OLS-calibrations using B lines

Calibration	Set		Training			Test
			SDev	*RMSEC*	ΔR_{train}	ΔR_{test}
OLS-1	1 B	without subtracting	0.0887	0.1569	139	109
	2 B	the blank	0.1387	0.1916	198	420
	1 B	with subtracting the	0.0436	0.1201	68	114
	2 B	blank	0.0877	0.1543	108	67
OLS -2	1 B	without subtracting	0.097	0.1578	109	170
	2 B	the blank	0.1574	0.1960	260	494
	1 B	with subtracting the	0.0307	0.1042	76	147
	2 B	blank	0.0627	0.1276	145	102

The line Fe 246.887 nm (2Fe) was used as correction of interference of Fe 249.653 nm for boron line of B 249.677 nm. These iron lines are close in the type of electronic transition and spectral intensity. The intensity of the 2Fe line was included in the data table. In order to estimate the influence of interferences and degree of their modifications, the multivariate PCR-calibrations with different structure for identical training and test sets were constructed. For linear calibrations without or with subtracting the blank the following values ΔR_{train} and ΔR_{test} were received:

OLS-1 (1B without subtracting the blank)	139	109
OLS-1 (2B without subtracting the blank)	198	420
OLS-1 (1B-bl)	68	114
OLS-1 (2B-bl)	108	67
PCR (1B & 2B without subtracting the blank)	98	100
PCR (2B-bl; 2Fe)	192	113
PCR (1B-bl; 2B-bl)	35	43
PCR (1B- bl; 2B-bl; 2Fe)	51	80

The best RRB, as training and test sets, were obtained for model PCR (1B-bl; 2B-bl) or PCR (1B2B). As further for the data of the multivariate calibration the intensities of boron lines obtained after subtracting the blank are utilized only, this mark (B-bl) is not repeated in the titles of structure of models. The execution of the Eq. (5) makes it clear that SiO molecular band renders a greater influence to quality of analytical results, than the interference of iron line (the bolt name and numbers in above the data). Nevertheless, the accounting of both interferences is necessary, especially if boron is defined in samples of unknown composition. Using two lines of boron in the calibrations improves the quality of training and prediction.

The atoms of boron have a high potential of ionization (8,298 e*V*); therefore altering matrix composition causes variations in intensity of boron lines [1], [7]. The analyzed samples (quartz and high-pure metallurgical silicon) are multifarious in a total chemical composition and the impurity concentrations [8]. For example, the contents of major elements modify in training sampling: silicon – from 42.6 to 99.8 %, iron to 1 %, calcium to 0.68 %, aluminum to 1.78 % (Table 4). For these reasons we have included columns in the data table, which can mirror the influence of matrix effects. The intensities of lines Si 302.0005 nm (Si), Fe 299.77 nm (1Fe), Al 265.247 nm (Al), Ca 317.933 nm (Ca) were measured in these spectra. The intensities of the lines enumerated vary from 500 to 1000 times in comparison with the background value.

The structure of the experimental data of the training and test sets were modeled by adding one line of major element to boron lines. The RRBs of the training and test samplings were calculated for each structure. The values obtained ΔR_{train} and ΔR_{test} are positioned as follows:

PCR (1B2B)	35	43
PCR (1B2B Si)	37	42
PCR (1B2B1Fe)	31	51
PCR (1B2B Al)	59	62
PCR (1B2B Ca)	154	122

Thus, the RRB values indicate that the degree of matrix influence on the intensity of boron lines diminishes from silicon to calcium in their content ranges: Fe ~ Si > Al > Ca. The accounting of matrix influences several major elements are possible simultaneously in PCR-calibrations. Compare the values of ΔR_{train} and ΔR_{test} for boron in the following models:

PCR (1B2B Si 1Fe)	33	51
PCR (1B2B Si 1Fe 2Fe)	**27**	54
PCR (1B2B1Fe 2Fe Al)	70	68
PCR (1B2B Si 1Fe 2Fe Al)	70	68
PCR (1B2B Si 1Fe Ca)	147	163
PCR (1B2B Si 1Fe 2Fe Ca Al)	149	137

The accounting of matrix influence of aluminum and calcium brings in additional noise and it does not improve results of boron determination. The best calibration takes into account the Si and 1Fe matrix effects.

Table 6 Model structures of features and ranges of relative bias for training and test sets for boron determination using PCR-calibration

Model structure	ΔR_{train}	ΔR_{test}
PCR (1B2B without subtracting the blank)	98	100
PCR (1B2B)	35	43
PCR (2B 2Fe)	192	113
PCR (1B2B 2Fe)	51	80
PCR (1B2B Si)	37	42
PCR (1B2B 1Fe)	31	51
PCR (1B2B Ca)	154	122
PCR (1B2B Al)	59	62
PCR (1B2B 1Fe 2Fe)	**27**	55
PCR (1B2B Si 1Fe 2Fe)	**27**	54
PCR (1B2B Si 1Fe)	33	51
PCR (1B2B 1Fe 2 Fe Al)	70	68
PCR (1B2B Si 1Fe 2Fe Al)	70	68
PCR (1B2B Si 1Fe Ca)	147	163
PCR (1B2B Si 1Fe 2Fe Ca Al)	149	137

The Table 6 combines the different variants of PCR-calibrations for the model structures of the experimental data. The data of Table 6 illustrate that the calibration PCR (1B2BSi1Fe2Fe) is the best calibration as the conditions (4)-(5) is executing. The first place is assigned for iron after comparing models of PCR (1B2B2Fe1Fe), taking into account the overlaps of SiO, 2Fe and matrix influence of 1Fe iron; and PCR (1B2BSi1Fe2Fe), considering of additional matrix influence of silicon. In both models identical values ΔR_{train} and ΔR_{test}: 27/55 and 27/54 (train/test) were received, accordingly. These calibrations are assessed as the best, since Eqs. (5) is satisfied for them.

At the close of numerical experiment it is possible to make the following recommendations for boron definition in quartz and crystalline silicon when executing the analytical technique considered.

- PCR-calibration (1B2BSi1Fe2Fe) using Eqs. (4)-(5) is the best calibration. It takes into account of spectral and nonspectral influences for boron definition in samples.

- The table of the experimental data for the best PCR-calibration includes: (*i*) dependent variables, representing intensities of two boron lines with the subtracted blank, iron line as spectral interference, lines of iron and silicon as matrix effect; (*ii*) independent variables are contents of these elements in calibration samples.
- Formulated recommendations concern only with the investigated content ranges of elements indicated for training sampling (Table 4). Their variations necessitate a verification of sufficiency of the structure selected.

Conclusions

The spectral data structure, construction of *n*-dimension calibrations and their comparing by the estimation "range of relative bias" were modeled for two analytical techniques: the determination of gold and platinum in insoluble carbonaceous substance; and definition of boron in crystalline silicon and quartz.

The multivariate calibration is perspective for application in arc atomic emission spectrometry due to its ability to simultaneously consider of the spectral and nonspectral effects. For this purpose the individual variables, which are reflective these effects or taking into account of their, are included in the data structure of training set, such as: (*i*) intensities of several analyte lines, corrected on presence of molecular bands; (*ii*) lines of interfering elements; and (*iii*) lines of major elements providing the main matrix effects. The range of relative bias is sensitive enough to model changes in the data structure and allows bringing to light the enumerated lines. Execution of Eq. (5) ensures choice of a good calibration, which adequately describes the concrete analytical process.

The numerical experiment has shown that the multivariate PCR-calibrations converse noncalibrated influences the calibrated ones; and they have advantages compared to one-dimensional models for analyte ranges, not exceeding two orders.

Acknowledgments

We are very grateful Igor L. Vasyliev for creation and adaptation of the software PCR-Calibration for processing atomic-emission spectra recorded by the instrument MAES-10, and Prof. Kim H. Esbensen for valuable instructions given in the book "Multivariate data analysis – in practice."

References

[1] P. W.J. M. Boumans, *Theory Of Spectrochemical Excitation*. London, Hilger & Wats, 1966

[2] A.S.C. Ehrenburg, *Data Reduction – Analysing And Interpretating Statistical Data*, Wiley Publishers, 1978

[3] K.H. Esbensen, *Multivariate Data Analysis – In Practice*, Camo ASA, 2001

[4] I.R. Shelpakova, V.G. Garanin, V.A. Labusov,, *Ind. Lab. Diagn. Mater.*, 65, 3-16 (1999)

[5] H. Kucha, , *Econ. Geol.*, **77**, 1578-1591 (1982)

[6] E.A. Razvozzhaeva, A.M. Spiridonov, N.V. Vilor. *Et Al Russian Geology And Geophysics,* 40, 1302-1307, (1999)

[7] E.A. Razvozzhaeva, A.M. Spiridonov, V.D. Tsykhanskii, I.E. Vasilyeva, S.I. Prokopchuk, *Russian Geology And Geophysics*, 43, 274-283, (2002)

[8] E.V. Smirnova, I.E. Vasilyeva, L.L. Petrov, S.V. Lontsikh, *J. Anal. Chem. (Russia)*, 41, 1361-1370 (1986)

[9] I.E. Vasilyeva, E.V. Shabanova, Yu.V. Sokolnikova Et Al., *Analitika I Kontrol (Russia)*, 5, 24-34 (2001)

In: Progress in Chemometrics Research
Editor: Alexey L. Pomerantsev, pp. 151-189
ISBN 1-59454-257-0

Chapter 12

Chemometrics and System Analysis of Modern Problems in General Chemistry

E. M. Slyusarenko and V. T. Mosidze
Lomonosov Moscow State University, Russia, slusarenko@laincom.chem.msu.ru

Abstract

The fundamental problems of common chemistry are considered by the system analysis and some mathematical solutions are discussed.

Keywords: system analysis; chemometrics; chemistry; multi-component systems; diffusion; design

Introduction

The system analysis consists of four basic steps: 1) formulation of problem (technical task); 2) generation of principal idea (brainstorm); 3) divergence (analysis); 4) convergence (synthesis). Chemometrics is often used during the steps 3 and 4; this is an important part of system analysis. In principle "...our knowledge about chemical phenomenon has to be presented by number, geometric images and during a time. (Vernadsky V.I.)".

Practically in all of the principal sections of general chemistry we have the problems, which are unsettled and difficulty understanding for pupils and students. Some problems of periodicity, the theory of chemical bond, thermodynamics and kinetics of the chemical interaction in multi-component systems have no single solution. Some solutions obtained by classic system analysis are set forth below.

MATHEMATICAL CORRELATION BETWEEN THE ATOMIC MASSES AND THE NUCLEI CHARGE OF ELEMENTS IN MENDELEEV'S PERIODIC TABLE

Formulation of Problem

It is necessary to determinate a correlation between the atomic mass of element and its place in the Periodic system or the charge in its nucleus.

Generation of Principal Idea

Practically all mass of an atom is concentrated in its nucleus. Although quantum mechanical calculations of the structure of the nuclei lead to a model of nuclear shells [1], this did not settle the problem of periodicity [2]. We can suppose that since the atoms of noble gases have the full-filled shells in their nuclei that explain their chemical stability. Hence, it is necessary to analyze the periodic changes of the number of protons and neutrons in nuclei of the noble gases

Early investigation by J. Thomsen [3] is particularly interesting. Thomsen predicted the existence of noble elements and calculated their atomic masses (Table 1). Two of Thomsen's working hypotheses attract attention: (a) about the equal changes in atomic masses of elements in paired periods; (b) about the existence of particularly stable groups of particles consisting of original proto-matter in atoms. In 1897, I. Rydberg [4,5] proposed a numerical row for the ordering numbers of noble gases (Table 1)

Table 1. Calculated atomic masses and ordering numbers for the noble gases [3,4].

Element	He	Ne	Ar	Kr	Xe	Rn	ERn
Ordering number [4]	2	10	18	36	54	86	118
Atomic mass [3]	4	20	36	84	132	212	292

Divergence

The atomic masses of the most abundant isotopes of the noble gases have been analyzed (Table 2) [7]. A number of elements in each period has been calculated according to Rydberg and shown in row 5. The mass number of the most stable isotope is used for radon. The atomic mass for ERn was calculated from a regular change of the number of neutrons in paired periods (row 6). The number of added neutrons in paired periods differs in 4. Hence, the gain in mass number of element 118 is 62 due to neutrons. The mass number of the ERn isotope satisfying the above regularity is 316.

There is nothing more remarkable in the row 6. However, if we use Thomsen's hypothesis on the equal increase of atomic masses in paired periods (Table 2, row 7), we can obtain a very interesting result. Numbers 2, 10, 28, 60 are a sequence of sums of a well-known quantum mechanical row: 2, 8, 18, 32,..., $2n^2$; and so we have

2 = 2; 10 = 8 + 2; 28 = 18 + 8 + 2; 60 = 32 + 18 + 8 + 2.

Table 2. Analysis of the atomic mass of more abundant isotopes of the noble gases in periods of Periodic System

Isotope	He	Ne	Ar	Kr	Xe	Rn	Ern
Mass number of isotope	4	20	40	84	132	222	(316)
Number of protons in isotope	2	10	18	36	54	86	118
Number of neutrons in isotope	2	10	22	48	78	136	(198)
Mass added by protons in each period (Rydberg)	2	8	8	18	18	32	32
Mass added by neutrons in each period	2	8	12	26	30	58	(62)
Mass added by neutrons (Thomsen's hypothesis)	2	10	10	28	28	60	60

This leads to the conclusion that in each nuclear shell there is a system of nucleon orbitals occupied only by neutrons, and that change of atomic mass is linked with the change in the number of protons in the isotope nuclei of noble gases in each period (ΔP) by formula:

$$\Delta M = \Delta P + \sum_{n=0}^{n=\sqrt{\Delta P/2}} 2n^2 \qquad (1)$$

Equation (1) is the mathematical expression of periodic change of the atomic masses in the Periodic table. Calculated atomic masses of the isotopes of the noble gases correspond to the upper boundary of their stability range. In even periods (Ne, Kr, Rn), the mass of the most abundant isotopes is less by 2 units. This lack of neutrons is totally compensated in the next paired period (Ar, Xe, Ern).

Convergence

The second term of equation (1) reflects the gain in mass due to neutrons. As a result we can present the structure of nucleon shells for isotopes of the noble gases as it is shown in Table 3. From this table it is evident that in order to obtain a stable isotope, it is necessary to fill additional neutron levels in each nucleon shell. It is obvious that the electronic system of atoms is determined by their nuclear structure.

Table 3. Structure of nucleon orbitals in the isotopes of the noble gases

Isotope	Period (shell)													
	I		II		III		IV		V		VI		VII	
	Number of nucleons in a level of shells													
	p	*n*	*p*	*n*	*p*	*n*	*p*	*n*	*p*	*n*	*p*	*n*	*p*	*n*
He^{4}	*2*	*2*												
Ne^{22}	*2*	*2*	*8*	*8*										
				2										
Ar^{40}	*2*	*2*	*8*	*8*	*8*	*8*								
				2		*2*								
Kr^{86}	*2*	*2*	*8*	*8*	*8*	*8*	*18*	*18*						
				2		*2*		*8*						
								2						
Xe^{132}	*2*	*2*	*8*	*8*	*8*	*8*	*18*	*18 8*	*18*	*18*				
				2		*2*		*2*		*8*				
										2				
Rn^{224}	*2*	*2*	*8*	*8*	*8*	*8*	*18*	*18 8*	*18*	*18*	*32*	*32*		
				2		*2*		*2*		*8*		*18*		
										2		*8*		
												2		
ERa^{316}	*2*	*2*	*8*	*8*	*8*	*8*	*18*	*18*	*18*	*18*	*32*	*32*	*32*	*32*
				2		*2*		*8*		*8*		*18*		*18*
								2		*2*		*8*		*8*
												2		*2*

The division of electronic levels into s-, p-, d-, f-sublevels can be employed for nucleon levels. The thin structure of proton and neutron sublevels obtained is shown below:

$$2 = \mathbf{2}; \tag{2}$$

If such correlation exists, the average change of atomic masses calculated for series s-, d- and f-elements has to be the sum of the numbers from equation (2). Calculated data are presented in Table 4.

Table 4. The average change of atomic masses in series of s-, d- and f-elements

Series s–elements	ΔM	Series d–elements	ΔM	Series f–elements	ΔM
0 - He	4,0	Ca – Zn	25,3	Ba – Yb	35,7
He – Be	5,0	Sr – Cd	24,8		
Ne – Mg	4,1	Yb –Hg	27,5		
Ar – Ca	-0,1				
Kr – Sr	3,8				
Xe – Ba	6,0				
Rn – Ra	4,0				
Mean quantity	4	Mean quantity	26	Mean quantity	36 (38)

In fact, the obtained results correlate with the thin structure of nucleon sublevels [8]. The filling of s-proton sublevel (2 protons) is accompanied by the filling of a single s-neutron sublevel (2 neutrons); and the change of atomic mass in series of s-elements is 4 units. The filling of d-proton sublevel (10 protons) is accompanied by the filling of two the d-neutron sublevel (10 + 6 neutrons) and the change of atomic mass in series of d-elements is 26 units. The filling of f-proton sublevel (14 protons) is accompanied by the filling of two the f-neutron sublevel (14 + 10 neutrons) and the change of atomic mass in series f-elements is 38 units. In Table 4, for series of f-elements the change of the atomic mass is only 36 units. It is connected with the enrichment of barium by hard isotope. In series Xe-Yb (s- + p-elements) the change of mass constitutes 42 units (4 + 38) as it is demanded for the ideal case. The rest of neutron sublevels are filled in series of p-elements.

The periodic change of the atomic masses of elements in Mendeleev's Periodic Table is strictly regular and connected with the order of filling of the proton and the neutron levels and sublevels in the nucleon shells of nuclei. This correlation can be expressed by the number row. The regular changes of the atomic masses of elements is determined also for series of s-, p-, d- and f-elements in periods of Periodic Table. Calculated data for isotopes correspond to upper boundary of their stability range.

Design of Molecules and Crystal Structures by Using the Geometric Conception of Chemical Bond

Formulation of Problem

Kekule's system for molecule designing having been used up to our time, doesn't correspond to the modern demands in the science and education. This system doesn't allow to indicate the direction of chemical bond but only the number of elements for design of molecules and crystals. It is necessary to create a new system of design taking into account the geometric places of the atoms and direction of chemical bonds in molecules elements (atoms) and the number of chemical bond between atoms.

Quantum-mechanical theory of chemical bond can't be applied either as it has no "atom" concept. In this theory atom is represented as a system of differential equations, which changes from one molecule to another. As a result in this theory there are no elements for design of molecules and crystals. Thus, it is necessary to create a new system of design taking into account geometric places of the atoms and direction of chemical bonds in molecules and crystal structures in the real space.

Generation of Principal Idea

For creation any design system is necessary to have some collection of designing elements and some method of joining these elements in construction. In chemistry these elements are atoms and the method of joining is a chemical bond. In our analyses we we'll use three principal postulates only:

1. In formation of molecules and crystals the main tendency is a minimization of integral tension of electric field in space.
2. Electrons move around nucleus along the close surfaces (orbits) having constant radius of curvature (spheres or a few spheres).
3. Chemical bonds are formed when the valence spheres of reacting atoms are contiguous.

Using these postulates a new approach for describing chemical bonds in molecules and crystals – geometric conception of chemical bond (GCCB) – has been elaborated [9,10]. In GCCB, the orbits are surface with a constant radius of curvature (spheres). For s-orbits it is obvious. Orbits (p, d, and f) consist of a pivotal sphere and valence (outer) of equal dimensions spheres. Configurations of s-, p-, d- and f-orbits are shown in Fig. 1

3.3. Divergence

Firstly configurations of atoms in different valence states (electronic ground, electronic excitation and orbital excitation) are to be determined.

Electronic Ground States for Atoms

The number of electronic ground-states for atoms is not high. In electronic ground states the valence orbits lie orthogonal. Four such states are formed in s- and p-series of elements in periods, as it is shown in Fig. 2,a-d.

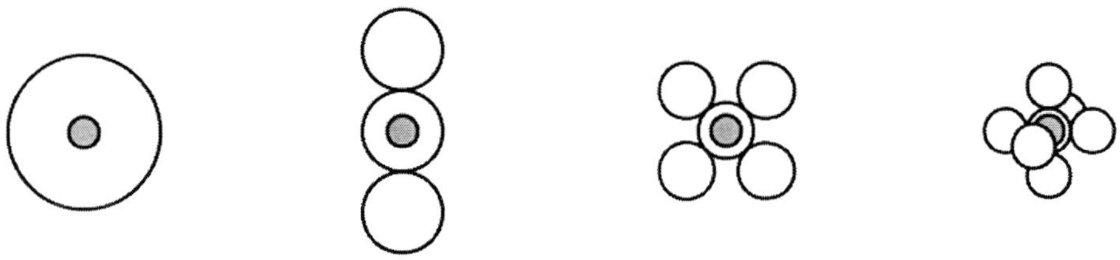

a) b) c) d)

Fig. 1. Configuration *s*- (a), *p*- (b), *d*- (c) and *f*- orbits (d) в GCCB

Other three ground states are added by d-elements (Fig. 2,e-g). These states consist of: single d-orbit (square disposing of four valence spheres); two d-orbits (tetragonal disposing of eight valence spheres) and three d-orbits (cubioctahedronal disposing of twelve valence spheres). More then twelve valence spheres (three d-orbits) can't be located around pivotal sphere. It is necessary to note that valence spheres of the same kind are not overlapped. Only a single f-orbit having six valence spheres (Fig, 2, h) is in electronic ground state. Eight configurations of atoms do not allow to design even such simple molecules as methane, water and so on.

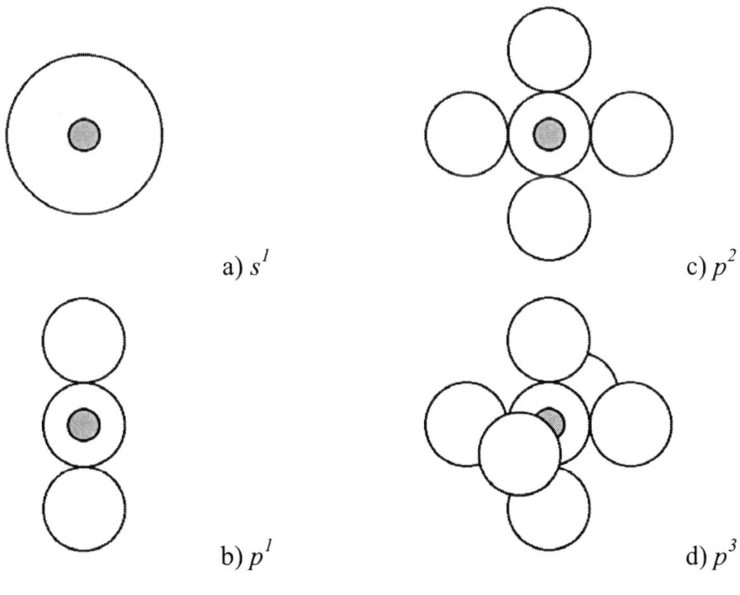

Fig. 2. Configuration of atoms in electronic ground state (Continued on next page)

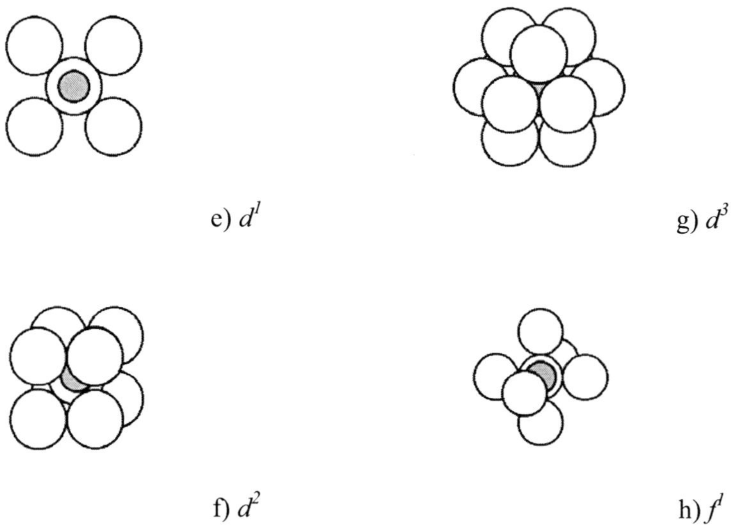

Fig. 2. Configuration of atoms in electronic ground state

In GCCB two kinds the excitation of atoms have been analysed — the electronic excitation and orbital excitation.

Electronic Excitation of Atoms

Having received energy any electron can transit from one sublevel to another sublevels. As a result the number of unpaired electrons participating in a formation of chemical bonds is increasing. In Fig 3 some electronic excited configurations of atoms are shown.

Orbital Excitation of Atoms

When the number of the valence spheres exceeds two (orbits with orthogonal disposition of the valence spheres), the orthogonal position of these spheres is broken and each sphere becomes independent generating new chemical bonds

It goes without saying that each orbital excited state has an even number of valence spheres. Some variants of orbital excited atoms (Fig. 4) are shown in two orthogonal projections. These configurations of atoms are used to forecast of chemical bonds in a molecules and crystal phases and to design their structures have been used.

Formation of Chemical Bond

Chemical bonds are formed when valence spheres of reacting atoms are contiguous. In the point of contact electron is indiscernible for both atoms, so electron exchange between atoms turn to be possible. Other kinds of interaction between atoms are just electromagnetic interaction without formation of the chemical bonds. If during formation of chemical bond centers of valence spheres lie on a line connected nuclei of the atoms, then s-bond is formed between atoms (Fig. 5,a, b). In other cases the π-bonds form (Fig. 5, c).

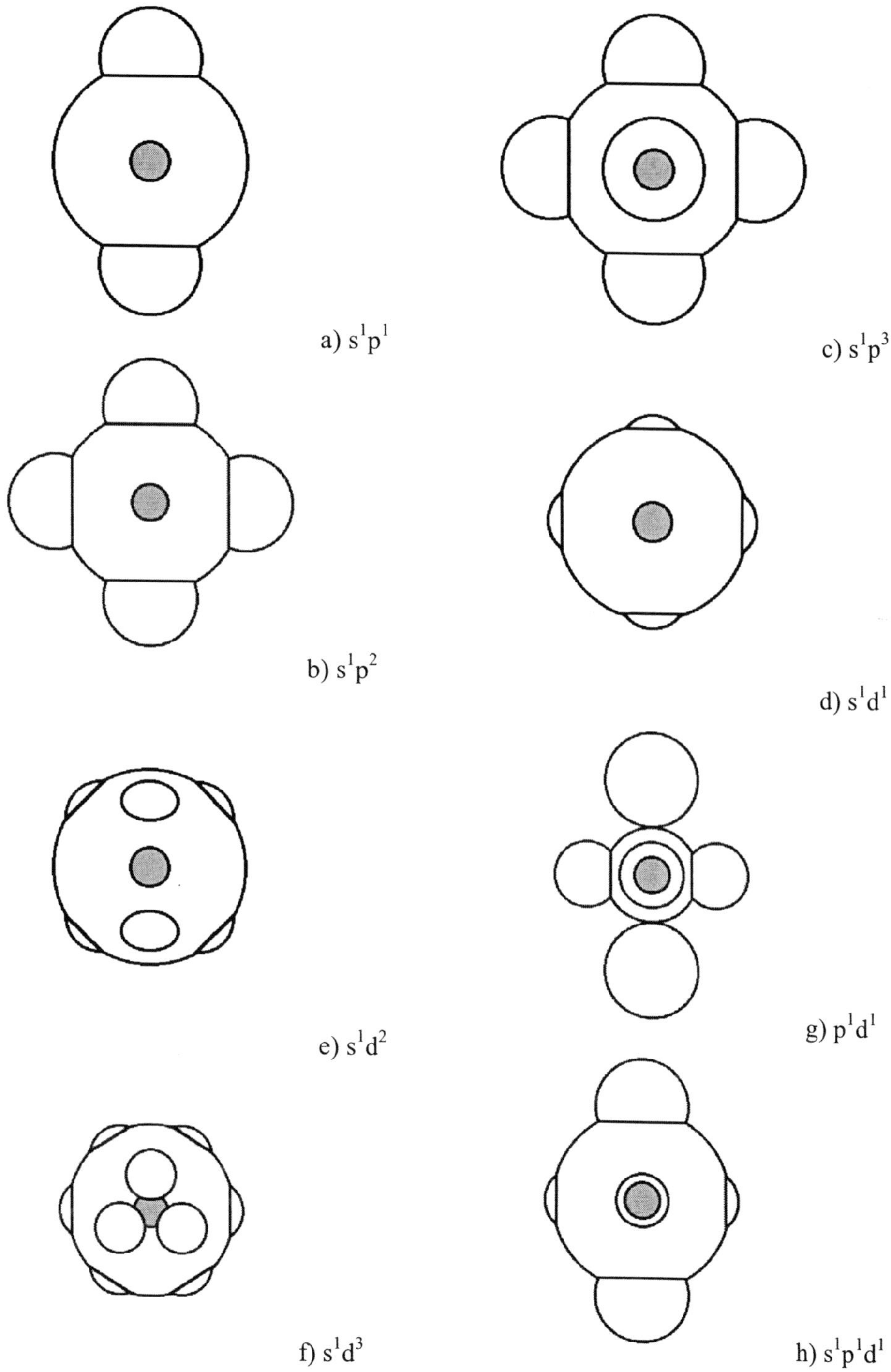

Fig. 3. Configurations of atoms in electronic exited state

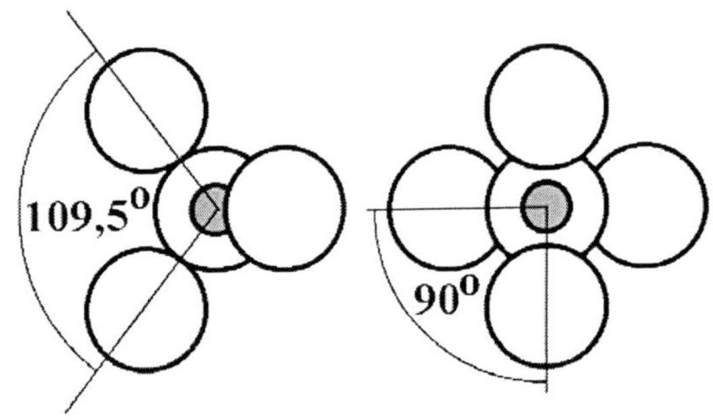

a) 4 spheres

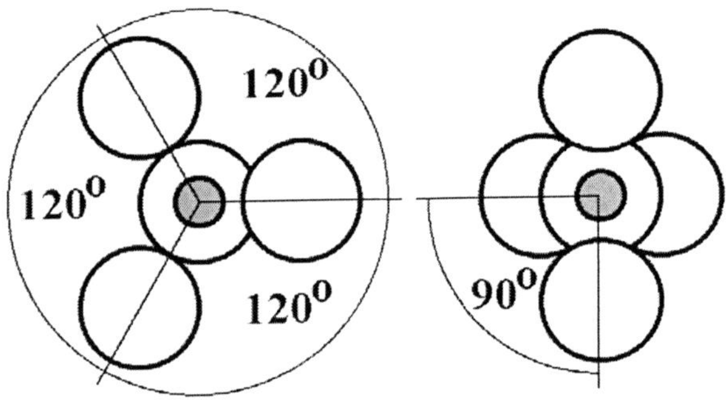

b) 4 spheres

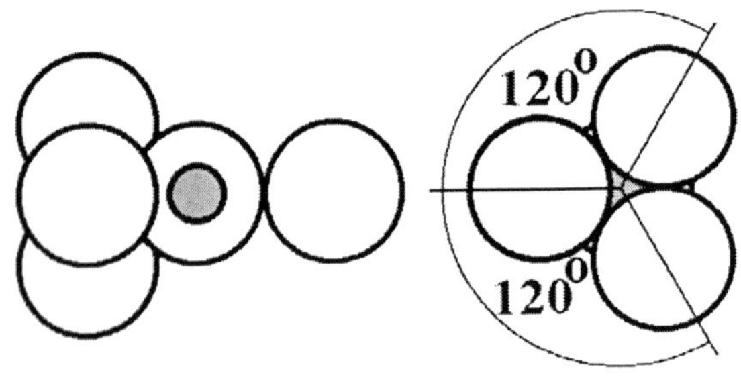

c) 4 spheres

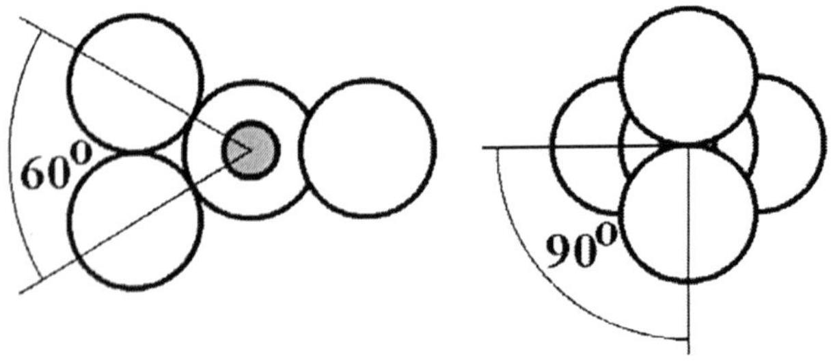

d) 4 spheres

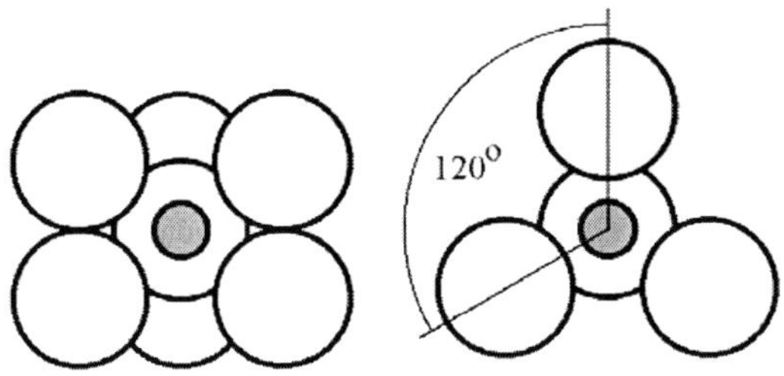

e) 6 spheres

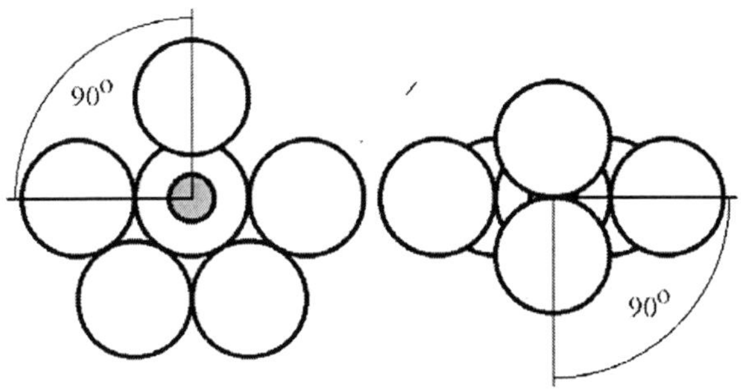

f) 6 spheres

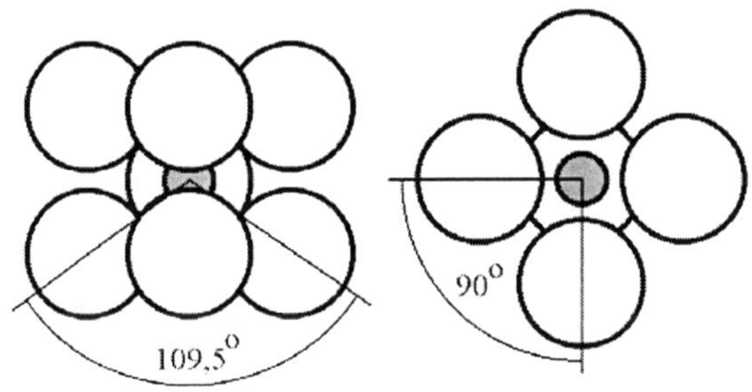

g) 8 spheres

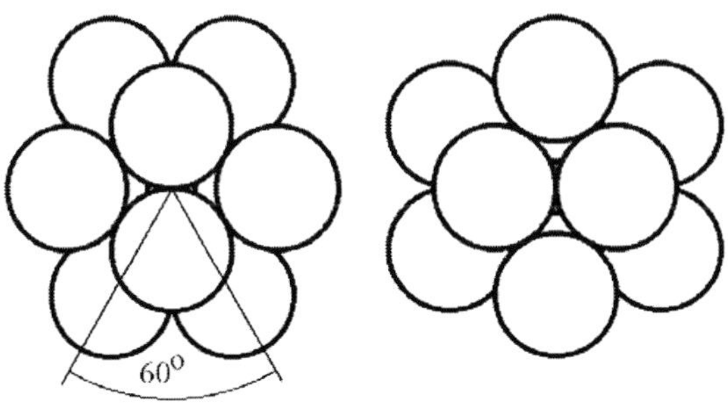

h) 12 spheres

Fig. 4. Configurations of atoms in the orbital excited states in vertical and horizontal projections: a) regular tetrahedron; b) irregular tetrahedron; c) irregular tetrahedron; d) triangular pyramid; e) triangular prism; f) cube; g) irregular octahedron; h) icosahedron

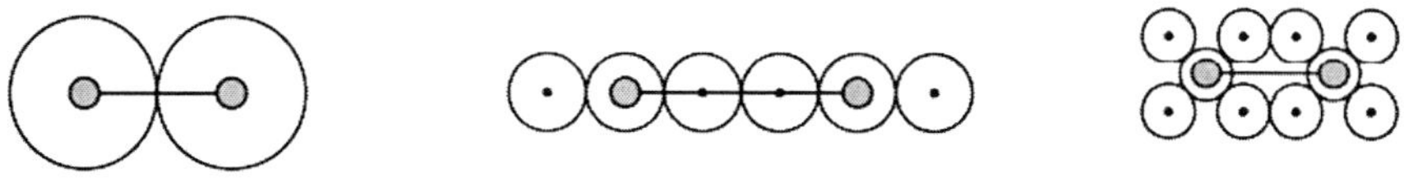

a) b) c)

Fig. 5. Formation σ-bonds between atoms: a) s —s bond ; b) p—p bond; c) d—d bond

Convergence

It is difficulty to say how many elements of design will be enough to give a full description of molecules and crystals, but it is evident that number of such elements is low. In this article we will design only a few molecules and crystals, which have some peculiarities in the formation of chemical bond.

Design of Hydrocarbons

The classic task that verifies any theory of chemical bond has already existed for more than hundred years. Correlation between the lengths and the orders of chemical bonds is evident, but the attempts to give an exact explanation of the phenomenon are still in progress. Vaunt-Hoff supposed that tetrahedronal atoms of carbon form a single chemical bond when their apexes are contiguous, a double chemical bond when their ribs are contiguous and a triple chemical bond when their facets are contiguous. Hence we can obtain a ratio of lengths C—C -bonds in hydrocarbons by simple geometric calculations. The calculated data are presented in Table 5. Vaunt-Hoff's data greatly differ from experimental data.

Table 5. Correlation lengths and orders of chemical bonds calculating by distinguish methods.

Method of calculation	Order of chemical bond			
	1.0(ethane)	1.5(benzene)	2.0(ethylene)	3.0(acetylene)
Experiment ratio [11]	1.000	0.907	0.877	0.782
Vaunt-Hoff method	1.000	—	0.707	0.544
GCCB method	1.000	0.911	0.911	0.816 (h) 0.878 (p)

In the atoms of carbon consisting of fore valence spheres all configurations of the orbital excited states are regular and irregular tetrahedrons. From regular tetrahedrons (109,7° — 109,7°) one can design practically all hydrocarbons and cyclohydrocarbons with the single bonds. By means of irregular tetrahedrons molecules consisting of double and triple bonds can be constructed. The ratio of lengths of C — C -bonds also obtained from a simple geometric calculation is presented in Table 5. These data are more close to the experimental data.

Ethane

Model of ethane (H_3C-CH_3), constructed by means of two tetrahedral elements of design (Fig. 4,a), is shown in Fig. 6 in two projections. Usually single s-bonds located at the angle of 109,7° are generated in the hydrocarbon chains. The experimental length of C—C -bond in ethane is 154 pm that is equal to three diameters of *p*-spheres of carbon [11]. From here we have $D_p(C)$=51.3 pm. Length of C—H in ethane is equal to 109 pm and $D_s(H)$ = 64 pm.

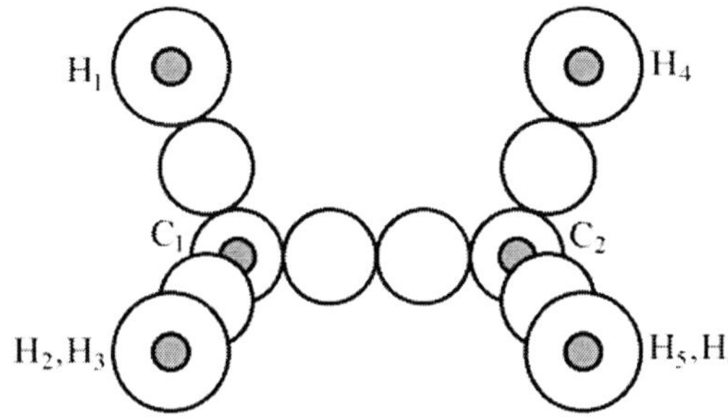

a)

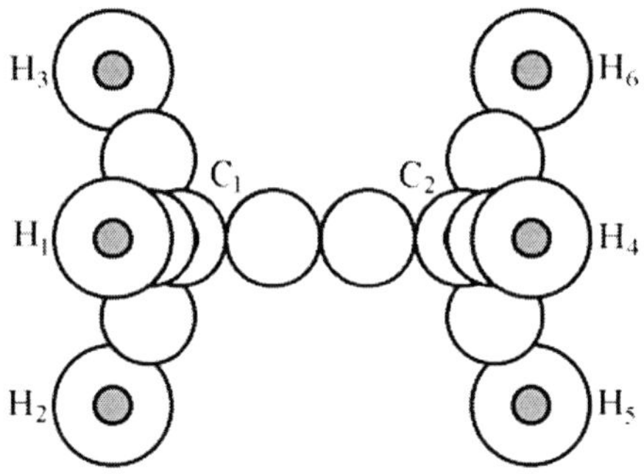

b)

Fig. 6. Model of ethane in vertical (a) and horizontal (b) projections

Ethylene

In the same way, the molecule of ethylene (H_2C=CH_2) is designed by means of two irregular (120° — 60°) tetrahedrons (Fig. 4, b) as it is shown in Fig. 7. Both bonds in ethylene are absolutely equal to π-bonds and all valence electrons participate in formation of usual double chemical bond.

The experimental length of the C—C - bond in ethylene is 135 pm [11]. By simple geometric calculations it is determined that experimental value of $D_p(C)$=49.4 pm. The length of the C—H - bond in ethylene is equal to 106 pm and D_s (H) = 64 pm. A small decrease (3-4%) of atomic dimension is connected with some polarization of bonds. If atomic dimension was not changed than the length of C—C in ethylene was equal 140 pm.

Acethylene

In Fig. 4,c the element for design of molecule of acethylene (HC≡CH) is presented. There are two variants for designing of this molecule. The combination of the valence spheres of elements like in ethylene gives a three-angle prism with three equal π-bonds (Fig. 8). Here each the valence sphere is combined with three other valence spheres. In the second case each of the valence spheres is combined with four valence spheres forming heptahedron (multi-center bond), and the distance C—C gets shorter.

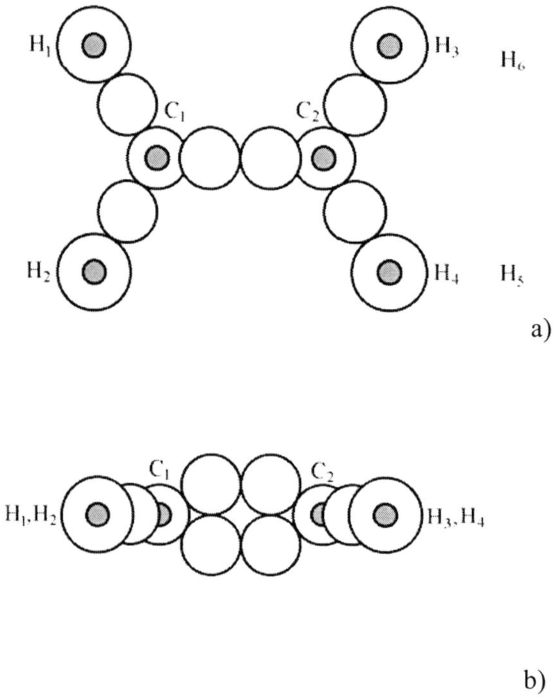

Fig. 7. Model of ethylene in vertical (a) and horizontal (b) projections

The experimental length of the C—C bond in acethylene is 120.5 pm [11]. Geometric calculations show that experimental D_p (C)=49.2 pm (the heptahedronal configuration of the valence spheres) and D_p (C)=45.8 pm (three-angle prism configuration of the valence spheres). The length of the C—H bond in acethylene equals 105 pm and D_s (H) = 31.2 pm (the heptahedronal configuration of the valence spheres) and D_s (H)=36.3 pm (the three-angle prism configuration of the valence spheres). If atomic dimension was not changed then the length C—C in acethylene was equal 125.7 pm (for the heptahedronal configuration of the valence spheres) and 135,2 pm (for the three-angle prism configuration of the valence spheres). The heptahedronal configuration of the valence spheres is performed in acetylene rather than the three-angle prism configuration, which gives a bad mistake for the length the C—C bond. In other compounds a realization of the three-angle prism configuration of the valence spheres is possible as well.

Allene

Molecule of allene (H_2C=C=CH_2) consists of one element (Fig. 4, d) and two elements (Fig. 4,b). In the molecule there are three atoms of carbon connected by two pairs of equal π-bonds. In this molecule the central carbon atom has different orbital exited configuration and is less than its neighbours.

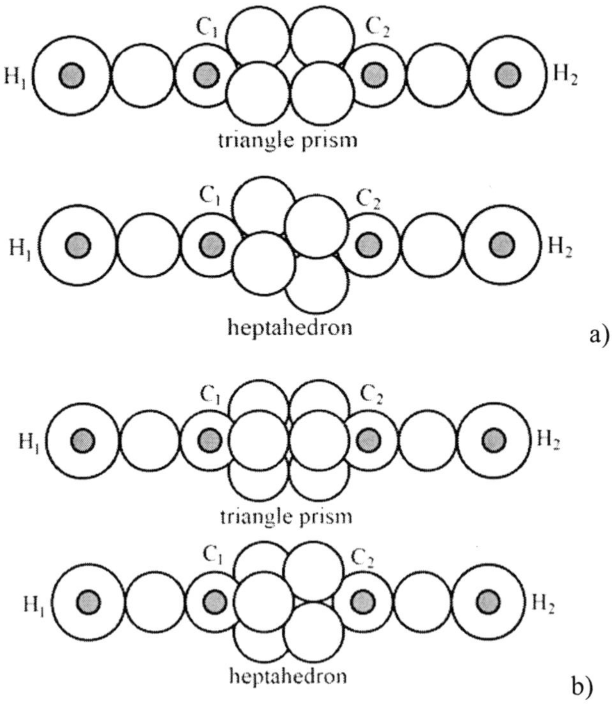

Fig. 8. Two variants model of acethilene in vertical (a) and horizontal (b) projections

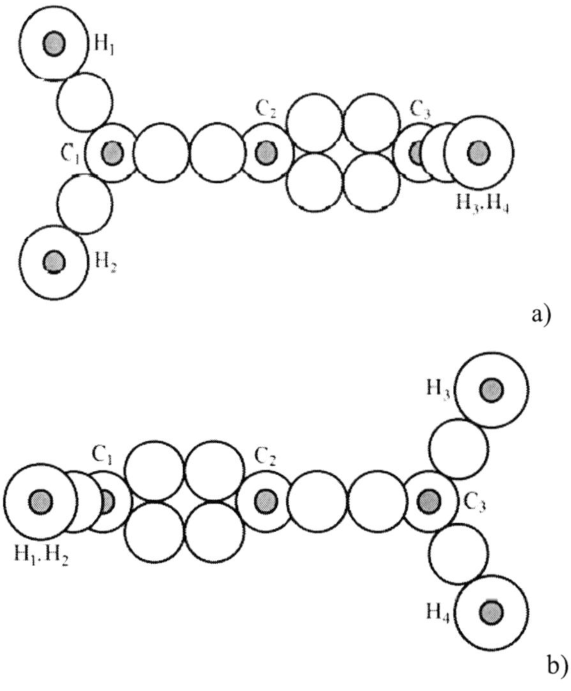

Fig. 9. Model of allene in vertical (a) and horizontal (b) projections

The experimental length of the C—C - bond in allene is 130 pm [11] though double length varies from 135 pm to 142 pm in other molecules. The length of the C—H –bond in allene equals 107 pm and in ethylene it equals 106 pm. It is evident that the C—C - bonds in allene are polar and there is a small positive charge on the central atom of carbon. Geometric calculations show that if the border atoms are the same as in ethylene (D_p (C)=49.4 pm) then the central atom of carbon has D_p (C)=46.6 pm.

Cyclopropane

In the three-angle molecule of cyclopropane (Fig. 10) the angle H–C–H is 120^0 [11]. Others cycloalcanes have the tetrahonal angle H–C–H. Angle 120^0 shows that cyclopropane consists of three elements of design (Fig. 4,b), which have been used for modeling of ethylene.

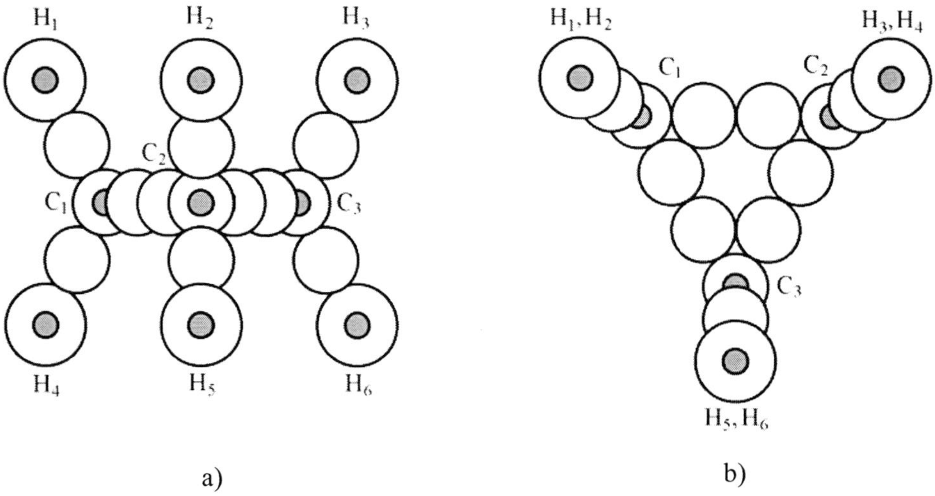

Fig.10. Model of cyclopropane in vertical (a) and horizontal (b) projections

In cyclopropane the C — C - bonds are σ-bonds and have the length of C—C 153 pm like in other cycloalcanes. As a result this molecule is not under tension.

Of One and Half Bonds between Atoms

Above shown examples are the well-known cases of single, double and triple bonds between atoms. It is more interesting that using elements of design shown in Fig. 4, b we can design a new kind of chemical bond – of one and half chemical bond. Such bond is obtained if two elements of design in Fig. 4,b are combined in such a way that centers of the valence spheres form a regular triangle (Fig. 11). The ideal example of such compound with one and half chemical bonds is benzene (Fig. 12, 6). One and half chemical bond and double chemical bond are very similar.

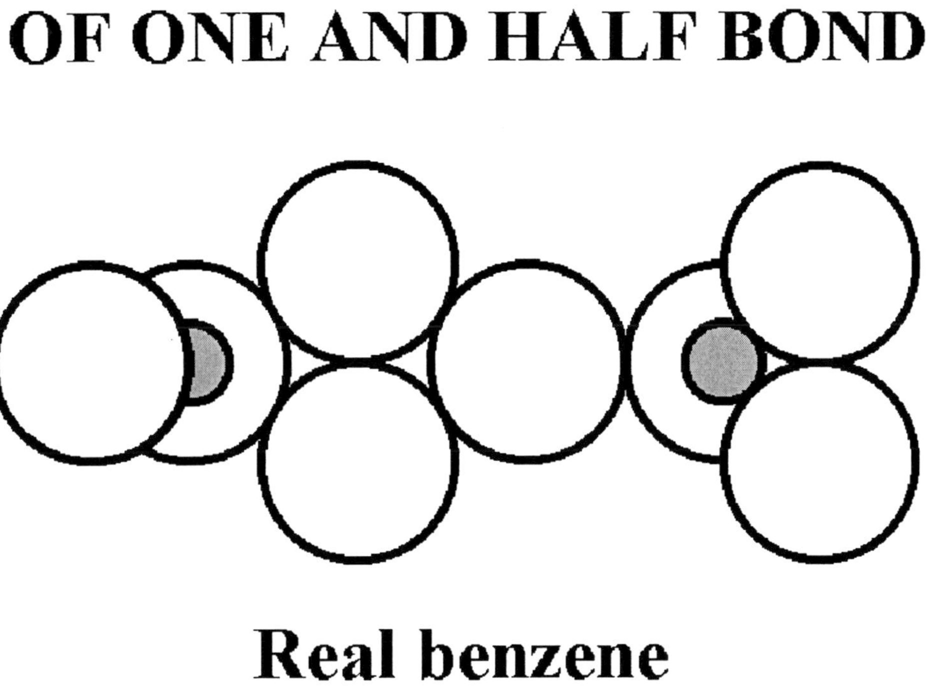

Fig. 11. Combination elements of design in formation of one and half chemical bond

Geometrically, both kinds chemical bond have equal lengths and equal valence angles bonding – 120°. As a matter of fact the length of double bond is shorter as compared with one and half bond because of polarization of bonds. Benzene is a member of homologous series of cyclohydrocarbons with conjugated double bonds and the same formula $(CH)_n$. Other members of this series pentadienil (C_5H_5) and heptatrienyl (C_7H_7) are shown in Fig. 6, b,d. Earlier these compounds were considered as ions of pentadien and heptatrien that is a mistake. More complicated compounds from this homologous series have been obtained as well: e.g. cyclododekahexaen; annulen, and so on. It is necessary to note that cycles with one and half bonds may contain other atoms (nitrogen, boron).

Such heterocycles, especially stabilized by the atoms of metals are often included in the structure some important biologically active compounds such as porphins, chlorophyll, haemoglobin, some vitamines, and so on.

Design Molecules with Bridge Bonds

Molecules with lack of electrons are not rare. In such molecules the number of electrons is less than it is necessary for formation of chemical bond by an electronic pair.

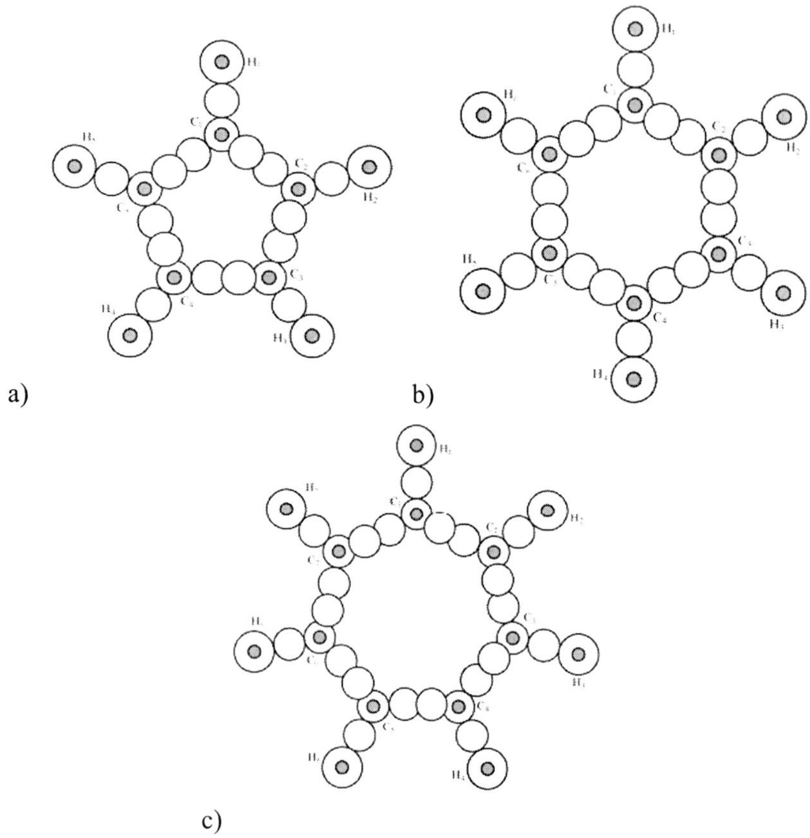

Fig. 12. Compounds with of one and half bonds: a) model of pentadienyl molecule C5H5; b) model of benzenemolecule C_6H_6 ; c) model of heptatrienyl molecule C_7H_7.

Diborane

Diborane (B_2H_6) is a good example of such molecule. During the long time diborane was considered as analogue of ethane. Later by the means of spectral investigations, it was demonstrated that two groups BH_2 lie in the same plane and two atoms H lay in the orthogonal plane with a larger length of the B—H bond [12]. In addition, an experimental investigation of the chemical reaction of diborane and others compounds with an electronic deficiency show exactly that a hydride ion H^- exists in diborane [13].

The designed molecule of diborane is presented in Fig. 13. Distance between B and bridge H is 133 pm and with other atoms H is 119 pm. The angle H–B–H in groups BH_2 with bridge H is 97^0 [12].

The dimensions of valence spheres in the molecule of diborane have been calculated using the experimental data. Atoms H from groups BH_2 have a small dimension (D_s (H)= 44.4 pm) and bridge atoms H have a considerably larger dimension (D_s (H)= 84.8 pm). Atom B has the diameter of *p*-sphere 64.5 pm (r_B = 96.8 pm), which is more than the covalence radius of boron. From this it follows that B and bridge H are the acceptors of electrons and have a negative charge. Atoms H from groups BH_2 are the donors of electrons and have a positive charge.

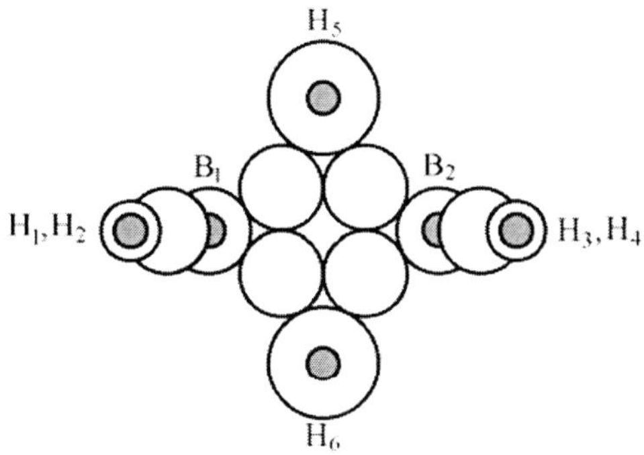

a)

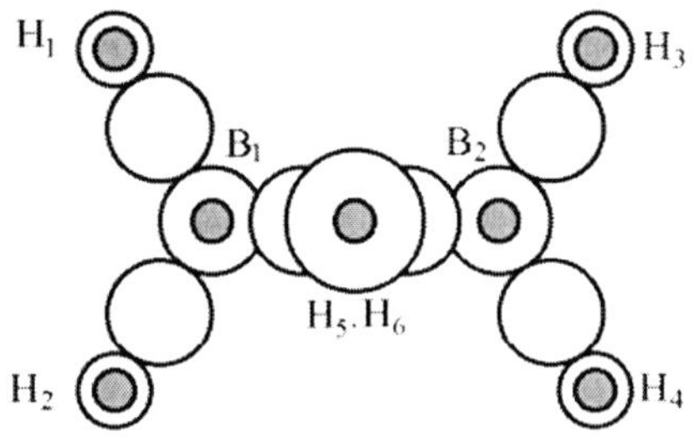

b)

Fig.13. Model of diborane in vertical (a) and horizontal (b) projections

Molecule $[Be(CH_3)_2]_n$ The polymeric chain in molecule $[Be(CH_3)_2]_n$ consists of alternating in two orthogonal projections fragments, which contain two atoms Be and two bridge groups CH_3. The distance between Be and bridge C is 193 pm. The angle C–Be–C in fragment of polymeric chain is 114^0 [14]. Designed structure of some fragment of polymeric molecule $[Be(CH_3)_2]_n$ is shown in Fig. 14. The position of atoms H is not determined. Atoms Be have diameter of *p*-sphere 76.9 pm (r_{Be} = 115.4 pm) and atoms C have diameter of *p*-sphere 64.2 pm (r_C = 96.3 pm) It is larger than covalence radius of both beryllium and carbon. Thus, both Be and C are acceptors of electrons, so are B and bridge H in the molecule of diborane. Atoms H are the donors of electrons.

Design of Two Atom Molecules N_2, O_2 and CO.

Molecules N_2, O_2 and CO (Fig. 15) have a bonding structure like in acethilene (triple bond). Molecule O_2 has two unpaired electrons in s-orbit of both atoms O and, therefore, is paramagnet. The two-atom molecule N_2 has to be the most stable, but such a molecule is CO. It is connected with transition of nonbonding electron from *s*-orbit of oxygen to valence *p*-spheres of carbon. It leads to appearance of charges on both atoms and extra electrostatic interaction between atoms. This interaction gives a small contribution of because strong polarization of all C—O - bonds. Therefore the atom of carbon has a small negative charge.

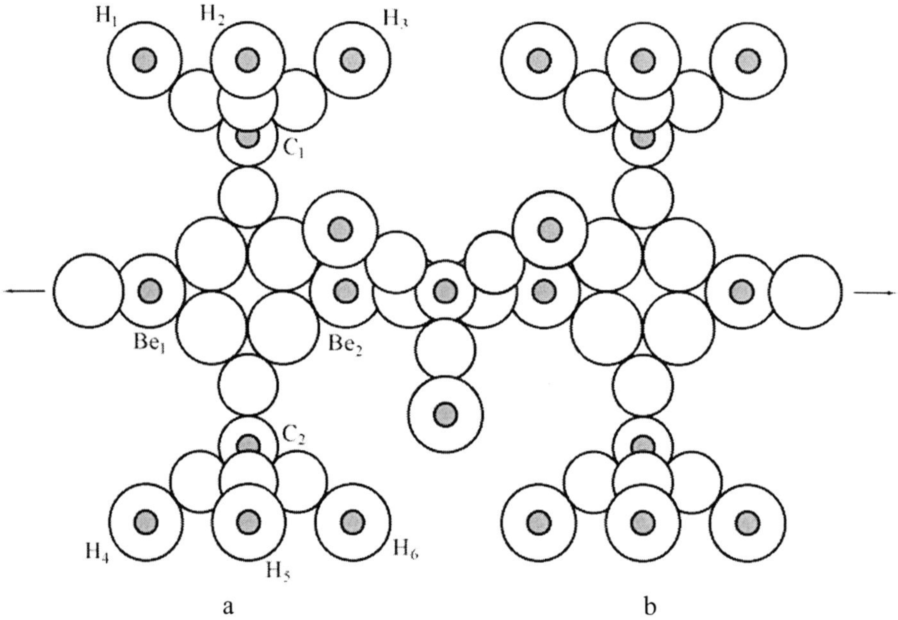

Fig. 14. Model one of fragments of polymer chain $[Be(CH_3)_2]_n$ in vertical (a) and horizontal (b) projections. (Atoms are lettered only for a single segment of polymeric chain. Arrows show an extension of the polymeric chain).

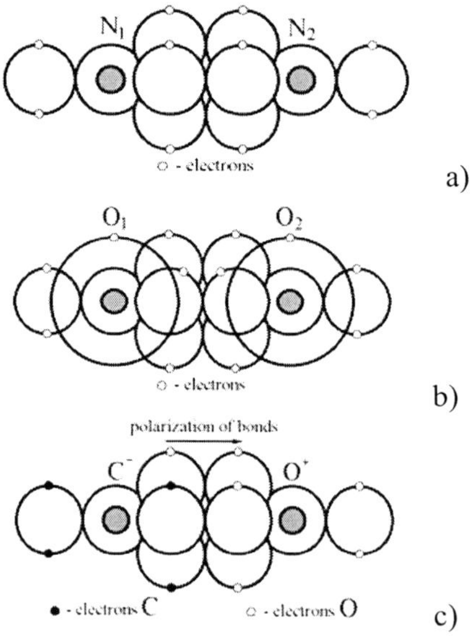

Fig. 15. Models of molecules: N_2 (a), O_2 (b) and CO (c).

Design of the Chemical Bonds of Beryllium

Hexagonal close packed crystal structure (hcp) $P6_3/mmc$ consists of atoms so that any atom has six neighbours in obtained hexagonal layer and three neighbours in upper and lower layers as it is shown in Fig. 16,a. Atoms from each following layer occupy only two equivalence positions named B and C (Fig. 16,b,c). Such crystal structure with alternated layers (ABAB or ACAC) obtained from ideal spheres and has maximum packed structure.

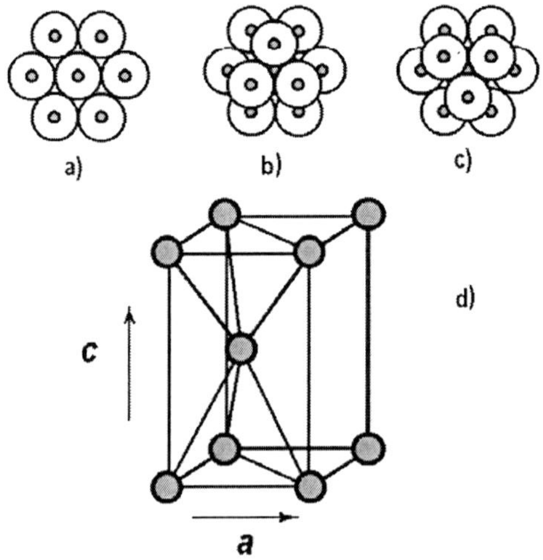

Fig 16. Hexagonal close-packed crystal structure (hcp) $P6_3/mmc$ in layers A (a), B (b), C (c) and elementary cell of structure (d).

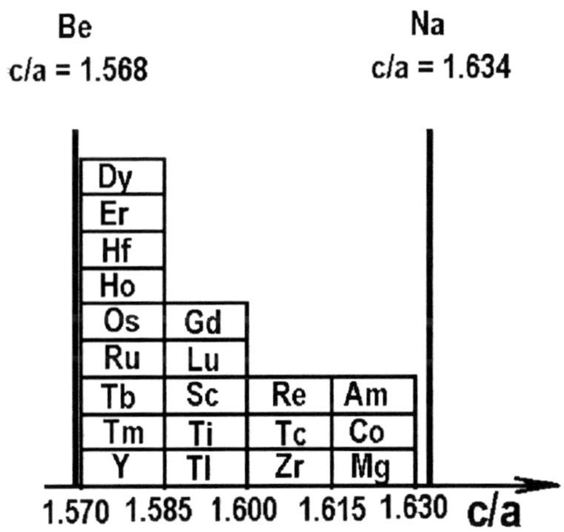

Fig. 17. Histogram of distribution elemental metals depending on axial ratio *c/a* in hcp structures [15].

The axial ratio *c/a* is easily calculated by the means of geometry as the ratio of double height to side rectilinear tetrahedron and it is equal to 1.6330 with packed density $\phi = 0.7405$ (for spherical atoms). From histogram (Fig. 17) it follows that axial ratio *c/a* for hexagonal metals does not correspond to the ideal ratio 1.6330, but rather to the axial ratio for beryllium, which is 1.568. It is a great difference for conception of ideal spheres. Repeated attempts to explain this deviation have failed [16].

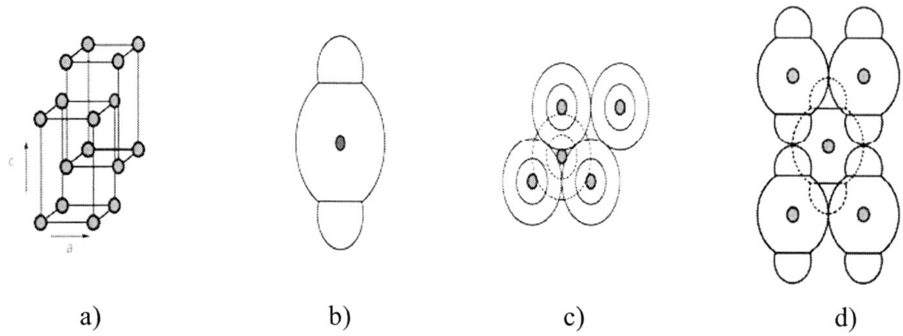

Fig 18. Structure hcp (P63/mmc), obtained by a two simple hexagonal structures (a); element of design of the structure –atom in electronic excited state slpl (b); system of chemical bond of hcp structure Be in two orthogonal projections (c,d).

Hexagonal close packed crystal structure may be designed by two simple hexagonal frames with the same axial ratio (Fig. 18,a). Atoms of one of the frame are put in the vacancies of another frame. Atoms of beryllium do not react in the ground state s^2p^0 so far as there are no unpaired electrons in the valence electronic sublevels of the atom. In electronic excited state (s^1p^1) atoms of beryllium are very asymmetric (Fig. 18,b). By the use of two such simple hexagonal frame is easily designed as it is shown in Fig.18,c,d.

As a result, each atom forms six the *s—s* - bonds in hexagonal plane and two the *p—p* - bonds along axis *c*. The structure of beryllium is generated by combination of two such simple hexagonal structures. In this case the axial ratio in such structure is equal to the ratio of the length of the *p—p* -bond to the length of the *s—s* - bond.

Therefore the ratio of the dimensions of spheres of the *p*- and the *s*-orbits may be calculated. The length of the *p*-orbit that consisting of three spheres to the length of the *s*-sphere is 1.568. From here diameter of *p*- sphere is

$$d_p = 1.5678\ d_s /3 = 0.5226\ d_s$$

The ratio of volumes of these spheres is more important.

$$V_s \,/\, V_p = d_s^3 \,/\, d_p^3 = 7.00635$$

The fact that electronic system of the *p*-orbits consists of seven spheres appears from second postulate of the GCCB theory, but it is surprising that the volume of electronic system of the *p*-orbits and the volume of the *s*-orbit are equal. This fact may be generalized to electronic systems of the *d*-orbits and the *f*-orbits. By such method it is possible to calculate the ratio of diameter sphere in the *d*- and the *f*-electronic systems to the diameter of the *s*-sphere as it is shown below in Eq. (3-7)

$$V_s = 7V_p = 21V_d = 43V_f \tag{3}$$

$$V_p = 1/7V_s\ V_d = 1/21V_s\ V_f = 1/43V_s \tag{4}$$

$$d_p^3 \,/\, d_s^3 = 1/7\ d_d^3 \,/\, d_s^3 = 1/21\ d_f^3 \,/\, d_s^3 = 1/43 \tag{5}$$

$$d_p / d_s = 1/\sqrt[3]{7} \quad d_d / d_s = 1/\sqrt[3]{21} \quad d_f / d_s = 1/\sqrt[3]{43} \tag{6}$$

$$d_s : d_p : d_d : d_f = 1 : 0{,}52276 : 0{,}36246 : 0{,}28544 \tag{7}$$

As a consequence we have correlation (7). Obtained correlation is fundamental and invariable within any changes the valence configuration and the dimensions of atoms.

Design of the Crystal Structure of Zinc and Cadmium

The second problem in hexagonal close packed crystal structure is the anomalous axial ratio *c/a* for zinc and cadmium (1.8564 for zinc and 1.8857 for cadmium).

Zinc and cadmium are disposed in second group of Periodic table, like beryllium, but they have different electronic structure. Atoms of beryllium have s^2p^0 configuration in electronic ground state and atoms of zinc and cadmium have configuration $s^2p^0d^{10}$. As a result, instead of the electronic excited state (s^1p^1) we have a possibility to excite the *d*-electron to the *p*-sublevel (d^1p^1). Simultaneously the *d*-orbit undergoes orbital excitation (regular tetrahedron), as it is shown in Fig.19,a.

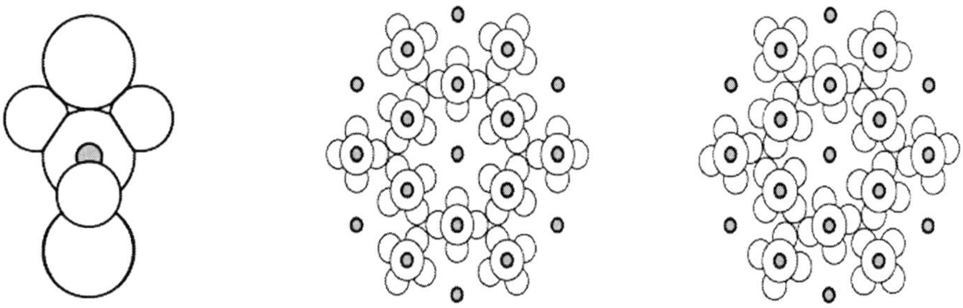

a) b) c)

Fig 19. Atom in electronic excitation d^1p^1 with orbital excitation the *d*-orbit (regular tetragon 109,7° — 109,7°) as an element for design of the crystal structure of cadmium and zinc (a). Formation Kagome's nets by three-central bonds in hexagonal plane: a) valence spheres taking part in three-central chemical bond dispose along the chains of atoms hexagonal plane; b) valence spheres taking part in three-central chemical bond turn round *p* — *p* bonds to angle 15°.

Two simple hexagonal frames as in beryllium construct the crystal structure of zinc and cadmium. As the chemical bonding along axis c of hexagonal close-packed crystal structure is realised by the *p* — *p* bonds, it is necessary to create a regular hexagonal layer with parameter $a = 1.5683 / 1.8856$ (Cd) $= 0.8317D_s$, using elements shown in Fig.19, a. Here D_s is the diameter of the *s*-orbit of cadmium.

Ideal hexagonal layer of such elements cannot be obtained but the elements form Kagome's net as it is shown in Fig.19, b, c. Two valence spheres of the *d*-orbit form two three-central bonds in Kagome's net above plane. Two valence spheres of the *d*-orbit below plane are disposed in the vacancies of Kagome's net. Obtained Kagome's net with maximum parameter $a = 0.8751\ D_s$ is shown in Fig. 19, b If the valence spheres of atom taking part in three-central chemical bond are disposed along the atomic chains of the hexagonal plane, then the axial ratio c/a = 1.7921 (Fig, 19,b).

Rotation of atoms around the p — p bonds to the angle 15° (Fig.19,c) gives the desired axial ratio c/a = 1.8840. The difference of the model axial ratio c/a and the experimental ratio for cadmium is less than 0.1%. The fourth part of atoms, which fill up the vacancies in Kagome's net, practically does not take part in formation of the chemical bond in the hexagonal plane. These atoms are connected with atoms of other planes disposed above and below of the p — p bonds as in beryllium and take part in a formation of the simple hexagonal structures.

Following combination of two simple hexagonal structures gives the hexagonal close packed crystal structure of cadmium and zinc. As a matter of fact, the packing parameter for this structure is less, than for other hexagonal close packed crystal structures with electronic excitation of atoms (s^1p^1) as the d-orbit has a volume less than the s-orbit.

Such structure is unstable (low fusion temperature, high pressure of saturated vapor, plasticity). Under high pressure only axial ratio is reduced (~ 1.73 and 180 Kb for Zn; 1.68 and 300 Kb for Cd). It is not surprising as the electronic excitation (d^1p^1) changes to the electronic excitation (s^1p^1) as in beryllium and magnesium.

New approach (the geometric conception of chemical bond) allows to create the system of design of molecules, crystal structures, polymeric chains, radicals it so on

POLYHEDRATION OF MULTI -COMPONENT PHASE DIAGRAM USING GRAPH METHOD

Formulation of Problem

Phase diagrams are the universal source of information about the interaction of the elements in multi-component systems and they serve as a basis for creation of new materials. However, the multi-component diagrams (with number of components more than four) have been researched insufficiently, because experimental work rapidly increases while one investigates many components.

Generation of Principal Idea

Constructing any phase diagrams can be presented by two stages:

1. All phases forming the system and the scheme of phase equilibria are determined. It is polyhedration of system.
2. The phase boundaries and the direction of tie line are defined. Then obtained information is presented in analytical form.

The performance of the second stage is a hard work and it is often impracticable. The first stage is qualitative and can be solved. The data obtained at this stage are useful for many users since they allow to choose a multi-component system fragment, which is perspective for creation of new materials with special properties.

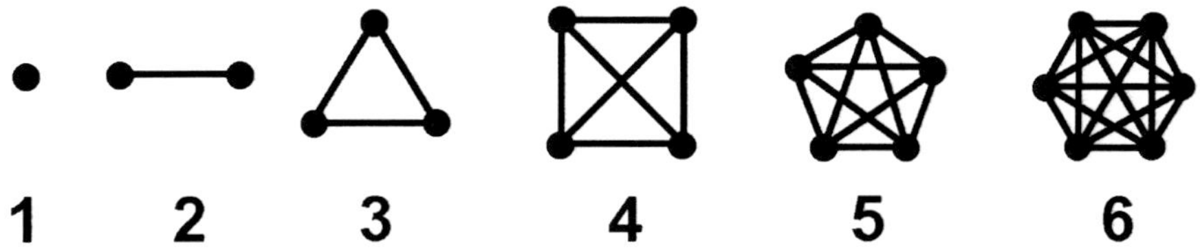

Fig. 20. Presentation of multiphase equilibria as the graphs

The graph method [17-19] can be used for polyhedration of the systems (determination of all phases and scheme of phase equilibria). It mainly consists of the following steps. Any the N-phase equilibrium may be presented as the complete graph with N vertices (Fig. 20). In this case each phase is marked as a point (independent from included components), and two vertices are connected with edge of graph (line among points), if equilibrium exists between these two phases.

Graph containing all equilibria in the system at the chosen temperature represents formalised isothermal cross-section of phase diagram and is named the total graph.

Divergence

Polyhedration of the n-component phase diagram using graph method can be realised on the basis of data of phase equilibria in the (n-1)-component systems by performing three consequent stages:

1. Presenting isothermal cross-sections of (n-1)-component systems as graphs. Graphs of the simplest isothermal cross-sections of ternary systems are given on Fig. 20. Combination of these graphs gives any formalised isothermal cross-section.
2. Building graph of n-component system (*total graph*) is realised by adding the graphs of the (n-1)-component systems. Number of vertices (points on plane) of the received graph is equal to number of phases in the n-component system. The edges of graph are received by copying all edges from the source graphs of (n-1)-component systems to the graph of the n-component system.
3. Decomposing the total graph of the n-component system on graphs of N- and (N-1)-phase equilibria. Moreover, it is possible to obtain several variants of such division. At decomposition of the graph it is necessary to note that the phase equilibria in the (n-1)-component systems do not give full information about all phase equilibria in the in component system.

Thus, the (n-1) phase equilibrium from the (n-1) component system in the n-component system can:

a. degenerates, moving over to the (N-2)-phase equilibrium (Fig.21,a);
b. recombines with such the (N-1)-phase equilibrium forming (N-1)-phase field (Fig.21,b);
c. participates in the formation of the N-phase equilibria (Fig.21,c).

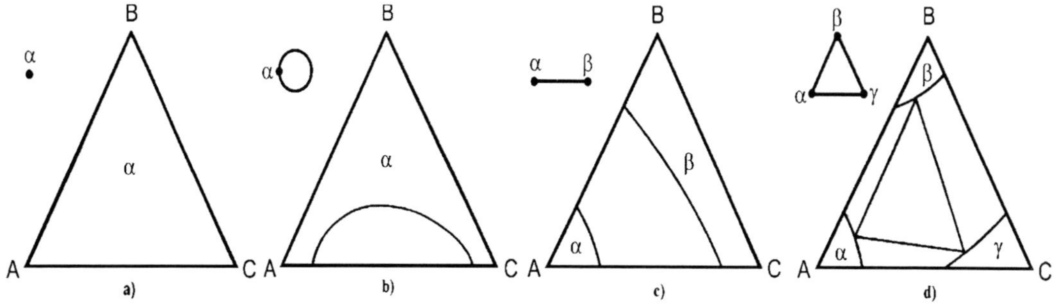

Fig. 21. Presentation of the simplest isothermal cross-sections of ternary systems as graphs: a) a single phase; b) degeneration of equilibrium; c) recombination of equilibrium; d) participate in formation of the in N-phase of equilibrium

Taking into account the foreside, the degenerated and recombined (N-1) equilibria should be separated from the total graph. The remaining graph will represent a collection of the N-phase equilibria, only if they do not form a general block. This is observed when several N-phase equilibria have similar the (N-1)-phase equilibria in polyhedron. In this case it is necessary to carry out an experimental investigation.

All steps of the total graph decomposition (Fig. 22), as example, have been shown for the Ni-V-Ta-Cr system. There are two of recombined three-phase equilibria λ_1-λ_2-β_1 and λ_1-λ_2-β_2 in this system (Fig. 22, b,c). Two four-phases equilibria μ-β-λ_2-$NiTa_2$ and α-μ-λ_2-Ni_2-Ta are isolated right away (Fig. 22,d,e). As the result of analysis and remaining graph consisting of five phases (σ, β, λ_2, α, γ) are obtained (Fig. 22,f).

After investigation of four alloys, melting in the field between these phases, there have been discovered the existence of two four-phases equilibria: σ-γ-α-β and β-σ-α-λ_2 (Fig. 22,g,h). Total graph of four-phases equilibria, preparing for analysis five-component systems, are shown in Fig. 22,i.

Convergence

Polyhedration of System Ni-V-Cr-Mo-Re

The system Ni-V-Cr-Mo-Re, which includes ten ternary systems V-Cr-Mo, Ni-Cr-Re, Ni-Mo-Re, Ni-V-Re, Ni-Cr-Mo, Ni-V-Cr , Ni-V-Mo, V-Cr-Re, V-Mo-Re, Cr-Mo-Re is investigated. The phase equilibria have been determined in the systems Ni-V-Cr-Mo, Ni-V-Cr-Re, Ni-Cr-Mo-Re, Ni-V-Mo-Re, V-Cr-Mo-Re and Ni-V-Cr-Mo-Re by using graph method. The obtained data have been confirmed by experimental investigation.

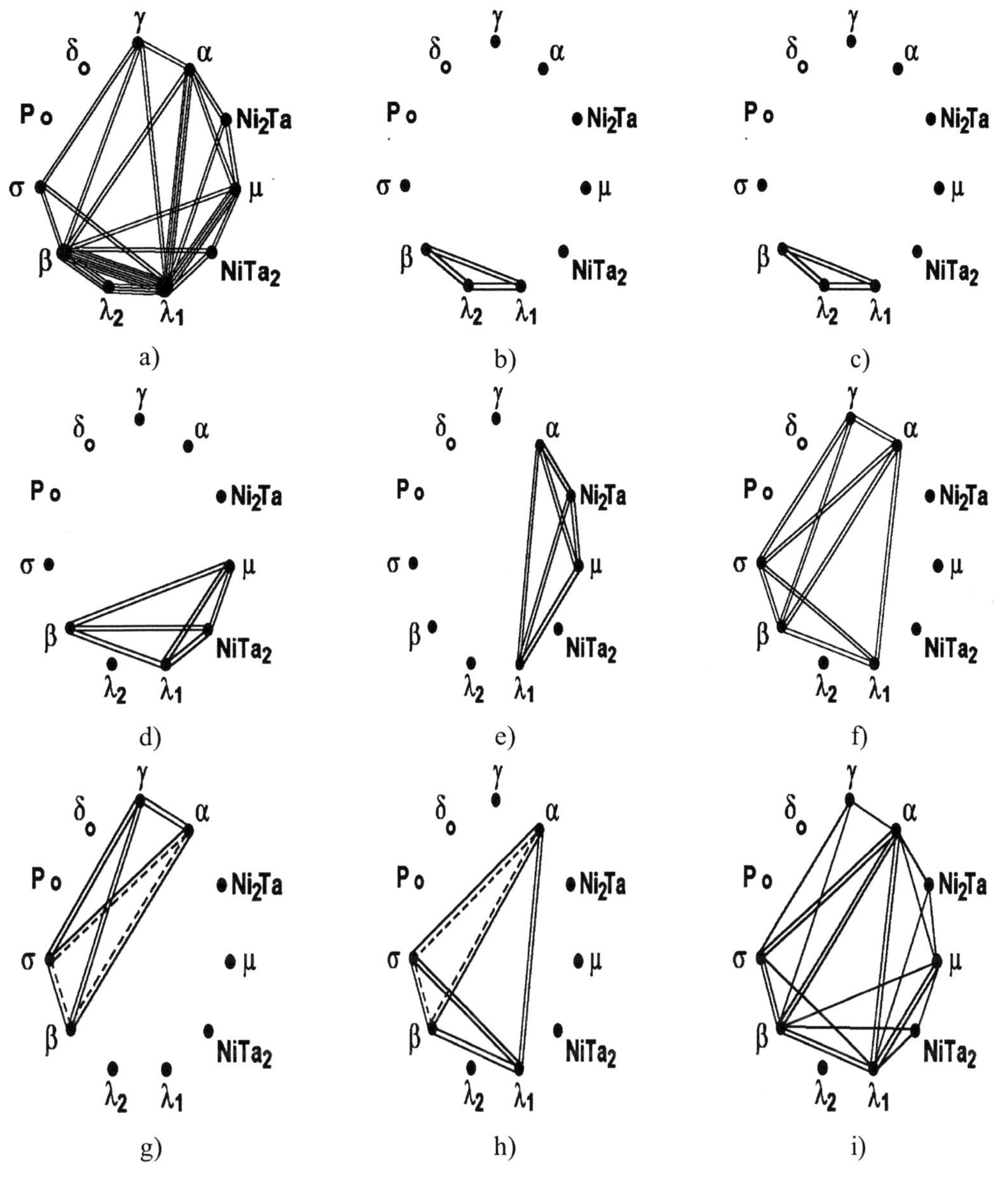

Fig. 22. The total graph three-phase eouilibria (a) of the Ni-V-Ta-Cr system, recombined equilibria (b,c), the remaining graphs of the zero rank (d,e), the remaining graphs of the first rank (f), its decomposition on the four-phase equilibria (g,h) and the graph of the four-phase eouilibria Ni-V-Ta-Cr system (i).

Thus, the following three four-phase equilibria occur in the five component Ni-V-Cr-Mo-Re system as it is shown in Table 6. For the formation of the five-phases equilibrium in quinary system it is necessary to have five different types of the four-phases equlibria. Therefore the five-phase equilibria in the five- component Ni-V-Cr-Mo-Re system are absent.

Table 6 Data polyhedration quaternary system in the quinary Ni-V-Cr-Mo-W-Re system.

System	Phases	Four-phases equilibria
Ni-V-Cr-Mo	β, γ, σ, δ, ρ	β+γ+σ+ρ; β+γ+δ+ρ
Ni-V-Cr-Re	β, γ, σ, Re σ	–
Ni-V-Mo-Re	β, γ, σ, δ, χ, ρ, Re	β+γ+σ+ρ; β+γ+δ+ρ; γ+σ+ρ+ Re
Ni-Cr-Mo-Re	β, γ, σ, δ, χ, ρ, Re	γ+σ+ρ+Pε
V-Cr-Mo-Re	β, σ, χ, Re	–

Table 7 Data of polyhedration quaternary system in the Ni-Mn-V-Cr-Mo-W system

	System	Phases	Four-phase equilibria
1	Mn-V-Cr-Mo	(β-Mn), β, σ	–
2	Mn-V-Cr-W	(β-Mn), β, σ	–
3	Mn-V-Mo-W	(β-Mn), β, σ	–
4	Mn-Cr-Mo-W	(β-Mn), β, σ	–
5	V-Cr-Mo-W	β	–
6	Ni-Mn-V-Cr	(β-Mn), β, γ, σ	–
7	Ni-Mn-V-Mo	(β-Mn), β, γ, δ, σ	–
8	Ni-Mn-V-W	(β-Mn), β, γ, σ, WNi_4	(β-Mn) – β –γ –σ
9	Ni-Mn-Cr-Mo	(β-Mn), β, γ, δ, σ, P	β–γ–σ–P, β–γ–δ–P
10	Ni-Mn-Cr-W	(β-Mn), β, γ, σ, WNi_4	(β-Mn) – β –γ –σ
11	Ni-Mn-Mo-W	(β-Mn), β, γ, σ, δ, WNi_4	(β-Mn) – β –γ –σ
12	Ni-V-Cr-Mo	β, γ, σ, P	β–γ–σ–P, β–γ–δ–P
13	Ni-V-Cr-W	β, γ, σ, WNi_4	–
14	Ni-Nb-M-Wo	β, γ, σ, δ, WNi_4	-
15	Ni-Cr-Mo-W	β, γ, σ, P, δ, WNi_4	β–γ–σ–P, β–γ–δ–P

Pholyhedration of the System Ni-V-Cr-Mo-W-Mn

Polyhedration of the six-component system Ni-V-Cr-Mo-W-Mn has been realized at 1250 K. Only in six systems the four-phase equilibria have been established. The equilibrium (β–Mn) – β –γ –σ is determined in the Ni-Mn-V-W, in the Ni-Mn-Cr-W, and in the Ni-Mn-Mo-W systems. The existence of remaining graph consisting of five phases: β, γ, δ, σ, P have been discovered in the Ni-Mn-Cr-Mo, in the Ni-V-Cr-Mo, and in the Ni-Cr-Mo-W. This graph was solved earlier the experimental investigation of the quinary system Ni-V-Cr-Mo-Re. It consists of two the four-phase equilibria: δ+β+P+γ and β+P+γ+σ.

Only three different types of the four-component equilibria exist in the Ni-V-Cr-Mo-W-Mn system. For the formation of the five-phase equlibrium it is necessary to have five different types of the four-phases equilibria. Consequently, in this system at 1250 K the five-phase equilibria are also absent.

Polyhedration of System Ni-V-Nb-Ta-Cr-Mo-W.

In seven-component system Ni-V-Nb-Ta-Cr-Mo-W the analysis of the phase equilibria in 35 quaternary systems, 21 quinari systems and seven six-component systems have been realized using the graph method.

All ternary systems (supposedly quaternary systems) contain only two phases: the bcc-solid solution and Laves λ_2-phase, which exists in the V-Ta, Cr-Nb and Cr-Ta systems. It follows, in the multi-component systems of the refractory metals only two-phase equilibria can exist. The Nb-Cr-Mo, Nb-Cr-W, Ta-Cr-Mo and Ta-Cr-W systems are exceptions, because they contain the three-phase equilibrium λ_1-β_1-β_2. In spite of the existence of only two phases in this systems three-phase equilibria are present because of the stratification of the β-solid solution.

At addition of the fourth component these regions form the three-phase field in the quaternary systems (in Nb-Ta-Cr-W system) or degenerate to the region of the two-phase equilibrium (in the rest systems). The review of the quaternary systems of the metal of the VB and VIB groups is given in Table 8.

Table 8. Equilibria in quaternary systems on the bases and refractory metals

Systems	Phases	Determined phase equilibria
V-Nb-Mo-W, Nb-Ta-Mo-W	β	-
V-Nb-Ta-Cr, V-Nb-Ta-Mo, V-Nb-Ta-W, V-Ta-Cr-Mo, V-Ta-Mo-W	λ_2, β	λ_2-β
V-Cr-Mo-W	β	Region of the stratification of β phase
V-Nb-Cr-W, V-Ta-Cr-W, V-Nb-Cr-Mo, Nb-Ta-Cr-Mo, Nb-Cr-Mo-W	λ_2, β	Degeneration of the three-phase equilibrium λ_2-β_1–β_2 to the two-phase equilibrium λ_2-β
Nb-Ta-Cr-W	λ_2, β	Recombination of the three-phase equilibria λ_2-β_1–β_2

In the systems on the basis of nickel and refractory metals there are many intermetallic compounds. For the polyhedration of the isothermal cross-sections of the quaternary phase diagrams the graphs method are used. Only in twelve diagrams the remaining graphs have been obtained.

Experimental decomposition of the remaining graphs carried out by investigations of the phase composition in 48 alloys and a few diffusion pairs, which are combined with using multi-phases alloys. The structure and the concentration distribution of the elements in the phases of the equilibrium alloys and the diffusion zones were investigated by scanning electron microscopy, microprobe analysis techniques and X-ray analysis. The established four-phase equilibria in the quaternary systems on the bases of nickel and refractory metals of the V-VI groups are given in Table 9.

Table 9. Four-phase equilibria in quaternary systems on the bases of nickel and refractory metals

	System	Phases	Determinated phase equilibria
1	Ni-V-Nb-Ta	α, β, γ, σ, μ, λ_1, λ_2, Ni_2Ta, $NiTa_2$	α–σ–γ–λ_1
2	Ni-V-Nb-Cr	α, β, γ, σ, μ, λ_1, λ_2	α–β–γ–λ_1, σ–β–γ–λ_1
3	Ni-V-Nb-Mo	α, β, γ, δ, σ, μ, λ1	α–β–γ–δ, α–β–μ–λ_1, β–γ–σ–λ_1, α–β–γ–λ_1
4	Ni-V-Nb-W	α, β, γ, σ, μ, λ_1	α–β–γ–λ_1, α–β–μ–λ_1, σ–β–γ–λ_1
5	Ni-V-Ta-Cr	α, β, γ, σ, μ, λ_1, λ_2, Ni_2Ta, $NiTa_2$	α–β–γ–σ, α–β–σ–λ_1, α–μ–λ_1–Ni_2Ta, β–μ–λ_1–$NiTa_2$
6	Ni-V-Ta-Mo	α, β, γ, δ, σ, μ, λ_1, λ_2, Ni_2Ta, $NiTa_2$	α–β–γ–δ, α–β–σ–λ_1, α–β–μ–λ_1, α–β–γ–σ, α–β–μ–Ni_2Ta
7	Ni-V-Ta-W	α, β, γ, σ, μ, λ_1, λ_2, Ni_2Ta, $NiTa_2$	α–β–γ–σ, α–β–σ–λ_1, α–β–μ–λ_1, α–β–μ–Ni_2Ta
8	Ni-V-Cr-Mo	β, γ, δ, σ, P	β–γ–σ–P, β–γ–δ–P
9	Ni-V-Cr-W	β, γ, σ	–
10	Ni-V-Mo-W	β, γ, δ, σ	–
11	Ni-Nb-Ta-Cr	α, β, γ, μ, λ_1, λ_2, Ni_2Ta, $NiTa_2$	α–μ–λ_1–Ni_2Ta, β–μ–λ_1–$NiTa_2$
12	Ni-Nb-Ta-Mo	α, β, γ, δ, μ, Ni_2Ta, $NiTa_2$	α–β–μ–Ni_2Ta
13	Ni-Nb-Ta-W	α, β, γ, μ, Ni_2Ta, $NiTa_2$	α–β–μ–Ni_2Ta
14	Ni-Nb-Cr-Mo	α, β, γ, δ, σ, μ, λ_1, λ_2, P	α–β–μ–λ_1, α–β–δ–P, α–γ–δ–P, α–β–σ–P, α–γ–σ–P, α–β–γ–σ, λ_1–λ_2–β_1–β_2
15	Ni-Nb-Cr-W	α, β, γ, σ, μ, λ_1, λ_2	α–β–μ–λ_1, λ_1–λ_2–β_1–β_2, α–λ_1–β_1–β_2, α–σ–β_1–β_2, α–σ–γ–β_1, α–σ–γ–β_2
16	Ni-Nb-Mo-W	α, β, γ, δ, μ	α–β–γ–δ
17	Ni-Ta-Cr-Mo	α, β, γ, δ, σ, μ, λ_1, λ_2, P, Ni_2Ta, $NiTa_2$	α–β–δ–P, α–γ–δ–P, α–β–σ–P, α–γ–σ–P, α–β–γ–σ, α–β–λ_1–Ni_2Ta, β–μ–λ_1–Ni_2Ta, β–μ–λ_1–$NiTa_2$, λ_1–λ_2–β_1–β_2,
18	Ni-Ta-Cr-W	α, β, γ, σ, μ, λ_1, λ_2, Ni_2Ta, $NiTa_2$	α–β–λ_1–Ni_2Ta, β–μ–λ_1–Ni_2Ta, β–μ–λ_1–$NiTa_2$, λ_1–λ_2–β_1–β_2, α–λ_1–β_1–β_2, α–σ–β_1–β_2, α–σ–γ–β_1, α–σ–γ–β_2
19	Ni-Ta-Mo-W	α, β, γ, δ, μ, Ni_2Ta, $NiTa_2$	α–β–γ–δ
20	Ni-Cr-Mo-W	β, γ, δ, σ, P	β–γ–δ–P, β–γ–σ–P

The five-phase equilibria in quinary systems which form in the seven-component system Ni-V-Nb-Ta-Cr-Mo-W can be obtained on the basis of the analysis of the blocks of the experimentally established four-phase equilibria of the quaternary systems by graphs method.

The single variant of the five- phase equilibria in the quinary systems is given in table 10. The number of systems in which the five-phase equilibria are realized is equal to 10. However, the number of types of the five-phase equilibria is equal to four. For the formation of the six-phase equilibrium it is necessary to have six different types of the five-phase equilibria. Consequently, in the Ni-V-Nb-Ta-Cr-Mo-W system at 1375 K the formation of the six-phase equilibria is not possible. Accordingly, the seven-phase equilibrium also does not exist.

In metallic systems, when phases have as a rule tightly packed structures the graphs method gives very good results.

Table 10. Five-phase equilibria in quinary systems on the bases of nickel and refractory metals

	System	Five-phases equilibria
1	Ni-V-Nb-Ta-Cr	$\alpha-\beta-\gamma-\sigma-\lambda_1$
2	Ni-V-Nb-Ta-Mo	$\alpha-\beta-\gamma-\sigma-\lambda_1$
3	Ni-V-Nb-Ta-W	$\alpha-\beta-\gamma-\sigma-\lambda_1$
4	Ni-V-Nb-Cr-Mo	$\alpha-\beta-\gamma-\delta-P$, $\alpha-\beta-\gamma-\sigma-P$
5	Ni-V-Nb-Cr-W	–
6	Ni-V-Nb-Mo-W	–
7	Ni-V-Ta-Cr-Mo	$\alpha-\beta-\gamma-\delta-P$, $\alpha-\beta-\gamma-\sigma-P$, $-\beta-\mu-\lambda_1-Ni_2Ta$
8	Ni-V-Ta-Cr-W	$\alpha-\beta-\mu-\lambda_1-Ni_2Ta$
9	Ni-V-Ta-Mo-W	–
10	Ni-V-Cr-Mo-W	–
11	Ni-Nb-Ta-Cr-Mo	$\alpha-\beta-\mu-\lambda_1-Ni_2Ta$
12	Ni-Nb-Ta-Cr-W	$\alpha-\beta-\mu-\lambda_1-Ni_2Ta$
13	Ni-Nb-Ta-Mo-W	–
14	Ni-Nb-Cr-Mo-W	$\alpha-\beta-\gamma-\delta-P$, $\alpha-\beta-\gamma-\sigma-P$
15	Ni-Ta-Cr-Mo-W	$\alpha-\beta-\gamma-\delta-P$, $\alpha-\beta-\gamma-\sigma-P$

MODELING OF INTERDIFFUSION PROCESSES IN MULTI-COMPONENT SYSTEM

Formulation of Problem

Description of diffusion processes in multi-component systems are based on the line non-equilibrium thermodynamic relationships introduced by Onsager [20].

$$J_1 = -D_{11}\frac{\partial C_1}{\partial x} - D_{12}\frac{\partial C_2}{\partial x} - D_{13}\frac{\partial C_3}{\partial x},$$

$$J_2 = -D_{21}\frac{\partial C_1}{\partial x} - D_{22}\frac{\partial C_2}{\partial x} - D_{23}\frac{\partial C_3}{\partial x}, \qquad (8)$$

$$J_3 = -D_{31}\frac{\partial C_1}{\partial x} - D_{32}\frac{\partial C_2}{\partial x} - D_{33}\frac{\partial C_3}{\partial x}.$$

For example, for the three-component system in the one-dimensional case, we have three such equations (8). For a quantitative description of interdiffusion in the n-component system n^2 parameters are necessary. This requires extremely complicated, and often unfeasible experimental techniques of determining these parameters. To solve this problem it is necessary to work out the method of simplification in the set of equations (8) in order to reduce the number of determined parameters.

Generation of Principal Idea

In the real systems, because of the inequality of the mobilities of components, the condition

$$\sum_{i=1}^{n} C_i = 1 \qquad (9)$$

is violated, which leads to the displacement of crystal planes in the diffusion zone i.e. the Kirkendall effect [21].

Darken assumed that this shift fully compensates deficit or excess of mass in a given plane and developed a fundamental theory of the observed phenomenon. According to Darken, we have equation (10)

$$\sum_{i=1}^{n} (J_i + VC_i) = 0, \qquad (10)$$

where V is the rate of the Kirkendall shift for the given plane. Introducing designation

$$\widetilde{J}_i = J_i + VC_i, \qquad (11)$$

we get the relation

$$\sum_{i=1}^{n} \widetilde{J}_i = -\sum_{i=1}^{n} D_i \frac{\partial C_i}{\partial x} = 0 \qquad (12)$$

In this work the convolution of the simultaneous equations (8) based on Darken's condition (10). For a three-component system we have (13)

$$-\sum_{i=1}^{3}(\sum_{k=1}^{3} D_{ik}\frac{\partial C_k}{\partial x}) + V(\sum_{i=1}^{3} C_i) = 0 \qquad (13)$$

Grouping the terms with the same derivatives we get

$$-\sum_{k=1}^{3}\frac{\partial C_k}{\partial x}(\sum_{i=1}^{3} D_{ik}) + V(\sum_{i=1}^{3} C_i) = 0 \qquad (14)$$

Using notation

$$D_k = \sum_{j=1}^{3} D_{jk} \qquad (15)$$

equation (14) can be rewritten as

$$-D_1\frac{\partial C_1}{\partial x} + VC_1 - D_2\frac{\partial C_2}{\partial x} + VC_2 - D_3\frac{\partial C_3}{\partial x} + VC_3 = 0, \qquad (16)$$

where $D_1 = D_{11}+D_{21}+D_{31}$, $D_2 = D_{12}+D_{22}+D_{32}$, $D_3 = D_{13}+D_{23}+D_{33}$ (*Dik* from the based set (8)) are the partial diffusion coefficients. From Eq. (16) it follows that

$$V = D_1\frac{\partial C_1}{\partial x} + D_2\frac{\partial C_2}{\partial x} + D_3\frac{\partial C_3}{\partial x}. \qquad (17)$$

Using the designations (11) one can pass on to the expression for the fluxes with the Kirkendall effect accounted for. Then

$$\tilde{D}_1\frac{\partial C_1}{\partial x} + \tilde{D}_2\frac{\partial C_2}{\partial x} + \tilde{D}_3\frac{\partial C_3}{\partial x} = 0, \qquad (18)$$

Where $\tilde{D}_i$ are the interdiffusion coefficients, which we can be determined experimentally.

Substituting Eq. (17) into Eq. (11) we get equation (19)

$$\tilde{J}_1 = -\tilde{D}_1\frac{\partial C_1}{\partial x} = -\left[D_1(1-C) + C_1 \times \frac{D_2\frac{\partial C_2}{\partial x} + D_3\frac{\partial C_3}{\partial x}}{\frac{\partial C_2}{\partial x} + \frac{\partial C_3}{\partial x}}\right] \times \frac{\partial C_1}{\partial x}. \qquad (19)$$

Hence

$$\widetilde{D}_1 = D_1(1 - C_1) + C_1 \times \frac{D_2 \dfrac{\partial C_2}{\partial x} + D_3 \dfrac{\partial C_3}{\partial x}}{\dfrac{\partial C_2}{\partial x} + \dfrac{\partial C_3}{\partial x}}. \tag{20, a}$$

Similar expressions are obtained for the components 2 and 3

$$\widetilde{D}_2 = D_2(1 - C_2) + C_2 \times \frac{D_1 \dfrac{\partial C_1}{\partial x} + D_3 \dfrac{\partial C_3}{\partial x}}{\dfrac{\partial C_1}{\partial x} + \dfrac{\partial C_3}{\partial x}}. \tag{20, b}$$

$$\widetilde{D}_3 = D_3(1 - C_3) + C_3 \times \frac{D_2 \dfrac{\partial C_2}{\partial x} + D_1 \dfrac{\partial C_1}{\partial x}}{\dfrac{\partial C_2}{\partial x} + \dfrac{\partial C_1}{\partial x}}. \tag{20, c}$$

The determination of the interdiffusion coefficients $\widetilde{D}_i$ in the model of independent fluxes is not difficult. The concentration curve for the i-th component is processed by the Matano-Boltzmann method as in binary systems (21).

$$\widetilde{D}_i = -\frac{1}{2t}\left(\frac{dx}{dC_i}\right)_{C_i^0} \int_{C_i^1}^{C_i^o} x(C_i) dC_i, \tag{21}$$

where C_i^1 is the concentration of the i-th component in the initial alloy, C_i^0 is the concentration of the i-th component in the point of composition where $\widetilde{D}_i$ is determined. The partial diffusion coefficients are calculated from a set of three equations (20). The full collection of the partial diffusion coefficients is calculated only in a point of crossing of two diffusion ways as one equation in the set of equations (20).

Divergence

Experimental examination idea was carried out by determination a partial diffusion coefficients iron, nickel and chromium in Fe-Ni-Cr system. This system has two phases with structure bcc (β) and fcc (γ), dividing the two-phase field (γ+β) as shown in Fig.22.

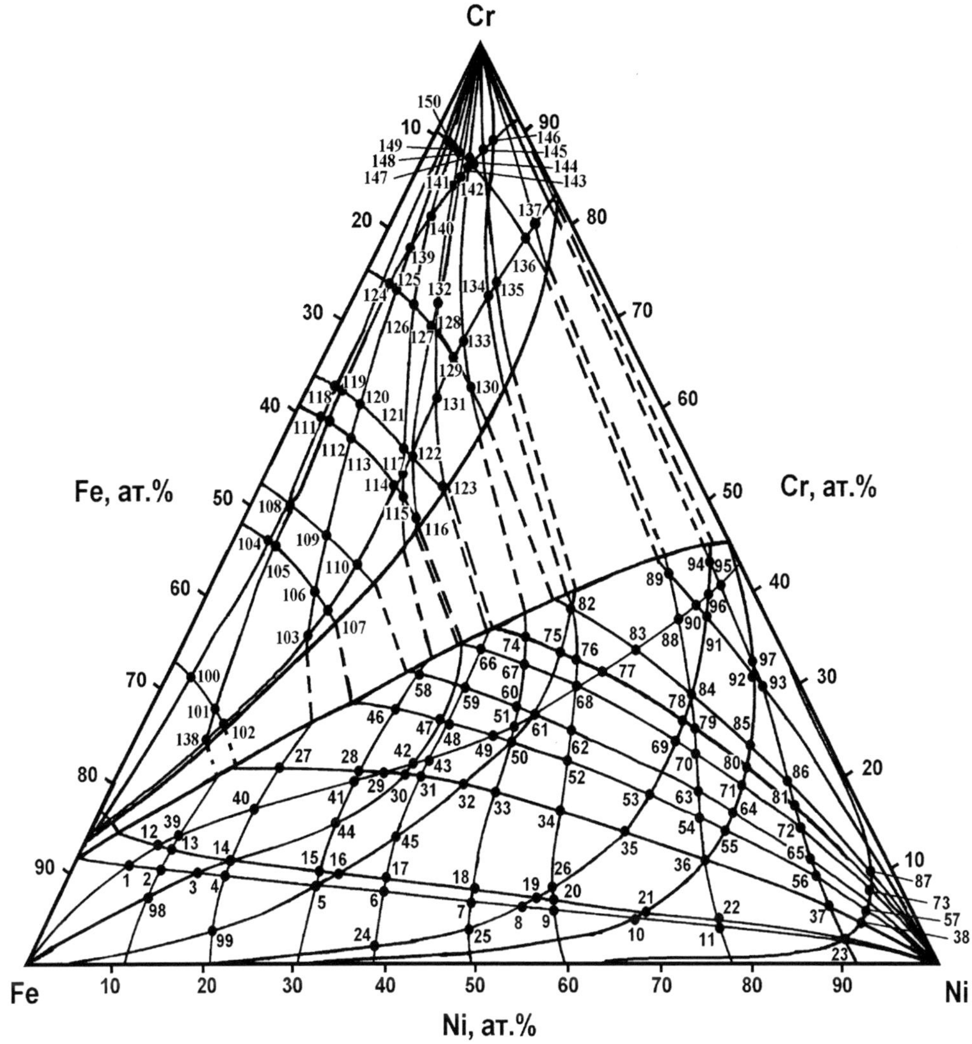

Fig.22. Diffusion ways in Fe-Ni-Cr systems at 1375 K

An analytical description Fe-Ni-Cr is realized. The next two equations are used for description of the boundary (γ+β) / β

$$C_{Ni} = -1.2C_{Fe}^2 + 0.448C_{Fe} + 0.1752, \text{ if } C_{Fe} \leq 0.3,$$

$$C_{Ni} = -0.28C_{Fe}^2 - 0.035C_{Fe} + 0.2373, \text{ if } C_{Fe} \geq 0.3,$$

Also two equations are used for description of the boundary (γ+β) / γ

$$C_{Ni} = 0.5C_{Fe}^2 - 0.85C_{Fe} + 0.545, \text{ if } C_{Fe} \leq 0.3,$$

$$C_{Ni} = 0.04C_{Fe}^2 - 0.618C_{Fe} + 0.5168, \text{ if } C_{Fe} \geq 0.3,$$

The equation for description the tie line in two-phase field (γ+β)

$$C_{Fe}(\gamma) = C_{Fe}(\beta) + 0.018 - 0.018C_{Fe}(\beta)$$

Nineteen diffusion couples (FeNi-alloys + Cr, FeCr-alloys + Ni, NiCr-alloys +Fe) and "sandwich" Fe + Ni (50 mkm) +Cr are under investigation at 1375 K. Partial diffusion coefficients Fe, Ni and Cr are determined in 150 points of the Fe-Ni-Cr system. The functional dependence of the partial diffusion coefficients Fe, Ni and Cr from their concentration is also determined. The obtained equations are shown below.

Convergence

A Discrete Model of Diffusion

In this investigation a discrete model of the classic diffusion pair [21] consisting of N shells with the same length ΔX (Fig. 23) are used

Each shell is filled with atoms with the concentration for the i-component in j-th shell as it is shown in equation (22).

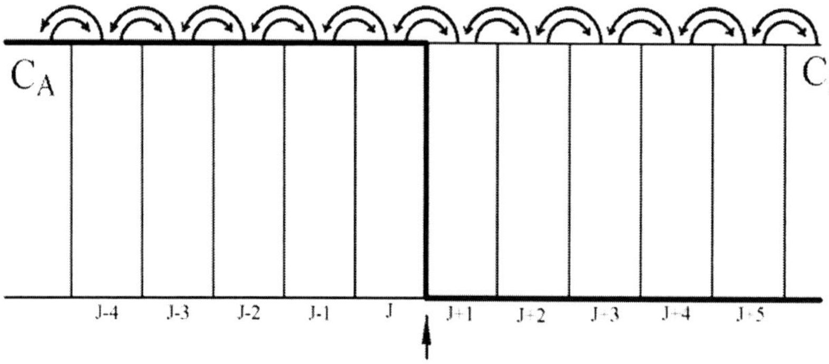

Fig. 23. Model the change of atoms between the micro-volumes of the diffusion zone

Partial diffusion coefficients D_i, represent the probability of the transition of atoms of i-th component from j-shell into neighbours during a unit of time and they are principal parameters for modeling

$$\overline{C}_i = \int_{X_{j-1}}^{X_j} C_i(X)dX \times \frac{1}{\Delta X}. \tag{22}$$

Shells change atoms of component according to equation (23)

$$\overline{C}_i(\tau = \tau_0 + 1) = \overline{C}_i(\tau_0) + D_i\overline{C}_j^{(j-1)} - 2D_i\overline{C}_j^{(j)} + D_i\overline{C}_j^{(j+1)}. \tag{23}$$

When $D_i = D_i(C)$, equation (22) is not applicable. In this case first general Fick's equation is used i.e, $J = \partial F(C)/\partial X$, when $F(C) = \int D_i(C)\partial C$, is a potential of diffusion. Then it is necessary to use equation)

$$\overline{C}_i(\tau = \tau_0 + 1) = \overline{C}_i(\tau_0) + F_{(j-1)} - 2F_{(j)} + F_{(j+1)}. \tag{24}$$

A single cycle of calculation in all shells of this model corresponds to a unit of time.

For modeling the concentration of component in diffusion couples Fe-Ni-Cr system we used the data of the analytical description of the Fe-Ni-Cr system (the equation of the phase boundary and the equation of tie lines) and the composition of alloys in diffusion couples as initial and limited conditions of modelling. Simple functions, positive in all fields Fe-Ni-Cr system, are used for interpolation of the experimental data of the partial diffusion coefficient iron, nickel and chromium (Fig. 24,a).

Modeling result are shown in Fig. 24,b-d. Practically full coincidence between the experimental results (points) and the modeling curves (line) is observed.

The discrete model of diffusion can be regarded as a generalization of the mathematical theory of diffusion, based on Fick's equations for multi-component systems. At the same time the basic equations of the model can be obtained within the framework of the thermodynamic theory. In principle the discrete model of diffusion combines both approaches.

For fcc field:

$$D_{Fe} = 12.4 - 0.096C_{Fe} - 0.132C_{Ni}$$

$$D_{Ni} = 6 + 37\exp(-C_{Fe}/42.2776) - 0.65C_{Ni}$$

$$D_{Cr} = 12 - 0.04C_{Fe}$$

For fcc field:

$$D_{Fe} = 1.7C_{Fe} + 1.1C_{Ni} + 2C_{Fe}^2$$

$$D_{Ni} = 1.0C_{Fe} + 2.2C_{Ni}$$

$$D_{Cr} = 2 - 0.2C_{Fe} - 0.05C_{Ni} + 0.01C_{Fe}^2 + 0.05C_{Ni}^2.$$

(a)

Figure 24 (a) and data of modeling of the concentration curves in diffusion couples (Continued on next page)

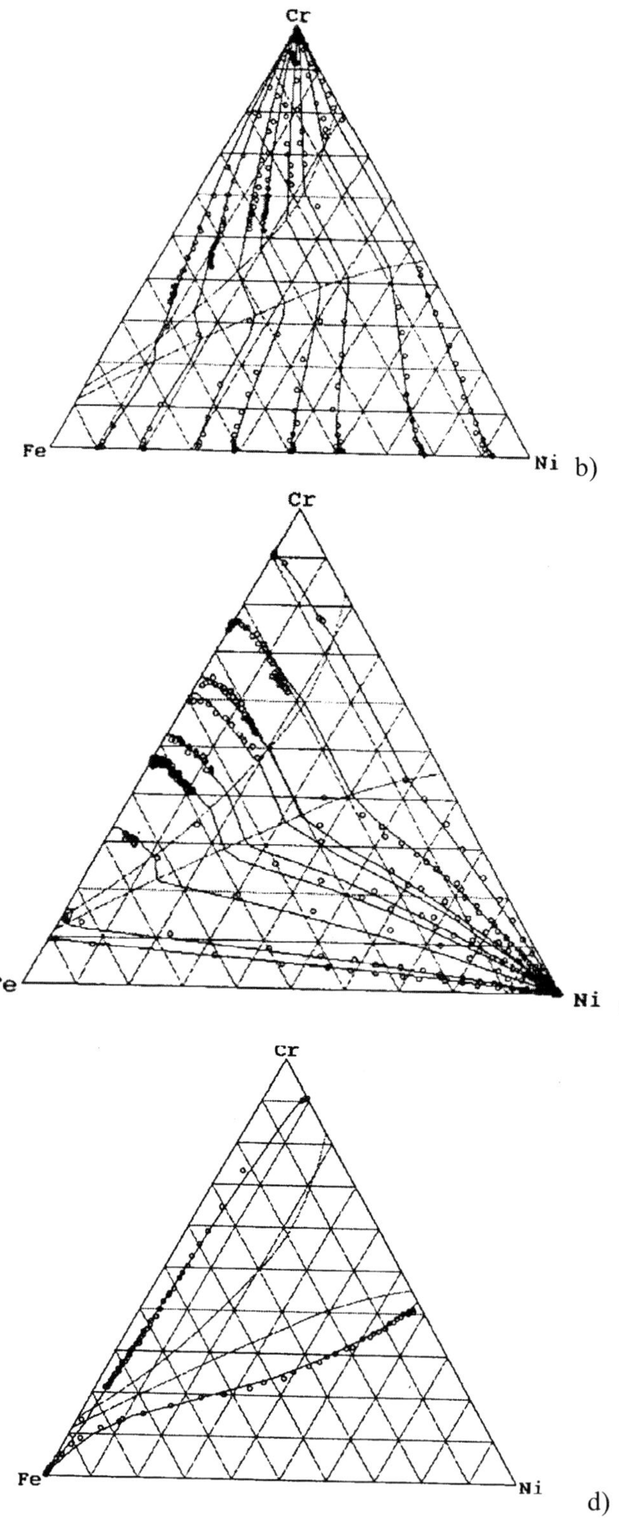

Fig.24. Interpolated equations for a partial diffusion coefficients in the phases γ and β in Fe-Ni-Cr systems FeNi-alloys + Cr (b); FeCr-alloys + Ni (c) and NiCr-alloys + Fe (d).

REFERENCES

[1] M. Goeppert-Mayer, J. Jensen, *Elementarnaya teoriya yadernyh obolochek*. Moscow, FL, 1958, (In Russian).
[2] T.B. Romanovskaja. *Istoriya kvantovomechanicheskoii interpretaczii periodichnosti*. Moscow, Science, 1986, (In Russian).
[3] J.Thomsen. *Z.anorg.Chem.*. 9, 285, (1895).
[4] Rydberg. *Swenska vet. Akad. Hand.*. 10, 2, (1885).
[5] Rydberg. *Z.anorg.Chem.*. 14, 66 (1897).
[6] D.N. Trifonov, *Kolichestvennaya interpretaziya periodicheskogo zakona*. Moscow, Gosa-tomizdat, 1963, (In Russian).
[7] E.M.Slyusarenko. *J. Mendeleev Communications*. 2, 78, (1996).
[8] E.M.Slyusarenko. *Istoriya chimii: oblast nauki i uchebnaya discziplina*. Moscow, MSU, 2001, (In Russian).
[9] E.M.Slyusarenko. *Geometric conception of chemical bond in crystalline structures of elements and intermetallic compounds*. Lvov, LSU (1999) (In Russian).
[10] E.M.Slyusarenko. *Mendeleev's Congress in Common and Applied Chemistry*. 2, 263 (2003).
[11] *Spravochnik chimika*. Ed. B.P.Nikol'sky, Moscow, Goschimizdat, 1963 (In Russian).
[12] *Razvitie ucheniya o valentnosti*. Ed. V.I.Kuznezov, Moscow, Nauka, 1977(In Russian).
[13] C.A. Coulson. *Valentnost*. Moscow, "Mir", 1965 (In Russian).
[14] R.J.Gillespie, I. Hargittai. *Model ottalkivaniya electronnyh par valentnoi obolochki b stroenie molekul*. Moscow, 'Mir', 1992 (In Russian).
[15] J. Emsly (Ed), *The Elements*. 2nd ed, Clarendon Press, Oxford, 1991.
[16] W.B. Pearson, *The crystal chemistry and physics of metals and alloys*. Wiley-Interscience, New York-London-Sydney-Toronto, 1972.
[17] E.M. Slyusarenko, V.A. Borisov, M.V. Sofin, E.Yu. Kerimov, A.E.Chastukhin. *J. Alloys and Compounds*. 284, 171, (1999).
[18] E.M. Slyusarenko, M.V. Sofin, E.Yu. Kerimov. *J. Mendeleev Communication*. 2, 56, (1999).
[19] M.V. Sofin, E.Yu. Kerimov, N.A. Bazhanova, Yu.V. Balykova, A.E.Chastukhin, E.M. Slyusarenko. *J. Alloys and Compounds*. 321, 102, (2001).
[20] S. De-Groot, P. Mazur. *Neravnovesnaya termodinamika*. Moscow, Nauka, 1964 (In Russian).
[21] E.M. Slyusarenko. *Kolichestvennye raschety parametrov vzaimnoi diffuzii v mnogokomponentnyh sistemah*. Moscow, MSU, 1987 (In Russian).

PART III. INDUSTRY

In: Progress in Chemometrics Research
Editor: Alexey L. Pomerantsev, pp. 193-208
ISBN 1-59454-257-0

Chapter 13

CHARACTERIZATION OF BIOFUEL MIXTURES BY ONLINE NEAR INFRARED SPECTROSCOPY

Paul Geladi[1], Tom Lillhonga[2] and Linda Reuter[2]

[1] Unit of Biomass Technology and Chemistry, Swedish University of Agricultural Sciences, Sweden

[2] Swedish Polytechnic, Finland, tom.lillhonga@syh.fi

ABSTRACT

Off-line measurements of NIR spectra of fuel mixtures (coal, peat, biofuel) were carried out. Eleven mixtures were made according to a mixture triangle. For each mixture ten replicate NIR measurements were made with an InGaAs spectrometer measuring between 911 and 1709 nm in 128 bands. The reference measurements were moisture, energy and ash, all carried out according to laboratory standards. All these reference measurements were done in triplicate. The data analysis consists of a PCA analysis to explain the structure of the spectra. Different PLS models for calibration were tested. Some important conclusions are reported. The replicates allow the determination of bias and standard deviation for the obtained predictions. Energy and moisture content can be measured to a reasonable precision. Ash content can also be measured, but with larger errors.

Keywords: process chemometrics, principal component analysis, partial least squares regression, process analytical chemistry, near infrared spectroscopy, mixture triangle, validation

INTRODUCTION

Biofuels are becoming more and more important as sources of energy production because they form part of a recycling chain for carbon dioxide. Forest and agricultural biofuels are part of the large scale production of energy where they are the starting material of many processes e.g. direct burning, burning as mixtures, production of gaseous and liquid fuels or pressing of fuel pellets and briquettes. The present material is part of a feasibility study carried out for a large energy production facility that uses forest biofuels, peat and coal for the production of electricity, district heat and process steam. Especially the forest biofuels have a large variability in chemical composition and physical nature. They can be branches, sawdust, needles, leaves, chips, part of roots etc. A previous study reported in [1] showed how it was possible to get a rough estimate of humidity from the near infrared spectra of the materials. The study used at-line measurement with a moving grating instrument giving 1050 wavelengths between 400 and 2500 nm. The results for the classical laboratory analysis and for the PLS prediction from NIR spectra are given in Table 1.

Table 1: Variation for the reference methodand PLS predictions from NIR spectra for replicate samples (5 replicates)

Replicate series	Reference Method	PLS Prediction
1	0.86	0.95
2	0.99	3.52
3	1.07	3.17
4	1.14	*
5	1.84	*
6	2.25	*

* means not measured because of equipment error

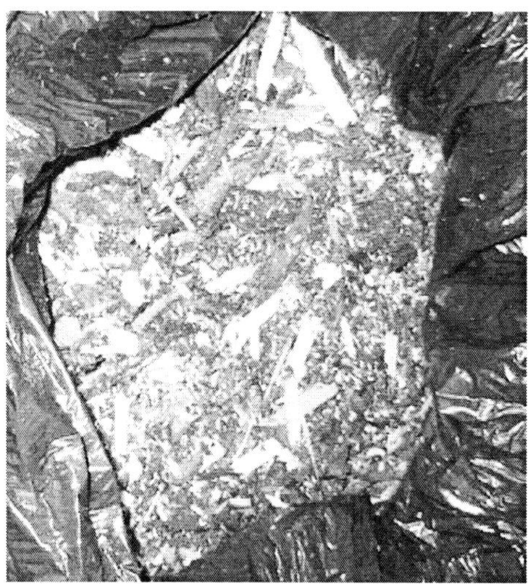

Fig.1. Very heterogeneous biofuel used in the measurements

It is clear from this table that an attempt has to be made to improve the precision of the method. Much improvement can be made by sampling and sample presentation modifications.

The reference method gives standard deviations close to 1.0 % in the best cases and up to 2.3 % for less homogeneous material. The PLS prediction gives standard deviations of 3.0 % or more in many instances. The reason for this is that the laboratory standard measures the bulk of the samples (about 500 g) while the NIR radiation only penetrates the surface for a few tenths of a mm.

The present example consists of laboratory measurements of mixtures. These are made according to a mixture triangle. Measurements of the mixtures are done with near infrared spectroscopy (NIR) using an InGaAs-array instrument giving 128 wavelengths in the region 911-1709 nm and with classical methods for moisture, ash and energy content determination. The goal is to be able to analyze the mixture triangle but also to be able to predict moisture, ash and energy and composition from the NIR spectra. The mixtures can also be used for classification purposes, but this is not included.

Process Chemometrics

The use of analytical instruments close to the process is called Process Analytical Chemistry (PAC) [2]. Off-line, at-line and on-line are used for defining measurement situations. Statistical Process Control (SPC) is the use of statistical methods to make and analyze control charts [3,4]. By combining multivariate analytical instruments and statistical process control, multivariate statistical process control (MSPC), also called process chemometrics, [5-9] is obtained.

The present study was done off-line in a laboratory environment in order to determine the feasibility of future at-line or on-line measurements. The line is actually under pressure and in an atmosphere containing much CO, making true at-line or online measurements difficult. The data analysis is a good illustration of the classical methods from chemometrics [10]. The goal is to replace slow and destructive laboratory methods with fast non-destructive spectroscopic methods. It is of interest to find out if moisture, energy and ash content can be predicted and how well this works

Principal Component Analysis

Data sets with many variables can be easier understood when they are simplified by data reduction. Principal component analysis (PCA) is a good reduction technique:

$$\mathbf{X} = \mathbf{TP}' + \mathbf{E} \qquad (1)$$

X: the spectral data (size I×K), sometimes mean-centered and/or scaled

T: a matrix of score vectors (size I×A)

P: a matrix of loading vectors (size K×A)

E: the residual (size I×K)

A: the number of components calculated

In the example, a reduction from 128 NIR wavelengths to 3 principal components is used. The vectors in T can be used to make score plots and the vectors in P can be used to make loading plots.

Calibration and Regression

The data require a calibration model to relate the spectra and reference values. This model is

$$\mathbf{y} = \mathbf{X}\mathbf{b} + \mathbf{f} \quad (2)$$

y: the mean-centered response data (size I)
X: the mean-centered spectral data (size I×K)
b: the regression coefficients (size K)
f: the residual (size I)

The high number of wavelengths requires a latent variable based method for calculating b. Partial least squares regression (PLS) was used. Building a regression model alone is not sufficient. The model also has to be tested. This is done with a test set and diagnostics are calculated. A simple but good diagnostic is simply looking at a plot of calculated against measured data. Root Mean Squared Error of Cross Validation is defined as

$$\text{RMSECV} = [\, \mathbf{f}'\mathbf{f} / \text{I}\,]^{1/2} \quad (3)$$

f is calculated by cross validation: some objects as left out of the calibration model and then predicted by it. This is done repeatedly until all I objects have been left out once. This gives a prediction diagnostic without using a separate prediction set. Where the curve of the RMSECV against number of components shows a minimum, the number of components that gives the best possible prediction results is situated.

EXPERIMENTAL

Raw Materials

Bags of coal, peat and forest biofuel were obtained from the plant. The coal and peat had particle size of less than 2 mm and were very homogeneous. The forest biofuel had much larger particles, up to a few cm, and was very heterogeneous; containing wood chips, bark chips and fine dust. The raw materials dried very quickly in storage and therefore water was added artificially during mixing. Figure 1 is a photograph of the biofuel material. It is very heterogeneous.

Mixture Triangle

The three ingredients were mixed according to a mixture triangle experimental design. Table 2 gives the design. Some runs were repeated.

Table2: The mixture triangle experiments

Mixture nr	Peat	Biofuel	Coal
1*	1	0	0
2*	1/2	1/2	0
3	1	0	0
4	0	1	0
5	1/2	1/2	0
6	1/4	3/4	0
7	3/4	1/4	0
8	1/3	1/3	1/3
9	0	1/2	1/2
10	1/2	0	1/2
11	0	0	1

* different biofuel / peat used

The mixtures were made on a weight % base with rather dry raw materials. The mixtures were then humidified by adding water. Some points in the mixture triangle were repeated. Figure 2 gives a schematic overview of the experiments: the raw materials are mixed according to the mixture triangle, water is added, they are measured by NIR and remixed and at the end the moisture content, ash and energy are measured by destructive laboratory methods. The mixture triangle is shown in figure 3.

One kilogram of each mixture was prepared. This amount of material was later split into three times approximately 300 g for destructive analysis. During the experiments new biofuel and peat were obtained, so some mixtures were made with a different biofuel. The mixtures were made in weight percent for wet materials, so the mixture proportions in table 2 are not completely exact on a dry weight basis

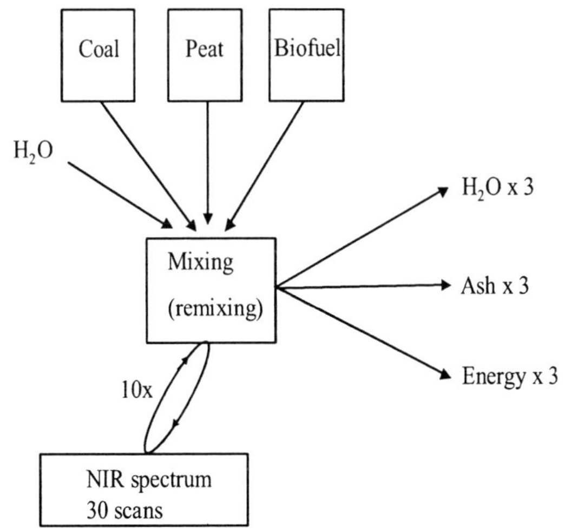

Fig 2. The experimental sequence. NIR spectra measured on 11 fuel mixtures with 10 replicates each and 3 reference values measured by industrial standards for each response variable

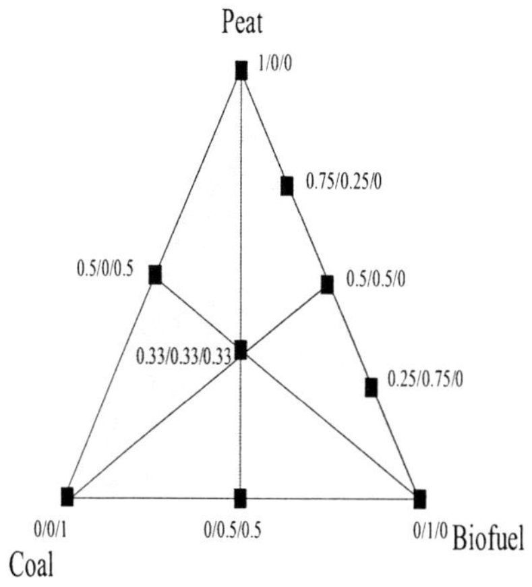

Fig.3. Mixture triangle with approximate proportions of each fuel type in the mixtures

Replicates

Of each run of the experimental design, ten replicates were measured by remixing the 1 kg of material and spreading the material out on the glass plate again.

NIR Spectrometer

The CDINIR128 spectrometer is based on an InGaAs detector array having 128 elements. The wavelengths form a sequence from 911 nm to 1709 nm with approximately 6 nm spacing. The spectrometer is connected to an integrating chamber with a glare trap by a fiberoptic cable. The material was spread out on a glass plate of 20 cm x 45 cm fitted to a frame structure with wheels. The measurements were made by adding 30 scans while moving the material over the integrating chamber. Integration time was approximately 25-30 ms. The setup is shown in figure 4. The reference spectrum was taken from a Spectralon standard in the same setup. With this kind of instrumentation it is important to have regular measurements of the reference standard.

Reference Measurements

The interesting parameters for the industrial process are humidity, energy content and ash content. These were measured by the classical laboratory standards (Swedish Standards, conforming with ISO standards). Moisture was measured gravimetrically by weighing the 300 g samples before and after drying at 105° Celsius (383 K) for 17 hours [SS 18 71 70 equals ISO 589-1981]. Energy content was measured by burning 0.7-1.0 g of dried and ground material as a pellet in a Parr bomb calorimeter. The energy content is given for the dry material. Ash content was measured gravimetrically by ashing about 1 g of sample in a Nabertherm oven at 550° Celsius (823 Kelvin) during at least 1 hour [SS 18 71 71 equals ISO 1171-1981].

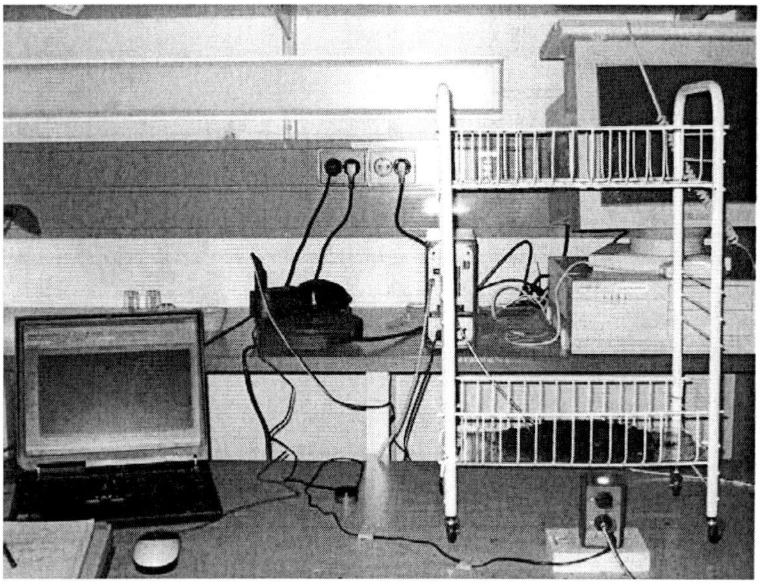

Fig. 4. Experimental setup (InGaAs spectrometer with fiber optic cable, integrating sphere with internal light source and sample trolley)

Moisture and ash content can be expressed in two different ways. Moisture percentages can be given as 100% for the dry material or 100% for the wet material. It was found that differences were small, so 100% for the dry material was used. The ash percentage can also be expressed based on the dry or wet material. Found differences were small and the percentage ash in dry material was used. For the energy content, there are large differences between wet and dry, but only dry energy content was tested in the calibration models. The compositions in table 2 can be added to the reference values, but this is not reported in this paper. The pooled standard deviation values are: 1.1% for moisture, 0.61 MJ/kg for energy and 1.26% for ash.

The Data

After removing obvious and explainable outliers, the obtained data form a matrix of 110x128 The reference measurements moisture, energy and ash were used as averages of the three replicates and as individual results. The reference measurements give either a 11x3 matrix, that can be expanded to 110x3, or a 33x3 matrix. All calculations were done in MATLAB and the PLS_Toolbox from eigenvector.com.

3. Results and Discussion

Principal Component Analysis

The data matrix was simplified by averaging over the replicates, giving an 11x128 matrix. The average spectra are shown in figure 5.

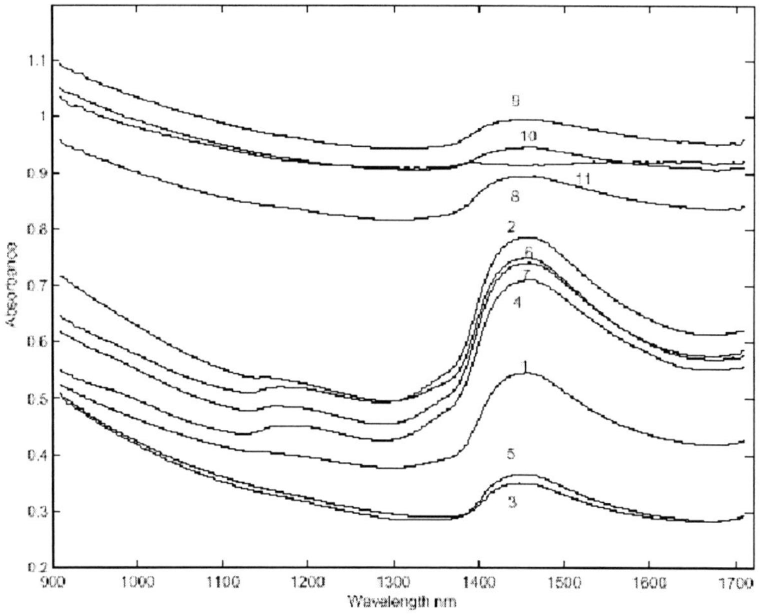

Fig. 5. Average spectra with fuel mixture reference number

They show more absorbance for the coal-containing samples. There is also a noticeable water peak. Mixture 11 with very little moisture does not show this peak. A more detailed analysis of the 11×128 matrix is given by principal component analysis [10,11]. This was done without any centering or scaling. The calculation gave 3 major components explaining 99.54, 0.38 and 0.08 % of the total sum of squares. Figure 6 gives a score plot for components 1 and 2. Figure 7 gives a score plot for components 2 and 3.

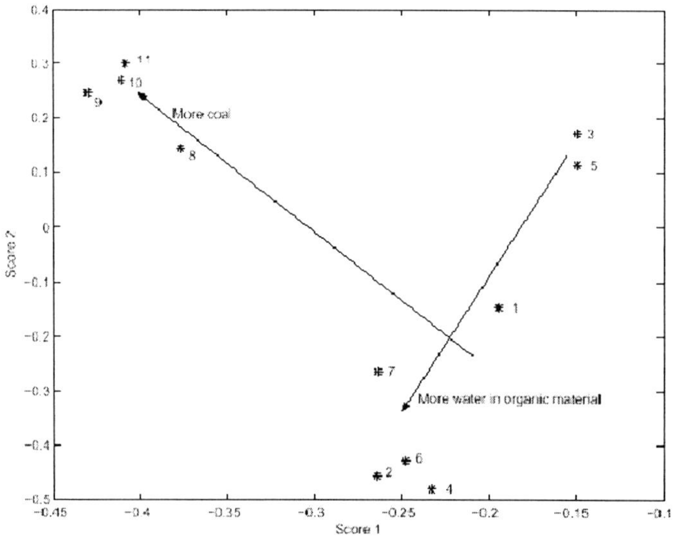

Fig. 6. Score plot of the two first major components in PCA model based on average spectra. Clustering and changes in water and coal content can be observed

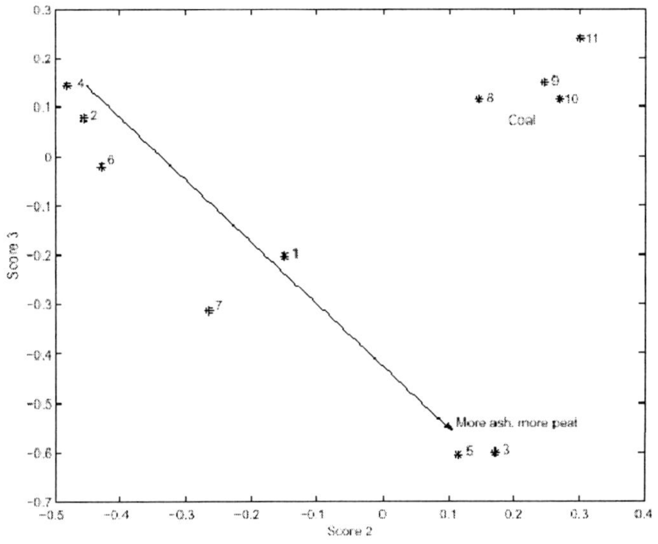

Fig. 7. Score plot of the second and third components in PCA model based on average spectra. Clustering and changes in ash and peat content can be observed

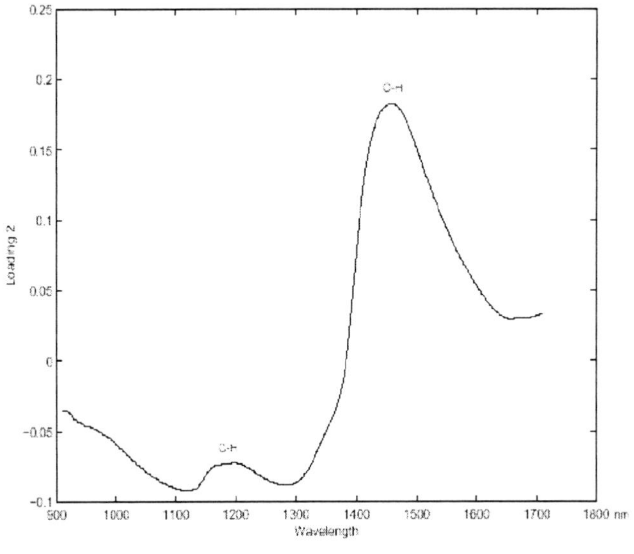

Fig. 8. Loading plot of the third component in PCA model based on average spectra. The peak at 1460 . nm is due to absorption by O-H bonds in water molecules and the lower peak at 1200 nm is due to absorption by C-H bonds in the organic material

It is possible to find structure in the score plots. The coal-containing samples cluster together in both scoreplots. In figure 6 a strong direction of moisture content for the organic samples is seen, mostly in the direction of component 2. A direction of increasing ash content (increasing peat content) is also seen in the third component in figure 7. This gives evidence that the latent structure of the data contains predictive information. The loadings can also be studied. Figure 8 gives the loadings for component 2 (with a sign reversal) and figure 9 gives the loadings for component 3 (also with a sign reversal). The second component loading shows the typical strong water peak at 1460 nm. A smaller peak is from the C-H bonds of the organic material. Loading 3 shows a smaller water peak at 1460 nm. It also shows a slope at short wavelengths, indicating absorption in the visual region.

A rough identification of the components would be:

Component 1: offset in absorbance from the much more absorbing coal-rich samples
Component 2: water content differences in the organic samples
Component 3: contrast between peat and biofuel, also correlated with ash content

The large dataset of all the 110 spectra can be used to detect outliers among the replicates. No extreme outliers were found and outlier detection was saved for regression models.

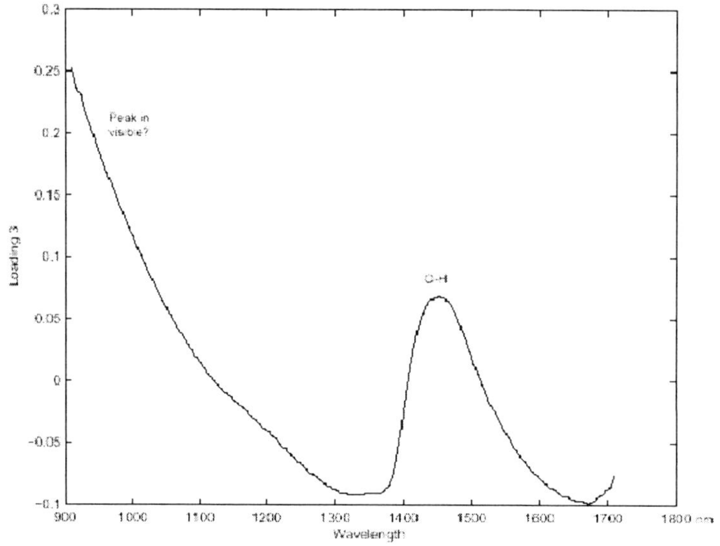

Fig. 9. Loading plot of the third component in PCA model based on average spectra. The peak at 1460 nm is again due to water absorption and an assumed absorption peak in the visible region can also be seen

Calibration

There are many ways of calculating calibration models, because of the replicates. They are shown in figure 10. The data sets are called 110S, 11S and 33S. A first calibration was done on the 110x128 data matrix called 110S. The spectral data were mean-centered and scaled and partial least squares regression [12,13] was used. It was found that leaving out scaling and using derivatives made little difference in the models, so these transformations are not reported here.

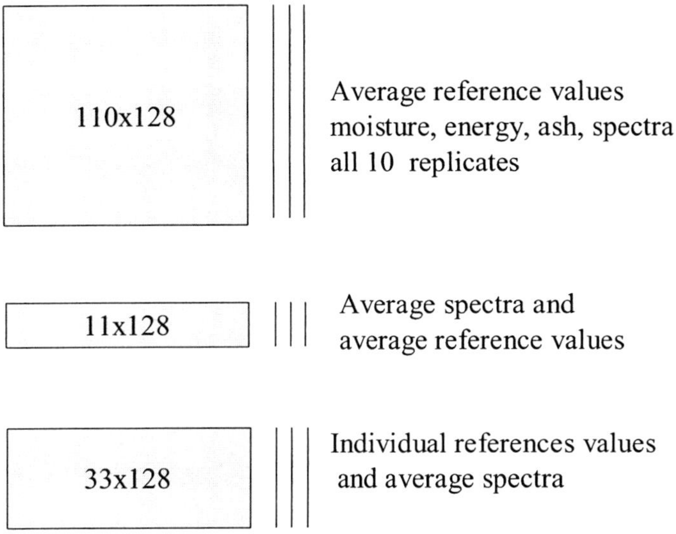

Fig 10. Three different data setups are evaluated. First setup is with the original dataset containing 110 spectra (11 mixtures and 10 replicates each) called 110S and second setup with one average spectrum for each mixture, with averaged reference values, called 11S. Third setup 33S with triplicate average spectra and replicate reference values.

Results for the 110S data set are shown in figure 11 for moisture, energy and ash content. These results are for PLS2, but match those for PLS1 closely. The cross-validation was by random subsets. According to figure 11, RMSECV values are approximately 0.94 % for moisture with 14 components, 0.39 MJ/Kg for energy with 8 components and 2.1 % for ash content with 12 components. These values look good, but they are misleading because there are only 11 different mixtures, so the number of components is too high and the RMSECV is too low. Cross-validation is far from ideal and in this case with many replicates it is clearly totally wrong.

Table 3: RMSECV results (in parentheses number of components used)

Data set	Moisture %	Energy MJ/kg	Ash %
110S	0.94 (14)	0.39 (8)	2.1 (12)
11S	2.3 (5)	0.63 (4)	5.6 (5)
33S	1.8 (7)	0.83 (6)	2.6 (8)

An alternative calibration is using the 11x128 data set (11S) of the average spectra. This makes the number of samples very small. Validation is not easy for such a small data set. Leave-one-out cross validation gives RMSECV values of 2.3 % for moisture with 5 components, 0.63 MJ/kg for energy at 4 components and 5.6 % for ash at 5 components. This is shown in figure 12. The data set is really too small to get more information. It is also possible to have a 33x128 data set called 33S. In this data set each average spectrum is present three times. In the responses, the replicates of the reference values are used. This gives an RMSECV of 1.8 % for moisture with 7 components, 0.83 MJ/kg for energy with 6 components and 2.6 % for ash with 8 components. These values seem more realistic for ash content. The results are also shown in table 3.

Table 3 shows the dangers of doing naive cross validation. The values for the 110S data set are based on overfitting. Taking enough components to explain 95% of the total sum of squares of the mean-centered response data gives equally good or better results.

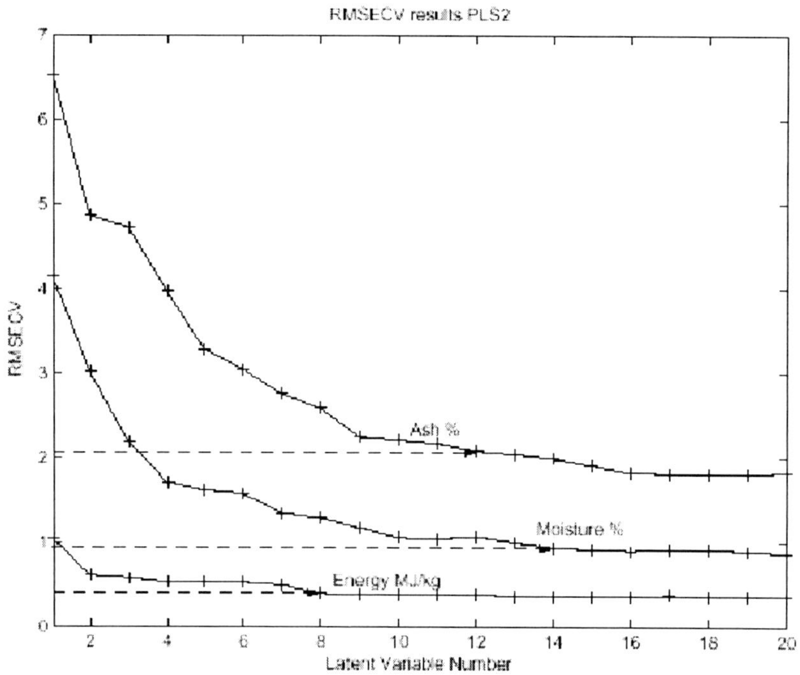

Fig. 11. PLS2 models calculated for 110S. The number of significant components is too high compared to this small dataset with only 11 average spectra.

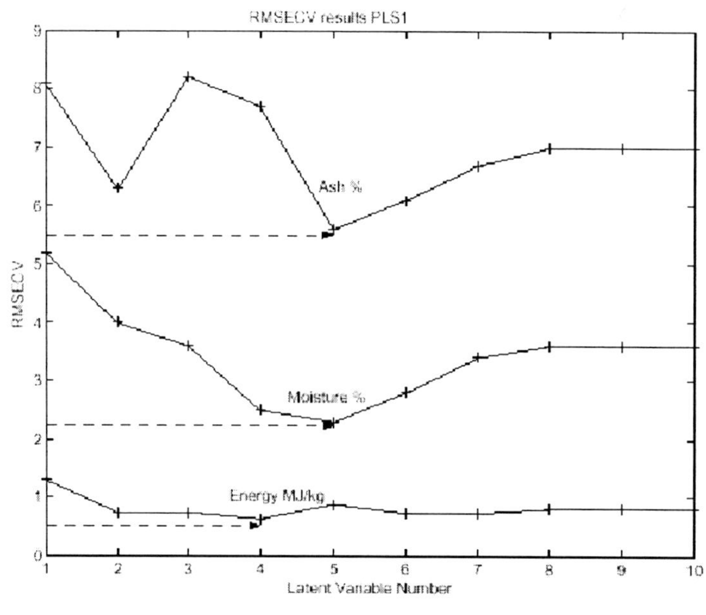

Fig. 12. PLS1 models calculated for 11S. The calibration errors seem more realistic than those in Fig. 11

The conclusion is that the smaller data sets 11S and 33S gave more realistic results. Replicate spectra are not very useful for model building if they have the same reference values.

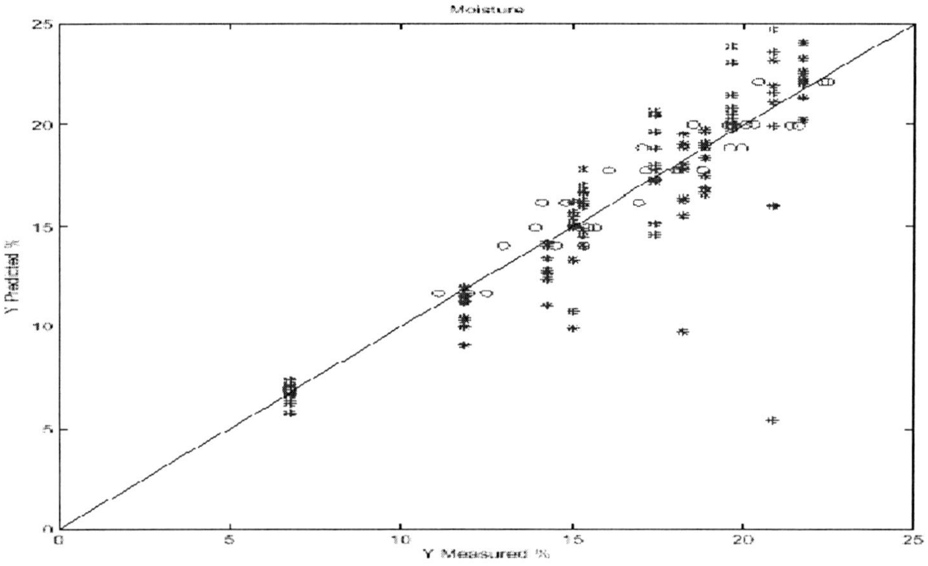

Fig.13. Moisture predictions versus measured.

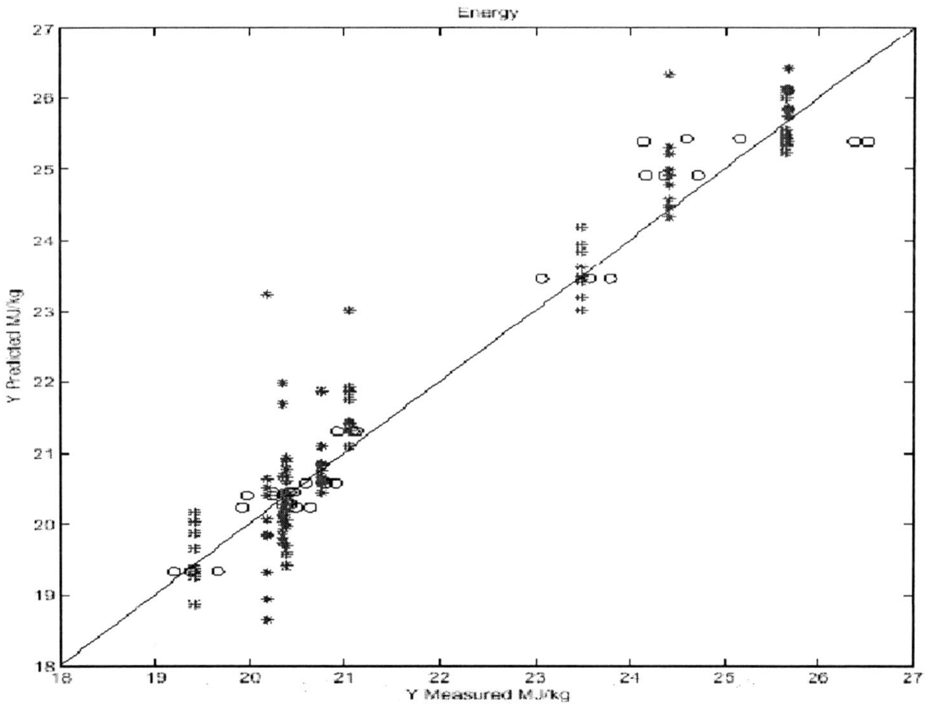

Fig.14. Energy predictions versus measured

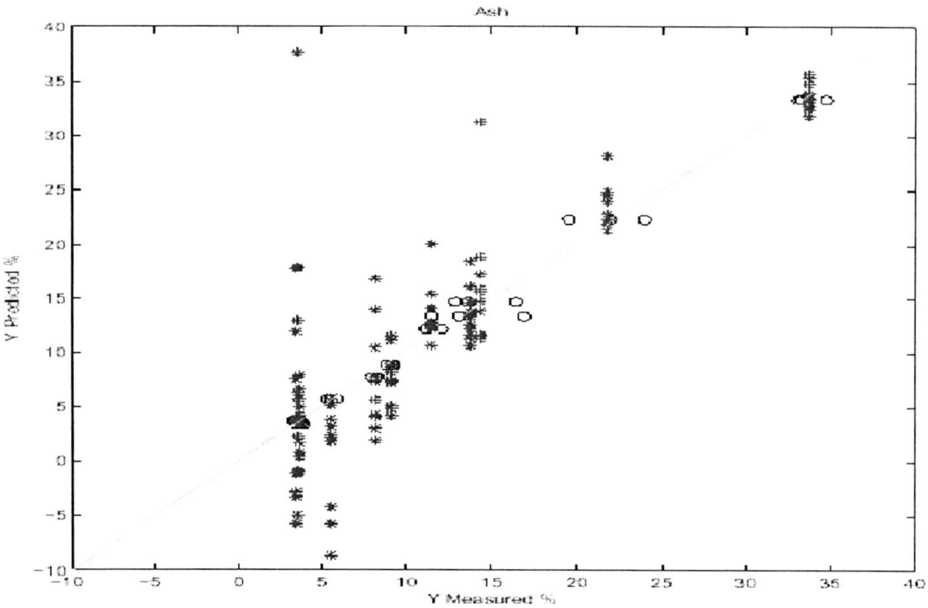

Fig. 15. Ash predictions versus measured.

The model should also be used for prediction. In this paper, the models made on 33S as in table 3 were used to predict 110S containing all the replicate spectra. The predictions are used to get an idea of bias and variance. These are not real prediction results, but they provide an indication of how good the calibrations are. Figures 13-15 give prediction results for respectively moisture, energy and ash content. The figures show a number of outliers, but also good linearity of the predictions, even though the spread around the diagonal is sometimes large. Some extreme outliers are noticeable. Moisture and energy are easy to predict from NIR spectra, while ash is more difficult.

It is important to look at the bias and standard deviation for each mixture. These values are given in figures 16 and 17. For mixtures 2 and 9 one obvious outlier was removed. In figure 17 a high standard deviation is seen mainly for mixtures with much biofuel. Ash has a higher standard deviation than water. Energy has a really low standard deviation.

The averages are given in Table 4. They are reasonably low, indicating that online measurements would be feasible.

Table 4: Averages for absolute bias and standard deviation

Parameter	Moisture%	Energy MJ/kg	Ash %
Bias (abs.)	0.7	0.2	1.2
St.dev.	1.5	0.5	1.7

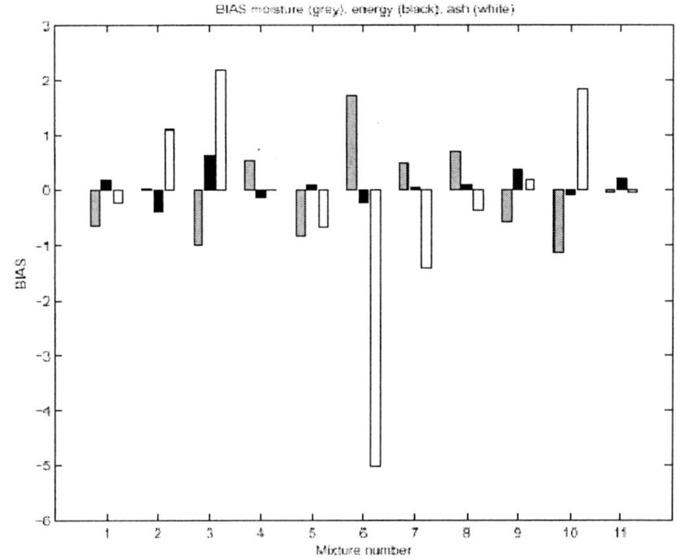

Fig. 16. BIAS of predictions for each mixture: moisture (gray), energy (black), and ash (white)

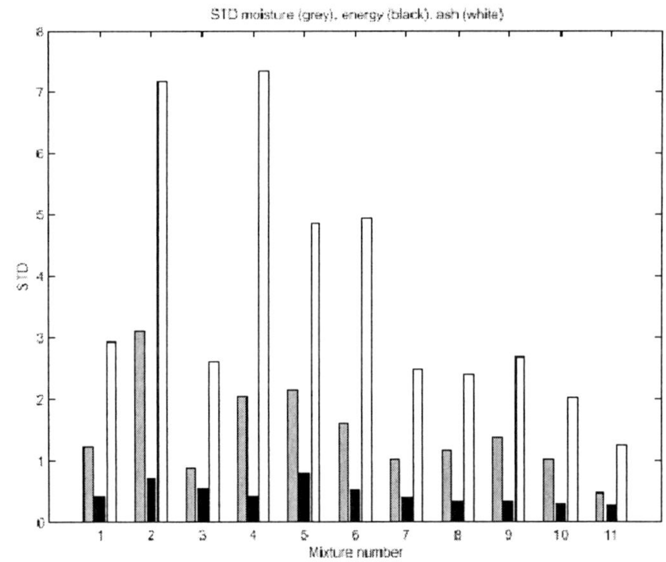

Fig. 17. Standard deviation of predictions for each mixture: moisture (gray), energy (black), and ash (white)

CONCLUSIONS

It is possible to measure NIR spectra of biofuel mixtures with the InGaAs spectrometer with only 128 wavelength bands. The mixtures in the mixture triangle are extreme, but still PCA on the data allows meaningful conclusions. Three components were explained by studying score and loading plots. For the calibration, cross validation gives misleading results if the replicates are used. It was found that it is possible to use average spectra to make a good calibration model with few PLS components. The spectra replicates can then be used as a test set for doing predictions. The predictions allow the calculation of bias and standard deviation values. These are really good for moisture and energy. For ash content large deviations are possible. Samples with a high biofuel content showed larger errors, because the biofuel was very heterogeneous.

ACKNOWLEDGEMENTS

The EU Unizon-Kvarken project, subproject NIRCE is acknowledged for financial support.

Stig Nickull of Alholmen's Kraft is acknowledged for supplying the samples.

REFERENCES

[1] P. Geladi, B. Sethson, J. Nyström, T. Lillhonga, T. Lestander , J. Burger, Chemometrics in spectroscopy. Part 2. Examples, *Spectrochimica Acta B*, submitted

[2] F. McLennan & B. Kowalski (eds.), *Process Analytical Chemistry*, Blackie Academic & Professional, London, 1995

[3] G. Box & A. Luceño, *Statistical Control: By Monitoring and Feedback Adjustment*, Wiley-Interscience, New York, 1997

[4] S. Qin, *J. Chemometrics*, 17 (2003), 480-502

[5] R. Lee & J. Young, *Multivariate Statistical Process Control with Industrial Application*, Society for Industrial & Applied Mathematics, Philadelphia PA, 2001.

[6] T. Kourti & J. MacGregor, *Chemom. Intell. Lab. Systems*, 28 (1995) 3-21

[7] B. Skagerberg, J. MacGregor & C. Kiparissides, *Chemom. Intell. Lab. Systems*, 14 (1992) 341-356

[8] T. Kourti & J. MacGregor, *J. Qual. Technol.,* 28 (1996) 409-428

[9] B. Wise & N. Gallagher, *J. Process Control*, 6 (1996) 329-348

[10] P. Geladi, *Spectrochimica Acta* B, 58 (2003) 767-782

[11] J. Jackson, *A User's Guide to Principal Components*, Wiley, NY, 1991.

[12] P. Geladi, *Chemom. Intell. Lab. Systems*, 60 (2002) 211-224

[13] T. Næs, T. Isaksson, T. Fearn, & T. Davies, *A User-Friendly Guide to Multivariate Calibration and Classification*, NIR Publications, Chichester, 2002

In: Progress in Chemometrics Research
Editor: Alexey L. Pomerantsev, pp. 209-227
ISBN 1-59454-257-0

Chapter 14

MULTIVARIATE STATISTICAL PROCESS CONTROL AND OPTIMIZATION

A. L. Pomerantsev and O. Ye. Rodionova
Institute of Chemical Physics, Moscow, Russia, forecast@chph.ras.ru

ABSTRACT

The multivariate statistical process control (MSPC) is an approach that helps to run the real-world processes. The MSPC concept is to apply the historical process X-data for construction of linear model, which explains how the final results (i.e. quality, y) depend on X-variables. Apparently, studding this model, one can suggest the program of actions that could improve the performance in general. However, the most important thing in production is the local optimization, which means the immediate actions that are performed in the course of process (i.e., in-line) in order to improve the quality of the final product or/and to reduce the production costs.. The paper contains new ideas, which (we believe) are interesting in process control. The first one comes within the conventional MSPC approach. We suggest constructing a series of MSPC models that describe the process in order of its cycles, stage by stage. This Expanding MSPC (EMSPC) method can be used for prediction of future final product quality at the early process stages while production cycle is not over yet. This may be considered as a passive optimization as we do not suggest the correcting actions. The second idea implies the novel SIC method to find the best in-line corrections. This technique may be called the Multivariate Statistical Process Optimization (MSPO) and it also applies the idea of EMSPC. The MSPO approach may be considered as an active optimization as we always suggest the correcting actions in the course of the process.

Keywords: MSPC, optimization, Expanding MSPC, SIC approach

INTRODUCTION

Methods based on the philosophy of Statistical Process Control (SPC) for process monitoring and fault detection were proposed by W. Shewhart [1] in early 1930's. The goal of SPC is to investigate whether a process is in a state of statistical control. A state of statistical control exists when key process variables or product attributes remain close to their target values. This kind of assessment is accomplished through use of different kinds of control charts: Shewart [1], CUSUM [2] and EWMA [3]. However examining one variable at a time makes diagnostics extremely difficult.

At present different industrial process are often monitored at frequent time intervals, giving rise to data bases with many variables and many observations, and what is more, all process variables are not independent of one another. Ordinary few original events are driving a process at any time, and all these measurements are simply different reflection of these original events. This is the reason why the SPC was joined with modern multivariate data analysis techniques (MDA, [4]). Multivariate approach based on principal component analysis (PCA) and partial least squares (PLS) [5] offers a number of informative diagnostic tools, they help us to understand how and why a particular process event occurs. The first real-world example in this paper is classical application of MSPC for continuous chemical process.

No doubt that the main goal of any statistical analysis of the process is to improve the quality of the final product or/and to reduce the production costs. In other words there should be a method that could optimize the process. Now we have to explain what kind of optimization we mean. The main MSPC concept is to apply the historical data collected on process for construction a linear (calibration) model, which explains how the final results (i.e. quality, y) depend on the observed X-variables that characterize the state of the process. Apparently, exploring this model, one can suggest the program of actions that could improve the performance in general. However, we think that the most important aspect in production is the local optimization, which means the immediate actions that are made in the course of process (i.e., in-line) in order to correct its current state and to improve the quality of the future product.

The main problem in optimization is not to find a solution, but to restrict the area where this solution could be found. Optimization of a linear model on unreasonable region of predictors always gives the senseless decision, where optimized characteristic tends to infinity. The first, rough restrictions on the predictor area come from the physical or the technological meaning of the process variables. The area could be decreased more with the explanatory analysis of historical data collected on the process. The data analysis yields the boundaries for each x variable that was actually run in practice. However, this is not enough, as typically process variables are correlated. This issue has been frequently discussed in the literature on MSPC [6]; variables x_1 and x_2 cannot be near their maximums simultaneously, if there is a negative correlation between them. This reason converts the p-dimensional rectangle of possible predictor values into more complicated ellipsoidal area. Nevertheless, this is not enough again. Still limited predictor area could give the unreasonable solutions in optimization. These are not unlimited quality measures, but values that lie out of the real quality limits. For example, if the quality variable y is a kind of purity measure that cannot be more than 100%, the solution y=105% is out of sense.

Another problem is that we are looking for the optimizing technological solutions locally, being in the process line. Thus, these actions should take into account the actual pre-history of the current process realization and they should not be the actions, which are unavailable or inapplicable now, in the current situation. However, we cannot state it in advance, which action is forbidden in which situation, as it would be rather complicated problem to describe a process explicitly. Only experienced production engineer can explain (sometimes on a hunch), why this action cannot be implemented now, and why that correction would give the result. To find a way out of this difficulty, we suggest to use the main concept of the MSPC approach and pose that for each state of a process those actions are admissible that are similar to ones that sometimes happened in the historical data in the similar situation. Strictly speaking, this means that allowable values of process parameters should lie within the MSPC calibration model, which describes the process.

The optimization part of the paper contains two main concepts, which (we believe) are interesting and fruitful for application in process control. The first one comes within the conventional MSPC approach and has much in common with multiblock PLS method [7],[8]. We suggest constructing a series of MSPC models that describes the process in the order of its manufacture cycle, stage by stage. This Expanding MSPC (EMSPC) method can be used for prediction of the future final product quality at the early process stages while production cycle is not over yet. This may be considered as a passive optimization as we do not suggest the correcting actions, but just predict the expediency of the offered actions. The second idea is not usual in the standard MSPC approach as it implies the novel SIC method [12],[13] to find the best process corrections in line. This technique may be called the Multivariate Statistical Process Optimization (MSPO) and it also applies the idea of EMSPC. MSPO may be regarded as an active optimization as we always suggest the correcting actions in the course of the process.

METHODS

Different mathematical methods are implemented in this work. First class is projection methods. PCA is used in the first example for process monitoring. PLS regression is applied to build the various calibration models. It is used in a single response version (PLS1), as well as in the multi response version, known as PLS2.

The second technique is a new approach called the Simple Interval Calculation (SIC). This is a method of linear modeling that gives the result of prediction directly in the interval form. The SIC approach also provides wide possibilities for the leverage-type object status classification, which can reveal the most important (boundary) objects among the calibration samples as well as the insider/outsider/outlier objects among the prediction samples [13].

MONITORING THE CONTINUOUS PROCESS

We have a set of historical data that describes the state of the process at 54 time points (observations). For each observation we recorded signals from 17 sensors. Applying traditional SPC approach we can construct Shewart charts for each of the sensor (variable). Based on the historical data we can calculate target value, warning and control limits. In case operator has to control only several variable his/her work seems to be rather simple (Fig.1a). The inspection of raw signals in terms of 17 time series plots is not very fruitful, especially when they are plotted on top of one another in one single graph (Fig. 1b). Evident data transformation may help to overcome this problem. On the first step we center observations for each variable (sensor). On the second step we scale or weight data. In our case we divide all observations for each sensor (variable) to the corresponding control limit. After that transformation all time series have common target value, here it is zero, and the same control limit equals to 1 (Fig. 2a). The operator work becomes a little bit simpler but still not very effective due to overlapping of 17 curves. To make the representation of the process state more illustrative we can show the process in column diagram view (Fig. 2b).

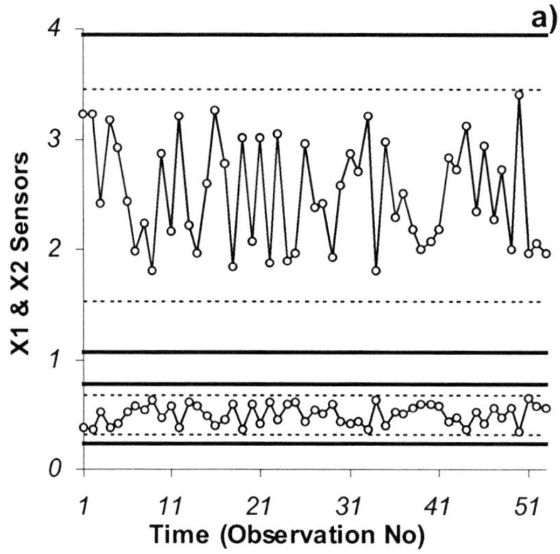

a) Two sensors.

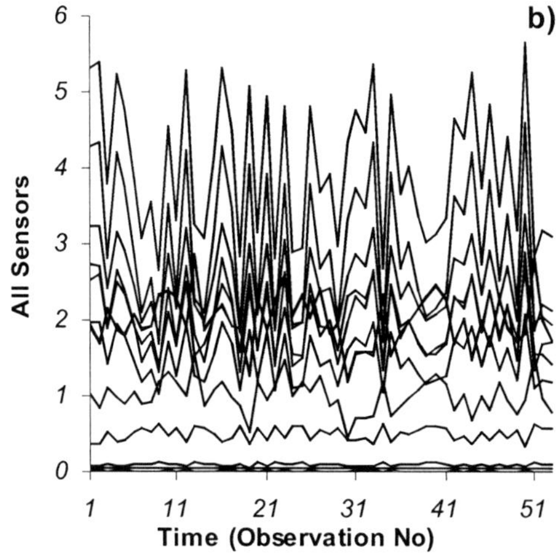

b) Seventeen sensors.

Fig.1 Continuous process. Raw historical data. Doted line – warning limits, solid line – control limits

Now each sensor is represented by the bar, warning and control levels are shown by the solid and dotted lines. For normal process state all bars should be inside the control limits. The SPC becomes simpler, but using process presentation in the form of Fig. 2b, we have lost information regarding changes of process in time, i.e. possible process drifts.

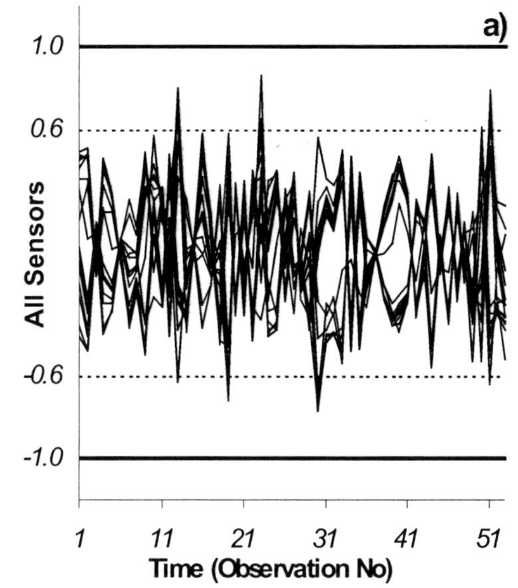

a) Time series Black curves are sensors values

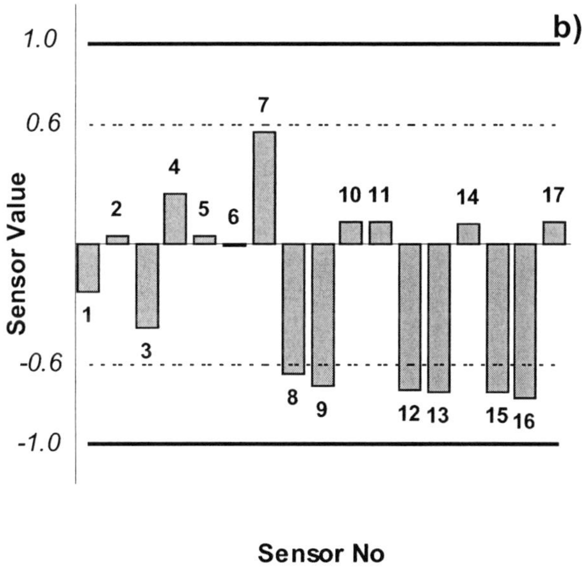

b) One time point t=30. Gray bars are sensor values

Fig.2 Continuous process. Scaled data, Doted line – warning limits, solid line – control limits

Another issue is 'out of control' situations. How to bring the process back into a state of control? Some of 17 measured characteristics are adjustable variables, which may be changed by operator. Evidently, it is necessary to decrease such a characteristic if it approaches the upper limit and increase in case it is near the lower limit. But what to do with uncontrolled variables? Ordinary all variables that characterize the technological process are highly correlated (i.e. the higher the temperature, the higher the pressure, *etc.*). Thus, operator can change the desired characteristic indirectly. Generally, for a complex system, variables cannot be manipulated independently. In ideal situation it would be preferable to change all controlled variable simultaneously. Practically this is impossibly and as a result there are many operator faults that cause a bad quality of the final products. The interconnections inside the system hinder us only if we stand on the old-fashion univariate point of view. At the same time high correlation between variables greatly help us if we use MSPC approach.

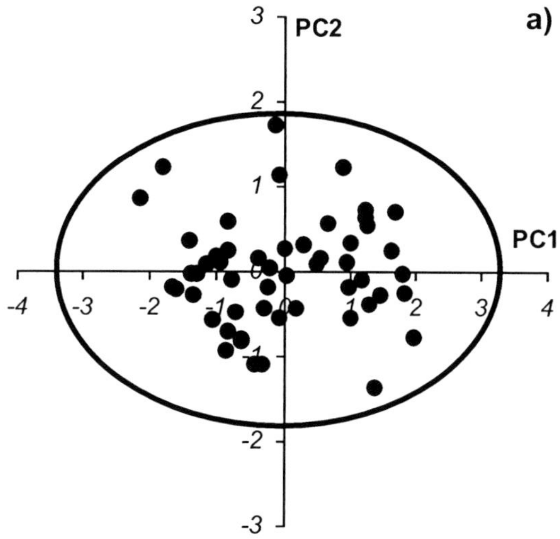

a). Score plot with Hotelling ellipse;● - observations

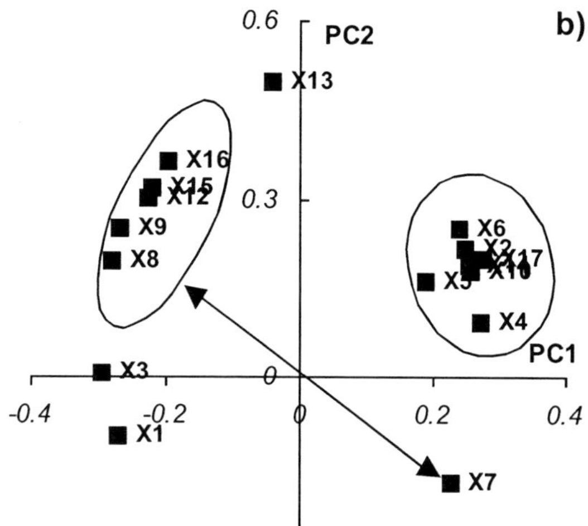

b). Loading plot. ■ – variables

Fig3. Continuous process. PCA analysis (PC1-PC2)

For these purposes we present all data as matrix ***X*** with 54 rows and 17 variables and apply PCA method. We need only two principal components (PC) for data description. The noise level is about 4%. Fig. 3a shows the score plot for all observations. We also managed to construct Hotelling's T_2 ellipse [9], which stands for in-state-of-control process limit. Now, for monitoring operator should trace only the position of the correspondent point on the score plot. Connecting the points on the score plot in time order we can trace the tendencies inside the monitoring process. Fig 3b shows all sensors in the loading plot. Several useful conclusions can be drawn from this plot.

1. We can clearly see two groups of sensors. This means that we can reduce the number of sensors in each group if this is necessary, i.e. instead of 5 sensors (X8, X9, X12, X15, and X16) we can select only one measuring device.
2. There is high negative correlation between sensor X7 and the first group. Also there is high negative correlation between sensor X1 and the second group (X2, X4, X5, X6, X10, X11, X14, and X17)
3. There is no correlation between sensors X3 and X13.
4. Sensor X13 plays a special role and brings essential contribution to the process modeling.

In this example application of the PCA method helps to present the numerous data clearly in the plane. Also, this approach helps to understand the dependences between different process variables and to compute the required influence upon *all variables* to keep the process inside the control ellipse, i.e. to compute the effective feedback. In the given example we have no outcome characteristics such as final product quality and all that we can do is to monitor and to bring the process back into a state of control in dangerous situations.

PROCESS OPTIMIZATION

Now, let us consider the case when we collect not only process variables, but also measure the final product quality.

Process Description

We consider a multi-stage continuous technological food process that is represented by 25 process variables ***x*** and one output variable ***y***, which is the final quality of the end-product. We prefer to look at the process more precisely. The production cycle (see Fig 4) is divided into seven stages numbered by the Roman numerals. Each stage may be described by input, current, and the further variables. Variables used on all previous stages are treated as the input fixed variables, current variables are the controlled ones, and variables that characterize next stages are out of scope on the current stage. Moving along the process, variables change their roles.

First stage (I) is represented by six input variables (W1, W2, W3 and S1, S2, S3) that stand for the properties of the raw components S and W. At the second stage (II) component W is refining and variables WR1 and WR2 characterize this process. Variables CW1, CW2, and CW3 (Stage III) represent the properties of the outcome product CW. The next stage (IV) is mixing of the raw component S and the refined component CW. The result M is characterized by variables M1, M2, and M3. Afterward, blend M is also refined (Stage V) with the process characteristics MR1 and MR2, and the properties of outcome CM are presented by variables CM1, CM2, and CM3 (Stage VI). The last stage (VII) stands for the ultimate amendments, which are done with additives A1,..., A6. The output variable (P=y) is the final product quality.

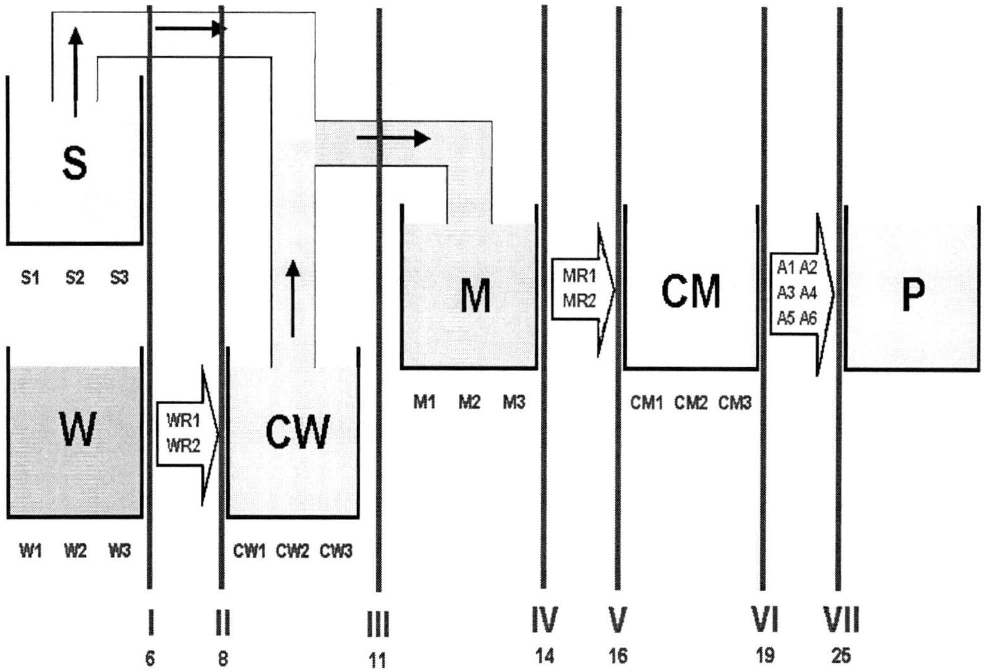

Fig 4 Production cycle

At the end of each stage, the production engineer could analyze the intermediate results and correct process parameters (variables) of the next stage. Both analysis and correction should be performed regarding the foreseeable output property y, and with respect to admissible range of correcting actions. Evidently it is unreasonable to suggest theoretically improving actions, which, however, cannot be implemented in practice. The MSPC approach based on the real production records is the appropriate tool that is able to give the reasonable solution of the problem.

Data and Notation

We have a collection of the actual historical data measured for 154 samples that characterize proper process performance. Each sample corresponds to the entire production cycle shown in Fig 4. The whole data set is divided horizontally (by samples) in two parts: the calibration set (102 objects) and the test set (52 object). Calibration data set is also divided vertically (by variables) into 7 blocks in conformity with the technological stages:

$\mathbf{X}=(\mathbf{X}_{I}, \mathbf{X}_{II},\ldots, \mathbf{X}_{VII})$

For example, block $\mathbf{X}_{IV}$ is the (102×3) matrix, which includes 3 variables: M1, M2, and M3, and 102 calibration samples. We also use notation $\mathbf{X}_{(M)}$ to represent the matrix that is composed with data from all previous stages. For example,

$\mathbf{X}_{(III)}=(\mathbf{X}_{I}, \mathbf{X}_{II}, \mathbf{X}_{III})$,

is the (102×11) matrix. In this notation we can write that for the M-th stage, matrix $\mathbf{X}_{(M-1)}$ presents the input data, matrix $\mathbf{X}_{M}$ represents the controlled current data. At stage I, we have no input data.

The data are centered and scaled in such a way that each variable, including quality measure *y*, varies within range (–1,+1) and that all values outside this interval are not valid. It is also assumed that the highest product quality corresponds to *y*=+1, as well as the lowest one corresponds to *y*=–1. Due to classical MSPC approach we take the whole calibration data set with 25 X-variables and construct the total PLS1 regression model

$$\mathbf{X}\boldsymbol{a} = \boldsymbol{y} \tag{1}$$

with 6 PLS components (PCs), which describes 98% variation in X and 96% variation in *y*. After that, we perform *additional data scaling*, multiplying some columns by factor –1, in order to make all regression coefficients $\boldsymbol{a}$ positive. This is done to standardize the process reply, because now increasing of any process variable x leads to the improvement of quality *y*. This is an important outcome showing how the process could be optimized *in general*. On the base of PLS-model we also construct the SIC-model of the entire process.

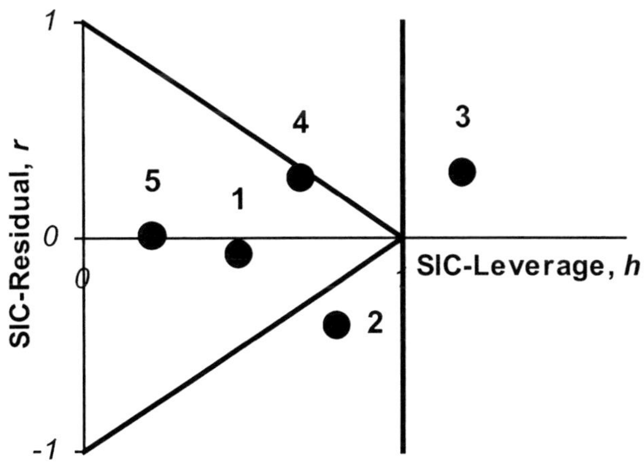

Fig 5 Status map of selected test samples

For illustration purposes we select five samples from the test set. They present the most typical cases with respect to the product quality (see Table 1) as well as to the SIC-status (Appendix) i.e. sample's relationship towards the process MSPC model (see Fig. 5).

Table 1 Selected test samples

Sample No	Quality status	SIC status
1	Normal	Insider
2	High	Outsider
3	Normal	Abs. outsider
4	Low	Outsider
5	Normal	Insider

Process Control

In this section we shall demonstrate how the process can be merely controlled without attempts to interfere in it, i.e. *passive optimization.* For this purpose we apply a method of *expanded MSPC*, which appears to be a new approach in this area.

It would be nice to have a chance to predict the future product quality being at the initial stages of its production. Using historical data we build a series of PLS1 regressions

$$\mathbf{X}_{(\mathrm{I})} \rightarrow \boldsymbol{y},\ \mathbf{X}_{(\mathrm{II})} \rightarrow \boldsymbol{y},\ \ldots,\ \mathbf{X}_{(\mathrm{VII})} \rightarrow \boldsymbol{y} \qquad (2)$$

where $\boldsymbol{y}$ are output data that characterize product quality and matrix $\mathbf{X}$ stands for input data expanded, moving together with a process. We can use these intermediate modeling for two different purposes. The first and the most obvious one is to predict and to monitor the possible output quality on the M-th stage of production process. The other way of expanded modeling application is much more interesting. After the process passed the M-1-th stage, the process engineer may vary the values of the forthcoming stage variables (X_M) in the mathematical model and analyze their influence on the future output quality, i.e. to use an "advisor" that helps to make a decision on the base of the previous experience expressed in mathematical model.

The results of expanded modeling (Eq. 2) and correspondent SIC-modeling are presented in Fig 6, where 5 plots correspond to five selected samples (see Table 1). In each plot we demonstrate the results of prediction of future quality y that are obtained on each process stage. They are: the predicted PLS point estimates (black dots) and the SIC-intervals (gray bars). At each stage we also show actual X-variables (white dots) that were used in the production of these samples. They can be presented in the same plot due to the uniform scaling performed before modeling. Open rhombus in the right part of the plots show the quality value, y that was actually obtained in the production. Each PLS1 regression (Eq. 2) is applied with the same number of PCs equals 6.

The following results are clearly seen in Fig 6.

1. The SIC-prediction intervals are decreasing along the process, but the reference value is always located inside them.
2. Sample 1 (insider) has a low predicted quality at initial stages, but further improvements have corrected it.
3. Sample 2 (outsider) has a high predicted quality at the beginning of the cycle and later actions improve it.

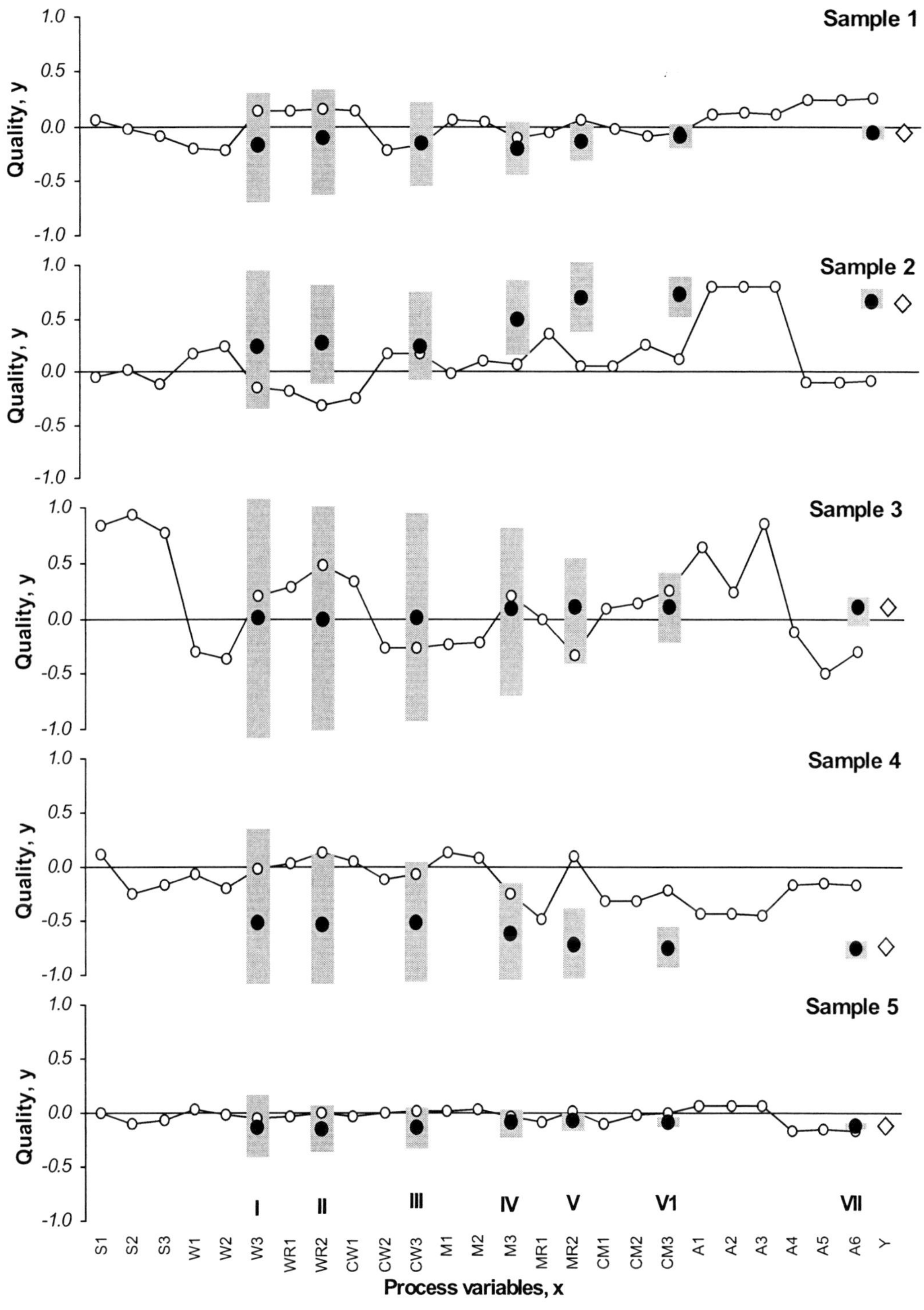

Fig 6. Passive control of the process, i.e. in-line prediction of quality

4. Sample 3 (absolute outsider) has the largest prediction intervals in the course of the process, but no corrections could improve its normal quality status.

5. Sample 4 (outsider) initially has a low predicted quality and all correction actions do not recover it, yet make it worse.

Process Optimization

In this section we shall consider how to find the optimal correcting actions that could be performed at the end of each stage, i.e. the *active optimization*. Actually, this means the proper choice of the controlled variables that become the input variables for the next coming stage. This choice should meets two crucial conditions. First, it must improve the quality of the end-use product. In our case this is maximization of y. Second, the choice must be performed within a range of acceptable bounds of controlled variables. Owing to the *additional scaling* procedure performed with raw data, we can state that increasing of each variable x leads to the growth of quality measure y. Therefore, the optimum is always reached at the upper boundary of the suitable variable region. Thus, all that have to be done is to find out which bounds are reasonable.

The rough limits come from the data scaling again. It is clear that each variable should lie within interval (–1,+1). However, these bounds are too large and optimization within this range gives the unrealistic result in predicted quality y, which becomes greater than 1, its upper scaled limit. This happens because there are evident correlations between X-variables, which make such combinations of parameters as (+1,+1,..+1) unavailable and impracticable.

To find the reasonable variable bounds we apply the Expanding MSPC method. For this purpose for each stage of the process we build partial calibration model

$$\mathbf{X}_{(\mathrm{I})} \rightarrow \mathbf{X}_{\mathrm{II}}, \mathbf{X}_{(\mathrm{II})} \rightarrow \mathbf{X}_{\mathrm{III}}, \ldots, \mathbf{X}_{(\mathrm{VI})} \rightarrow \mathbf{X}_{\mathrm{VII}}$$

that predicts the block of future coming X-variables using the known block of past X-variables as the predictor data. These models are constructed by the PLS2 method on the base of the historical process records. They have various complexities, the number of PCs, etc. For example, a model

$$\mathbf{X}_{(\mathrm{III})} \rightarrow \mathbf{X}_{\mathrm{IV}}$$

that predicts the forth block of variables (M1, M2 and M3) by block $\mathbf{X}_{(\mathrm{III})}$ of the first 11 variables, has the (102×11) X-matrix, the (102×3) Y-matrix, and it uses 4 PCs. This approach implies the general MDA concept: the difference between X and the Y variables are very problem dependent. It turns on our own choice and intention, which data block is the X or the Y matrix in the current task. In application to the problem in question, the expanding MSPC approach gives the most reasonable estimates of the future variables, which agree with the historical experience collected on the plant. Now, we can apply these models for any new process realization and obtain the prediction of the expected X-values.

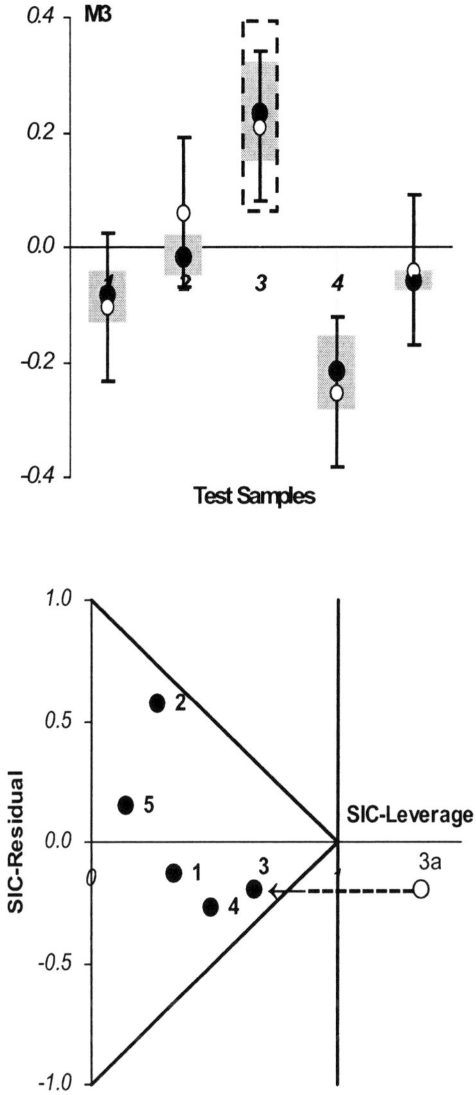

Fig 7 The SIC prediction (left plot) and the SIC object status classification map (right plot)

However, PLS model gives a point prediction only. This is not enough for solution of the process optimization problem, where some interval estimators should be involved. Such range could be calculated as a confidence interval using various methods suggested in [10],[11]. We propose to apply the SIC-method, which gives the result of prediction directly in the interval form. In Fig 7 (left plot) we present the SIC-intervals (wide gray bars) that are obtained in the regression prediction

$$\mathbf{X}_{(III)} \rightarrow x_{14} = \mathrm{M3}$$

for the selected test samples (Table 1). This plot also shows the PLS estimates (black dots), the known reference values (open dots) together with the error bounds (black error bars). The right plot in Fig 7 represents the object status map for these samples.

Actually, one can see two objects 3 and 3a in both plots. In the left plot, object 3a is shown by the dashed rectangle and in the right plot it is represented by the open dot. Object 3a is the primary solution found by the SIC method for sample 3 (see Table 1). In the right plot it can be seen that object 3a is the evident absolute outsider (Appendix) as it lies out of the insiders area (triangle) and even in the area of absolute outsider, which is bounded by the vertical line. Respectively, in the left plot the predicted interval for this sample (dashed rectangle) includes the error interval (black error bar). This means that prediction error for sample 3 is greater than calibration error (i.e. maximum error deviation β) so this sample is not reliable object in prediction (see Appendix). In other words, the predicted interval is not a reasonable range for optimization. As a matter of fact, this conclusion may be confirmed by the calculations of the optimal actions for the sample 3. If we apply the primary SIC intervals (like 3a) in this optimization, the predicted quality measure y becomes unreasonable (more than 1) even at the early process stages.

However, we have to suggest optimizing actions for outsiders and even more for them. The solution is to shrink the predicted interval until it will be less than error interval, or, in other words, until this sample becomes an insider. This operation is shown in Fig. 7 (right plot) by arrow and object 3 represents the result. In the left plot it corresponds to the gray bar, which is less than dashed rectangle, and is also less than black error bar. Such a 'shrinkage' of the predicted intervals is performed at the each stage of the process with every sample that appears to be 'out of the model', i.e. outsider. The adjusted intervals are used as a reasonable range of predictors, where optimization with respect to future quality is performed. In our case, the optimal x-point is always an upper border of the interval, so optimization is a very simple task.

We carried out this procedure for each selected test sample and demonstrate the results in Fig 8. The layout of these plots is similar to the plots shown in Fig. 6: large black dot and gray bars represent the PLS point and the SIC interval prediction of quality measure y respectively; the open rhombus at the end stands for the reference y value; and the doted kinked curves show the corresponding values of X variables in the course of production. We present two such curves: the open dot line represents the historical records (the same as in Fig. 6) and the black dot line shows the optimized run of the process. The initial stage cannot be optimized as it contains the given properties of raw components S and W, so the black dots coincide with the open dots. Stage II can already been optimized and one can see that the black dot lie above the open dots. Using these optimized X values as predictors, we can find the improved variables for stage III, and so forth.

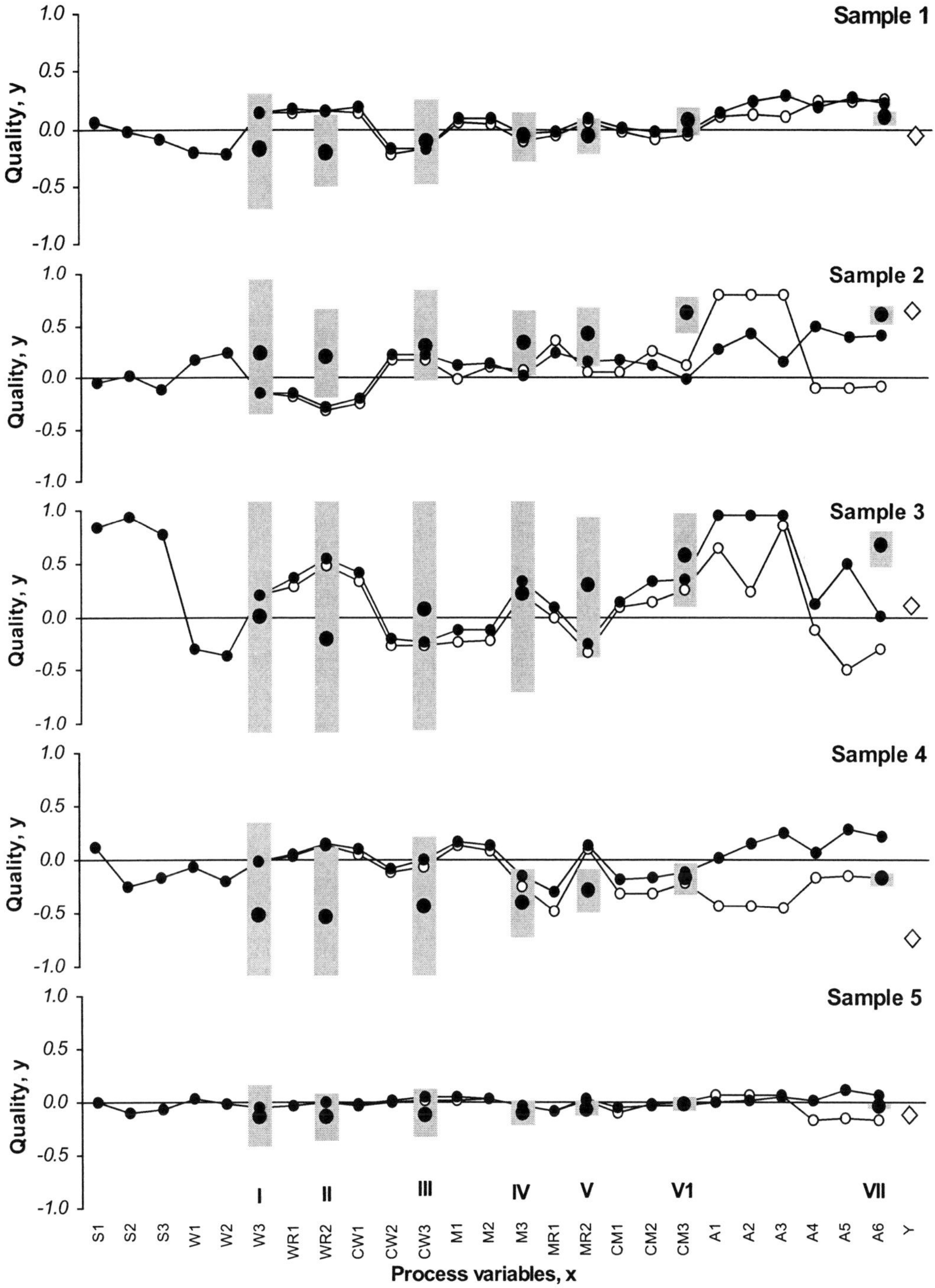

Fig. 8 Active optimization of the process, i.e. the in-line improvement of quality

The following results can be observed in Fig 8.

1. Sample 1 (insider of normal quality) became a little bit better.

2. Sample 2 (outsider of high quality) did not change its quality, but also did not became worse.
3. Sample 3 (absolute outsider of normal quality) turned into a high quality object
4. Sample 4 (outsider of low quality) became a normal quality object
5. Sample 5 (insider of normal quality) did not change its quality; it kept its normal status.

Discussion and Conclusions

The presented results appear to be interesting and even give some reasons for general conclusions. Both passive control (Fig 6) and active optimization (Fig 8) show that status of a sample regarding the calibration model determines its chances to be optimized. All insiders (test Samples 1 and 5) are the worst objects for improvements. These samples are so average, so normal, that no actions could enhance them. Apparently, this comes from the common technological wisdom that if something is going well let it go at that. This historical experience collected in the actual process records, constricts the area of available actions with x-variables to a very narrow range, which does not permit any enhancements. On the other hand, the outsiders, i.e. the outstanding objects that are far from the common zero level, give an excellent chance to improve or to worsen them. In the passive control of the process (Fig 6), Samples 2 and 4 have developed their natural inclinations: the good quality Sample 2 became even better, and the bad quality Sample 4 got worse. If, however, we undertake some correcting actions with such marginals, they immediately respond with improvements. This happens because the area of possible X-variables is much wider for these unusual, extreme samples. The special case is absolute outsider (Sample 3), which is an extraordinary object with abnormal initial properties of the raw components. The historical data contain no knowledge about such samples, so the area of possible corrections is always so wide that it requires some shrinkage to put it in the routine mainstream. Nevertheless, even restricted, such a sample demonstrates excellent possibilities to 'upbringing' and it could become a high quality object. Further generalization, association, and parallels to a human inclinations and their development are, unfortunately, out of scope of this paper.

In conclusion we discuss some special issues relating to the proposed optimization procedure. It is worth mentioning that it is possible to optimize the process with respect to other demands than maximization. For example, we could look for the actions that tend quality measure y to its mean, zero value, instead of maximum. No problems arise for that target. Another issue is the choice of optimizing rules, which are the global, or the local optimization. The first (i.e. the global) rule was presented in the paper and it says that each increasing of any X-variable leads to the improvement of quality. This follows from the general PLS model where all coefficients $\boldsymbol{a}$ are positive. Another rule is to optimize the quality response y with respect to each local model (Eq. 3), where some regression coefficients $\boldsymbol{a}$ could be negative. We have performed optimization with respect to the both rules and found no differences in the ultimate results. Now, we cannot explain this effect, but it actually demonstrates a stability of the proposed optimization procedure.

The exciting item is the selection of actual data records that represent the historical knowledge about the process. This data set gives the base for the MSPC regressions, imposes restrictions on correcting action, and, as a matter of fact, determines the ultimate result of optimization. We made this selection with respect to the following considerations. All failure realizations were declined, as well as, the records that were characterized by the experienced production engineer as atypical or incidental. Generally speaking, we made there nothing except the conventional data pretreatment including outlier detection and explanatory analysis. However, this issue is not properly investigated yet and demands further revision.

The last issue concerns the shrinkage algorithm of the predicted X-variables. In the paper, it was explained in case, where reference y-value is known. In reality, we do not know the future process parameter values, and, therefore, have no possibility to calculate the SIC-residual (Eq. 7) that is necessary for the SIC-object classification. However, we can calculate the SIC-leverage (Eq 8), which does not depend on the reference and determines whether the sample is an absolute outsider. If it has this status, the prediction interval has to be decreased until its SIC-leverage becomes smaller than 1, and the sample lose the status of absolute outsider. We have performed the process optimization with respect to this algorithm, without using information about known y-values and the obtained outcomes were very close to the results shown in Fig 8. This demonstrates that the MSPO approach can be applied in practice without any problem.

The following general conclusion can be drawn from the paper.

The Expanded Multivariate Statistical Process Control (EMSPC) approach has possibilities to support the in-line improvements of production. It helps to predict what effect on the product quality could be caused by the planned actions, and thus, it assists in the passive quality control.

The Multivariate Statistical Process Optimization is a technique that proposes the real actions for the quality improvement in the course of production. This is an active optimization of quality, which takes into account the actual history of the process and suggests the available corrections while production process is not over yet.

APPENDIX: SIC DEFINITIONS

The detailed description of Simple Interval Calculation (SIC) may be found in this book [13]. Here we will present only main definitions, necessary for this work.

The SIC approach is based on the single assumption that all errors involved in the calibration problem

$$\mathbf{X}\boldsymbol{a}=\boldsymbol{y} \tag{3}$$

are *limited* (sampling errors, measurement errors, modeling errors, *etc*.). This assumption means that there exists a positive, β value (unknown, as a matter of fact), which limits the difference between the predicted response $\hat{y}$ and the true response value y

$$|\hat{y}-y| \le \beta \tag{4}$$

This value, β, is called the maximum error deviation (MED).

Rely on this assumption, and using the given calibration data $\{\mathbf{X}, \boldsymbol{y}\}$, it is possible to build the system of inequalities regarding the unknown regression parameters $\boldsymbol{a}$ (Eq. 3), which determines a closed convex set A – the region of possible (parameter) values (RPV). The region A is a volumetric analogue of the conventional parameter estimates $\hat{\boldsymbol{a}}$ that could be obtained by some traditional regression method (i.e. PLS).

Using the obtained RPV, we can solve the prediction problem for any given predictor vector $\boldsymbol{x}$. The result of prediction is also presented as an interval for response y

$$V = [v^-, v^+] \tag{5}$$

where

$$v^- = \min_{\boldsymbol{a}\in A}(\boldsymbol{x}^t\boldsymbol{a}), \qquad v^+ = \max_{\boldsymbol{a}\in A}(\boldsymbol{x}^t\boldsymbol{a}) \tag{6}$$

The size and the position of interval V (Eq.5) regarding the known reference value y and with respect to the MED value β, characterize the quality of prediction. To quantify this quality, we introduce the following measures.

SIC-residual is defined as –

$$r(\boldsymbol{x}, y) = \frac{1}{\beta}\left(y - \frac{v^+(\boldsymbol{x}) + v^-(\boldsymbol{x})}{2}\right) \tag{7}$$

SIC-leverage is defined as –

$$h(\boldsymbol{x}, y) = \frac{1}{\beta}\left(\frac{v^+(\boldsymbol{x}) - v^-(\boldsymbol{x})}{2}\right) \tag{8}$$

Definitions (7) and (8) help to apply the following object status classification.

Statement 1 An object $(\boldsymbol{x},y)$ is an insider, iff $|r(\boldsymbol{x},y)| \le 1-h(\boldsymbol{x})$
Statement 2 The calibration object $(\boldsymbol{x}_i,y_i)$ is boundary, iff $|r(\boldsymbol{x}_i,y_i)| = 1-h(\boldsymbol{x}_i)$
Statement 3 An object $(\boldsymbol{x},y)$ is an outlier, iff $|r(\boldsymbol{x},y)| > 1+h(\boldsymbol{x})$
Statement 4 An object $(\boldsymbol{x},y)$ is an absolute outsider for any y, iff $h(\boldsymbol{x}) > 1$.

REFERENCES

[1] Shewhart W.A. *Economic Control of Quality of Manufactured Product*, Van Nostrand, Princeton, NJ, 1931
[2] Woodward R.H., Goldsmith P.L., *Cumalative Sum Techniques*, Oliver and Boyd, London, 1964
[3] Hunter J.S., *J. Quality Technology*, 18, 203-210, 1998
[4] Nomikos P., MacGregor J.F. *Technometrics*, 37 (1) 41-59, 1995.

[5] Hoskuldsson A. *Prediction Methods in science and Technology*. Thor Publishing, Copenhagen, Denmark, 1996

[6] Yacoub F, MacGregor J.F. *Chemom. Intell. Lab. Systems*, 70, 63-74, 2004.

[7] Wold H. Soft modelling. The basic design and some extensions, in Joreskog K.G, World, H. (Editor), *Systems under Indirect Observations*, Vol. 2, North-Holland, Amsterdam, 1982

[8] Wagen L.E., Kowalski B. *J. Chemometrics*, 3, 3-20, 1998

[9] Hotelling H. Multivariate quality control, illustrated by the air testing of samples bombsights, in Eisenhart C., Hastay M.W., and Wallis W.A. (Editors), *Techniques of Statistical Analysis*, McGraw-Hi, New York, 113-184,1947

[10] Efron B., Tibshirani R. *Statistical Science*, 1, 54-75, 1986

[11] Faber K., Kowalski B. *J. Chemometrics* , 11,181-238, 1997

[12] Pomerantsev A.L., Rodionova O.Ye. Prediction of Antioxidants Activity Using DSC Measurements. A Feasibility Study. In *Aging of polymers, polymer blends and polymer composites*, Nova science Publishers, NY, v.2, 19-29, 2002

[13] Rodionova O.Ye., Pomerantsev A.L. *This issue,* 40-60

In: Progress in Chemometrics Research
Editor: Alexey L. Pomerantsev, pp. 229-236
ISBN 1-59454-257-0

Chapter 15

Adaptation of Calibration Models while Transferring within a Series of IR-Spectrometers

***L. A. Rusinov*[1], *K. A. Zharinov*[2], *A. L. Sulima*[2] and *V. A. Zubkov*[2]**
[1] St. Petersburg Technological Institute (Technical University), Russia, rusinov@tu.spb.ru
[2] Lumex Ltd., St. Petersburg, Russia, zharinov@lumex.ru

Abstract

A number of methods for the transfer of calibration models of quantitative analysis from a master instrument to slave instruments as applied to near-IR spectrometers are considered. A direct standardization method and calibration transfer method involving piecewise regression are compared.

Keywords: NIR-spectroscopy, calibration models transfer

Introduction

The problem of enhancing the accuracy of analysis of complex organic compounds gains great importance in view of the necessity of ensuring ecological and foodstuff safety. Methods based on spectral analysis in the near-infrared (NIR) range play an important role in coping with this problem. However, large labor content of calibrations and requirements for individual calibrations of analyzers over the whole spectrum impose essential restriction on the wide use of this kind of analysis.

Spectral data processing is complicated by substantial overlap of the NIR spectral lines of the constituents in question. Therefore, full-spectra methods allowing for account of the complete spectral data were developed, among which the partial least square (PLS) method has gained wide use [1-6]. This method involves expansion of the spectral data in the principal components (that are, factors explaining the components variance), which implies transfer from an optical density – wave number basis to a formal basis that provides orthogonality of readouts. This makes it possible to avoid inversion of a variance matrix, which is badly conditioned in the case of a NIR analysis, whereas the inversion has to be carried out in the conventional method of least squares.

The processes of calibration of NIR-spectrometers are complex and laborious, requiring a large amount of work not only for performing chemical analyses during preparation and certification of calibration samples, but also for performing calibration procedures, subsequent data processing, and designing calibration models.

The state of metrological support in NIR spectroscopy and imperfect manufacturing technologies of optical and electronic subsystems of the spectrometers give rise to the fact that spectra taken on two identical spectrometers may significantly differ from each other (Fig. 1). Therefore, calibration models obtained on one instrument may bring about significant errors in analysis when used on the other instrument of the same model and same manufacturer.

In addition to differences in parameters of the instrument subsystems, successful transfer of calibration models from one instrument to another can be complicated by the fact that quantitative determination of the components of a material analyzed by NIR spectroscopy is commonly based on a very small variation in spectral data. For example, the largest span of variation in diffuse reflectance spectra of 95 samples of wheat grains does not exceed 0.13 optical density units [7]. Taking into account that such a small variation is caused by differences in all physical and chemical factors that affect the spectra, it appears that each of the factors shares extremely small values in the variation magnitude, and reliable determination of these values calls for a very high measurement accuracy

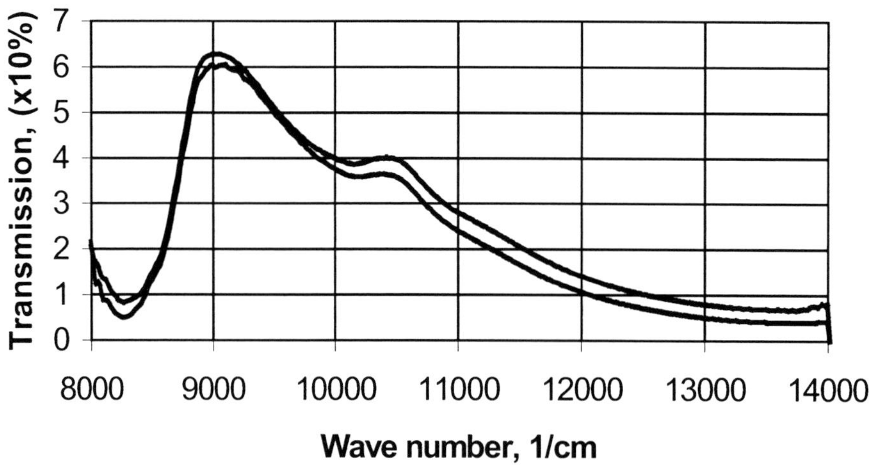

Fig. 1 Spectra of wheat registered on two instruments.

At present, this problem is evaded by making individual calibrations of analyzers. At the same time, the development of methods enabling the calibration model transfer within instruments of the same type would substantially reduce the labor input and thereby would allow extending application areas for NIR analyzers.

METHODS FOR TRANSFER CALIBRATION MODELS

For the transfer of calibration equations from one instrument to another various approaches to standardization of spectrometer responses can be used. These methods have been considered in a number of papers [8-13], the review of the methods was done by Fearn [14]. Some of the methods are based on the adjustment of a calibration model of the master instrument so that it could be applied to slave instrument, for example, a method using a classical calibration model [8]. However, methods involving adjustment of spectra measured on a slave instrument so that those conform to spectra measured on a master instrument are conceived to be more promising [8, 10, 13]. In this case, it becomes possible to use a calibration model of the master instrument. For example, these are the direct standardization method and the piecewise direct standardization method. In all of the transfer methods, the adjustment is made by a partial set of calibration samples, which is much smaller than in a conventional calibration procedure and thus the expenditures of labor on calibration decrease.

METHOD OF DIRECT STANDARDIZATION

In the direct standardization method, spectra obtained on a slave instrument are adjusted so that those conform to spectra taken on a master instrument whereas the calibration model remains invariable. Thus the possible shifts of spectrums on wave numbers axes are not taken into account. The response matrices of the two instruments are interrelated via the matrix of coefficients F [8]:

$$Xm = Xs\ F \tag{1}$$

where F [mxm] is the square matrix of coefficients β, m being the number of readouts in the spectrum (the number of wavelengths at which the spectrum is detected). This matrix brings a spectrum of the slave instrument in correspondence with the spectra taken on the master instrument:

$$F = (Xs)^{-}\ Xm \tag{2}$$

where $(Xs)^{-}$ - is the generalized inverse of Xs.

The vector of response of an unknown sample measured on a slave instrument, $x_{s.un}$, is recalculated to the response vector $x^*_{s.un}$, which would be expected if that sample will be analyzed at the master instrument:

$$(x^*_{s.un})^T = (x_{s.un})^T\ F \tag{3}$$

From vector $x^*_{s.un}$ by using masters full calibration model, the unknown concentration of an analyzed component is found. In this case, no information on concentrations is necessary for the transfer of calibration models. This means that any samples can be used for calculation of matrix F.

It should be noted that matrix F is calculated from a relatively small sample set and Eq. (1) is commonly undefined with respect to the matrix F that may give rise to large errors in evaluation of concentrations. One more shortcoming of the method is its inability to compensate for possible shifts of spectra along the wave number axis. Therefore, a method based on projecting a segment of the spectrum taken on a slave instrument onto a point in the spectrum taken on a master instrument is of interest.

METHOD OF PIECEWISE DIRECT STANDARDIZATION

Piecewise direct standardization method. In this method, a total calibration is performed on a master instrument and response matrix X1 [nxm] is obtained (where n is the number of samples and m is the number of readouts in the spectrum), then a calibration model is designed. Some small portion of samples (with response matrix Xm) is selected from the complete set of calibration samples and a partial calibration of a slave instrument is performed, whereby response matrix Xs is obtained.

Two spectrometers may differ for different reasons, but any difference is exhibited in a shift of digitization of the wavelength axis and in differences in intensities. In a particular case when there is no shift in digitization of the wavelength axis but the responses of the instruments differ only in intensities, matrix F in Eq. (1) is diagonal one. If a systematic shift, being constant over the spectrum, is present, the diagonal in the matrix is displaced aside.

In a general case, matrix F is the one with nonzero elements, which allow for possible shifts along the wavelength axis and differences in intensities. The number of matrix elements can be very large (typically, on the order of 700 [14]). To define this matrix, the number of transferred samples should be at least equal to rank Xs, which causes certain difficulties in practice. Fortunately, in most of spectroscopic techniques, including NIR spectroscopy, the intensity in spectra shows continuity with respect to wavelengths. Therefore, it can be assumed that the elements of matrix F lying far from a point under consideration in the spectrum do not noticeably contribute to parameters of the given point, and these elements can be considered to be zero. Hence, matrix F can be treated as a banded matrix, the bandwidth, which defines the width of the regression window, being subject of inquiry.

Thus, when the spectrum of a slave instrument is shifted relative to the wavelengths of a master instrument, the corresponding pairs of the wavelengths lie in contiguous segments of the spectrum. Since the shift is *a priori* unknown and is not constant over the spectrum, F can be found by determining the relation between the i-th point in the spectrum taken on master instrument x1.i and some zone (regression window) in the spectrum taken on slave instrument Ri in the vicinity of point x2.i:

$$Ri = [x2.(i-j), x2.(i-j+1), \ldots, x2.(i+k-1), x2.(i+k)] \quad (4)$$

In this case, the shift of the center of window Ri of width p=j+k relative to point x2.i can be asymmetrical, that is, j can be unequal to k. Then, assuming that variations in intensities are linear and a possible shift in digitization of the wavelength axis lies within this window, we seek the relation with point x1.i of the master instrument in the form

$$x1.i = Ri\ \beta i, \quad (5)$$

where βi is the row in matrix F in Eq. (1). Then, for the total spectrum X1 we obtain

$$X1=R\ F \tag{6}$$

where matrix R is formed by rows Ri, F= diag(β_1, β_2, ... , β_i, ... , β_p), and p is the width of window [i-j, i+k].

Equation (6) can provide approximation even for nonlinear variations in intensity over the spectrum. In this case, Rank(R)≤Rank(X1), which makes it possible to reduce the number of the necessary samples. The vector of coefficients βi can be calculated by the PLS method or a conventional multivariate regression method. For the extreme points in the spectrum taken on a master instrument, the corresponding zone cannot be formed in the spectrum taken on a slave instrument; hence, those are either not taken into account or extrapolation is done.

Having once defined matrix F from the responses of the sets x2 which analyzed on two instruments, we can recalculate any spectrum x2.un for an unknown sample measured on a slave instrument to spectrum x1.un, as if was it measured on a master instrument:

$$x1.un = x2.un\ F \tag{7}$$

Approaches to Forming the Set of Samples to Transfer Calibration Models

An adequate selection of samples for analysis on both instruments x2 is of great importance in this procedure. Many investigators note the difficulties in defining the minimum number of samples and their necessary characteristics (see, for example, [11, 12, 14]). It is suggested in [12] to define the sample set by the significance of the contribution of a sample to the Standard Error of Calibration (SEC) in the prediction from a model constructed for sample array x2 with and without the given sample. If at inclusion of the sample in the set SEC is diminished insignificantly, the sample should not be included in x2.

A different approach to the selection of samples for array x2 involves the selection of those samples for calibration on a slave instrument whose spectra differ most significantly from one another (in [11], the Euclidian distance between the spectra is used to evaluate the difference in spectra).

The main drawback of these selection methods is a large amount of calculations. In the given investigation the modification of the second method is offered. With due account of the fact that, both in the direct calibration transfer and in the piecewise regression method, it is supposed to use the PCA or PLS methods to seek the elements of matrix F, we suggest to select the spectra in the factor space rather than in the spectra space. This approach is considered in more detail in our second paper.

Comparative Analysis of the Calibration Model Transfer Methods

For a comparative analysis of the calibration transfer methods, we used spectral data arrays obtained in analysis of mix fodder (32 samples) and wheat (38 samples) on FT-NIR spectrometers InfraLum FT-10 manufactured by Lumex Ltd. Research and Production

Company. We measured concentrations of two characteristics of the analyzed substances, protein and moisture. Typical absorption spectra taken on two instruments are depicted in Fig. 1. Each spectrum comprised 290 points in a wave number range 8500 - 12600 cm-1.

Two arrays were formed from the initial data. The first one (22 for mix fodder and 28 for wheat samples) was used to construct calibration models and the second one (10 samples) to check their adequacy. Both arrays contained spectra of the samples covering the entire variation range of the concentrations.

Table 1 presents the results of calibrations of the master and slave instruments against the complete set of calibration samples, the slave instruments used for analysis of mixed feed and wheat being different.

Table 1. Results of the direct calibration

Analyzed product	Com-ponent	M*)		SEV error, wt. %		Component concentration range in the mix, wt. %
		Master instru-ment	Slave instru-ments	Master instru-ment	Slave instru-ments	
Mixed fodder	Protein	4	4	0,79	0,80	20-26
	Moisture	4	4	0,49	0,49	8-11
Wheat	Protein	3	4	0,38	0,40	11-15
	Moisture	3	3	0,25	0,22	10-14

*) M is the number of the principal components in the PLS method corresponding to the minimum Standard Error of Validation (SEV).

The direct use of the calibration model of the master instrument for two slave instruments brought about the following SEV errors in concentrations of the components: protein - 0,90 % and 0,62 %; moisture - 1,00 % and 0,44 % (the first value in the pair concerned to mix fodder, the second one – to wheat). A comparison with the data in Table 1 shows that, when the main calibration model is used in this manner, the error due to the nonidentity of the instruments significantly increases (by a factor of 1.2 – 2) as compared to the errors arising in individual calibrations.

We studied the latter two methods as most promising ones of all those considered here. Tables 2 and 3 present SEV errors as functions of the number of the principal components (iterations) when the algorithm of the PLS method is used for calculation of the vector of coefficients β in Eq. (5) involved in forming matrix F in Eq. (1). In this case, the width of the window around the points in the spectrum taken on the slave instrument, which are projected onto the i-th point in the spectrum taken on the master instrument, was varied from 1 (this corresponds to the absence of the shift along the wavelength axis in the spectra of both instruments, that is, to the direct standardization method) to 5 readouts (the piecewise direct standardization method).

Our study has demonstrated that better results in the calibration transfer, as compared to a conventional calibration, are obtained with the smaller number of the PLS iterations used for calculation of the main calibration model and subsequent calibration transfer (as can be seen from comparison of M numbers in Tables 1, 2, and 3).

Table 2. Transfer of the calibration models for mixed feed from the master instrument to the first slave instrument.

Characteristics	Window width						
	1	3			5		
	SEV	M*)	K**)	SEV	M*)	K**)	SEV
Protein	0.90	4	3	0.81	3	3	0.85
Moisture	0.58	2	2	0.57	3	3	0.64

*) M is the number of the principal components in the PLS method used in calculation of the main calibration model, which correspond to the minimum SEV in its transfer to the slave instrument.

**)K is the number of the principal components in the PLS method used in calculation of the vector of coefficients β in Eq. (5)

Table 3. Transfer of the calibration models for wheat from the master instrument to the second slave instrument

Characteristics	Window width						
	1	3			5		
	SEV	M*)	K**)	SEV	M*)	K**)	SEV
Protein	0,45	3	3	0,40	3	3	0,42
Moisture	0,36	3	3	0,32	3	3	0,35

*), **) See notes to Table 2.

It appeared possible to use conventional multivariate regression procedures for calculation of coefficients β in Eq. (5) and, hence, of adjustment matrix F.

Conclusions

Therefore, the following conclusions can be drawn from a comparison of the errors brought about in the calibration model transfer with those arising in the direct calibration.

1. The calibration transfer from one instrument to another without adjustment of possible shifts and variations in intensity results in a significant error increase and cannot be used.
2. The piecewise direct standardization method yields better qualitative results as compared to other methods. The direct standardization method does not yield satisfactory results, apparently, because of the nonlinear behavior of variations in intensity and inconstant shift along the wavelength axis in different instruments. So it can be recommended to compensate shifts as accurate as possible.
3. The coefficients in the calibration transfer equations of PDS can be calculated without loss of accuracy by the multivariate linear regression method. The PLS method is more laborious, and it is expedient to this effect if used for calculation of the main calibration model of a master instrument.

4. Due to the indeterminacy of the set of equations for calibration, it is expedient to increase the spectrum readout step (for example, by two-four times as the standard step in an InfraLum FT-10 instrument). This effect can cause decreasing the overall error (in our case, by a factor of 1.2).
5. The numbers of the principal components in the PLS method in the direct calibration, calculation of the coefficients in the calibration transfer equations, and for various analyzed characteristics are different, and the optimum number should be sought by varying any of these factors. Commonly, the number of the principal components used for calculation of the calibration model equals 3 – 7. In this case, the error in the calibration transfer increases by a factor of 1.2-1.5.

References

[1] L.A. Rusinov, K.A. Zharinov, A.V. Tolstoy, and V.A. Zubkov, *Vestnik Metrologitcheskoy Academi*, **7**, 17 –30, (2001), (in Russian).
[2] Hoskuldsson, *Prediction methods in science and technology*, Thor publising, vol.1, 1996.
[3] H. Martens, T. Naes, *Multivariate calibration*, John Willey & Sons, 1998.
[4] K. Esbensen, *Multivariate analysis in practice*, Camo, 2000.
[5] D. Haaland, Thomas V. *Anal. Chem.*, 60, 1193-1202, (1988).
[6] Standard practices for infrared multivariate quantitative analysis. *ASTM Standards. Designation* E 1655-99.
[7] V.P. Kritshenko. *Near range infrared spectroscopy,* Agroteh, 1997 (in Russian).
[8] Y. Wang, D. Veltkamp, B.R. Kowalski, *Anal. Chem.*, 63, 2750-2756, (1991).
[9] J.S. Shenk, M.O. Westerhaus, W.C. Templeton, *Crop Science,* 25, 159 –161, (1985).
[10] Y. Wang, B.R. Kowalski, *Appl. Spectrosc.*, 46, 764-771, (1992).
[11] E. Bouveresse, C. Hartmann, D.L. Massart, *et.al.*, *Anal. Chem.*, 68, 982-990, (1996).
[12] L. Zhang, G.W. Small, *Anal. Chem.*, 74, 4097-4108, (2002).
[13] Y. Wang, T. Dean, B.R. Kowalski, *Anal. Chem.*, 67, 2379-2385, (1995).
[14] T. Fearn, *J. Near Infrared Spectrosc.*, 9, 229-244, (2001).
[15] E. L. Sulima, K.A. Zharinov, V.A. Zubkov, and L.A. Rusinov, *This issue*, 212-218

In: Progress in Chemometrics Research
Editor: Alexey L. Pomerantsev, pp. 237-243
ISBN 1-59454-257-0

Chapter 16

SPECIFIC FEATURES OF PRACTICAL IMPLEMENTATION OF CALIBRATION MODEL TRANSFER FROM A MASTER INSTRUMENT TO SLAVE NIR ANALYZERS FOR ANALYSIS OF MAIN CHARACTERISTICS OF WHEAT

***E. L. Sulima*[1]*, K. A. Zharinov*[1]*, V. A. Zubkov*[1] *and L. A. Rusinov*[2]**

[1] Lumex Ltd., St-Petersburg, Russia; zharinov@lumex.ru

[2] St. Petersburg Technological Institute (Technical University), Russia; rusinov@tu.spb.ru

ABSTRACT

A technique for the transfer from a master instrument to slave NIR analyzers «INFRALUM FT-10» of an initial calibration model designed using the PLS method is considered. Sample selection methods for the calibration model transfer according to the values of reference data and using the score parameter of the first principal component are discussed. Results of the quality assessment of the model transfer from a master instrument to 14 slave NIR analyzers deployed at elevators in Krasnodar District in Russia are exemplified by an analysis of characteristics of wheat.

Keywords: NIR-spectroscopy, calibration models transfer.

INTRODUCTION

In many industries, including food industry, one of the main problems is the study of the composition of products. Methods based on spectral analysis in the near-infrared (NIR) range play an important role in solving this problem.

Calibration of instruments involves laborious and complex processes, which call for a large amount of work. To design calibration models, a set of samples whose properties cover the whole range of possible characteristics of samples under study is selected. The calibration samples are analyzed using conventional chemical techniques to determine concentrations of their characteristics or properties. Then the spectra of the calibration samples are measured on a spectrometer and the calibration models are constructed, which relate the spectral data to those of the chemical analysis [1].

Unfortunately, it is impossible to apply same calibration to several instruments because of specific design features of each spectrometer. Therefore, it is expedient to carry out complete calibration for a single instrument (master instrument) and then to transfer the calibration to slave instruments of the same series. Various techniques can be used for the transfer of calibration equations from one instrument to another. For example, spectra taken on a slave instrument can be adjusted so that they correspond to the spectra obtained on a master instrument, thus the calibration model remains invariable. The spectra of a slave instrument are adjusted according to the regression coefficients calculated by the method of least squares. Predictions of the adjusted spectra of the slave instrument are made using the master calibration. For example, the direct standardization method [2,3] and the piecewise direct standardization (PDS) method [2,3,4] are used to this effect.

In this paper, the subject of inquiry is food-grade wheat. We designed initial calibration models against 145 wheat samples on a master instrument for main characteristics, such as protein, gluten, glassiness, and moisture. The ranges of reference values were as follows: for protein - 10.5-17.5%, gluten - 14.8-26.9%, glassiness - 35-59%, and for moisture - 8.7-13.7%. The calibrations were transferred from an «INFRALUM FT-10» master instrument to 14 «INFRALUM FT-10» NIR analyzers using 10 samples selected from the complete master calibration set. To check the quality of the calibration transfer, we used a validation set of 10 samples that were not involved in the complete calibration of the master instrument. The algorithm and results of the calibration transfer are presented below.

EXPERIMENTAL

Selection of the Transfer Method

As was demonstrated in [3], the PDS method is preferable over the direct standardization method because the former method allows for possible shifts in intensities and wavelengths in spectra of a same sample measured on a master and slave instruments. For a series of instruments on which the experiment was carried out, we implemented automatic correction for the wavelengths and the necessary normalization of the spectra over the intensities was performed. As our study revealed, in such an approach to standardization of the instruments and normalization of the spectra, the use of the PDS method show no advantage over the direct standardization. This can be seen from Table 1, in which the values of the Standard Error of Validation (SEV – Eq.6 below) are listed for a master and slave instruments with the calibration being transferred from the master instrument by the direct standardization and the PDS methods.

Table 1. Results of the calibration model transfer by the direct standardization and the PDS methods (for protein).

	SEV	
Master instrument	Direct standardization (transferred calibration)	PDS (transferred calibration)
0.34 %	0.38 %	0.39 %

Thus, at execution of automatic correction on wavelengths SEV of the calibrations transferred by the methods of direct standardization for slave instruments are approximately identical. Besides matching SEV for the master instrument and slave of instruments also are in the close agreement. Therefore, in our further experiment we used the direct standardization.

Implemented Transfer Technique

We denote the spectra matrix for a master instrument as Xm and the same for a slave instrument as Xs. Relation between a matrix of a spectrum of the slave-instrument Xs and matrix Xm, appropriated to an expected spectrum of the same samples on the master instrument, at a selected method of direct standardization will be defined by expression [3]:

$$Xm=Xs\,F \tag{1}$$

At execution of automatic correction on wavelengths the matrix of coefficients F becomes diagonal and coefficients in Eq.(1) can be defined by the simple least squares method:

Spectra of a slave instrument are adjusted according to the calculated coefficients:

$$F=(Xs)^{-}\,Xm \tag{2}$$

where $(Xs)^{-}$ - is the generalized inverse of Xs.

The vector of response of an unknown sample measured on a slave instrument, $x_{s.un}$, is recalculated to the response vector $x^*_{s.un}$, which would be expected if that sample will be analyzed at the master instrument:

$$(x^*_{s.un})^T = (x_{s.un})^T F \tag{3}$$

Predictions of characteristics values of unknown substances can be made for the adjusted slave instrument spectra $x^*_{s.un}$ by using master calibration.

Selection of Samples for the Transfer

To implement the calibration transfer, set of samples was selected and their spectra were measured on slave instruments. The samples were selected from the complete calibration set of the master instrument according to the scores. We selected 28 samples uniformly distributed over the scores. The transferred calibration was verified by a validation set of 10 samples. To determine the minimum sample set for transfer, the number of samples was decreased, those whose scores fell in the middle of the range being eliminated. Each calibration transferred was verified by the validation set. Figure 1 depicts a graph of the Standard Error of Validation (SEV) of the transferred calibration versus the number of samples involved in the transfer. In this case, a calibration transfer from the master instrument to five slave instruments is shown for protein. When the number of samples is smaller than 12, SEV deteriorates.

Samples can be selected for transfer also by the reference data, whose values can be maximum and minimum for the content a certain characteristic, because their minimum values approximately correspond to maximum scores and vice versa. These values are listed in Table 2. The best transfer is attained when samples having extreme scores for other characteristics as well (e.g., moisture, gluten, and glassiness) are added to the selected samples with extreme scores for a single characteristic (e.g., protein). So the set from 10 samples (1, 2, 5, 7, 10, 23, 24, 26, 27, and 28 - Table 2) was formed. These samples are selected basing on the fact that they have extreme score values for all of the characteristics of wheat, that are, protein, gluten, glassiness, and moisture.

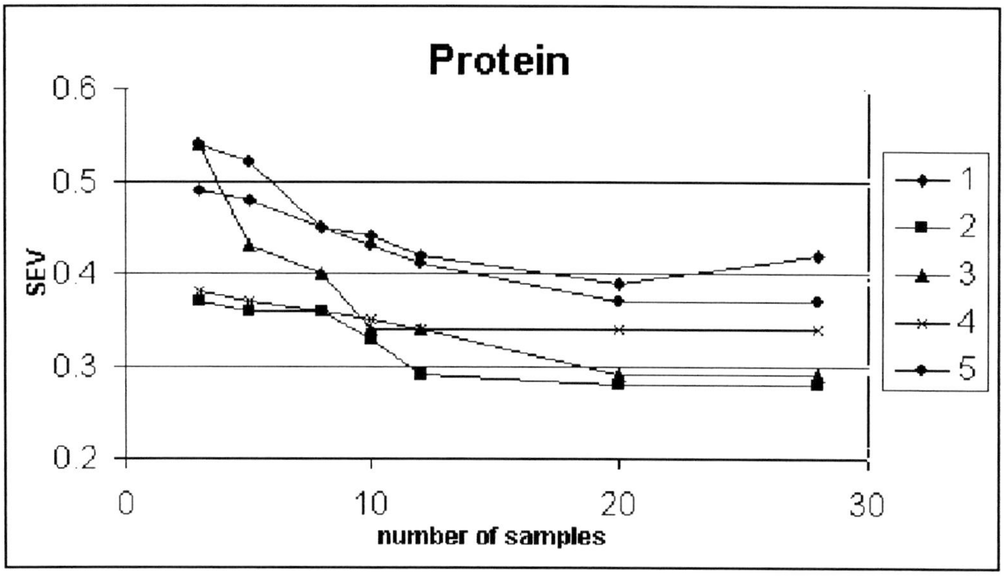

Fig.1. Curves 1, 2, 3, 4, and 5 correspond to slave instruments.

Results of the Calibration Transfer

The calibration equations were transferred to 14 instruments as per ten samples selected basing on the scores for several characteristics (protein, gluten, glassiness, and moisture) of food-grade wheat. The initial calibration of the master instrument was designed as per 145 samples whose main characteristics, namely, Standard Error of Calibration (SEC), Standard Error of Cross Validation (SECV) and Standard Error of Validation (SEV) were calculated by the following equations and listed in Table 3:

Table 2. Reference values for protein.

Sample number	Reference values, %	Scores	Sample number	Reference values, %	Scores
1	**10.6**	**0.0953**	15	13.4	-0.0845
2	**11.34**	**0.506**	16	13.52	0.0683
3	11.4	0.177	17	13.6	0.00784
4	11.4	0.125	18	14.41	0.23
5	**11.5**	**0.206**	19	14.5	-0.211
6	11.5	0.00599	20	14.54	-0.00834
7	**11.7**	**0.752**	21	14.6	-0.362
8	11.99	-0.482	22	14.6	-0.639
9	12.12	-0.568	**23**	**15.7**	**-0.543**
10	**12.2**	**0.322**	**24**	**16.3**	**-0.944**
11	12.2	-0.183	25	16.6	-1.07
12	12.3	0.209	**26**	**16.8**	**-1.18**
13	13.3	-0.571	**27**	**17.41**	**-1.25**
14	13.3	0.01	**28**	**17.42**	**-1.28**

$$SEC = \sqrt{\frac{\sum_{i=1}^{n} (\hat{y}_i - y_i)^2}{d}}, \tag{4}$$

where d is the number of the degrees of freedom in the calibration model, $d=n-k$, n being the number of calibration samples, and k is the number of the principal components used in the model; $\hat{y}$ denotes the values of the components being determined for the calibration samples, which were obtained with the use of this calibration, and y denotes the corresponding reference data;

$$SECV = \sqrt{\frac{\sum_{i=1}^{n} (\hat{y}_i - y_i)^2}{n}}, \tag{5}$$

where $\hat{y}$ is the vector comprising the estimates of the cross validation, y is the vector of the reference data, and n is the number of the calibration samples;

$$SEV = \sqrt{\frac{\sum_{i=1}^{n'} (\hat{y}_i - y_i)^2}{n'}}, \tag{6}$$

where n' is the total number of the validation samples, y denotes the reference values of the analyzed component for the i-th spectrum of the validation set, $\hat{y}$ denotes the predicted values of the analyzed component for the i-th spectrum of the validation set.

Table 3. Results of calibration of master instrument

Characteristic	SEC	SECV	SEV
Protein	0.41	0.41	0.34
Gluten	0.62	0.69	0.65
Glassiness	2.56	2.74	2.37
Moisture	0.19	0.20	0.25

Results of the calibration transfer are presented in Table 4.

Table 4. Results of calibration model transfer.

instruments	SEV			
	protein, %	gluten, %	glassiness, %	moisture, %
master	0.34	0.65	2.37	0.25
1	0.31	0.82	2.30	0.33
2	0.29	1.14	2.36	0.25
3	0.31	0.72	2.41	0.30
4	0.33	0.76	2.36	0.21
5	0.38	0.65	2.82	0.21
6	0.33	0.78	2.06	0.28
7	0.44	0.68	2.49	0.28
8	0.35	0.69	2.18	0.28
9	0.48	0.64	2.54	0.31
10	0.42	0.65	2.58	0.30
11	0.45	0.65	2.54	0.31
12	0.45	0.60	2.36	0.36
13	0.40	0.70	2.41	0.36
14	0.51	0.65	2.00	0.35
average SEV	0.38	0.72	2.39	0.30

CONCLUSIONS

Thus, the following conclusions can be drawn from analysis of the obtained results:

- Adjusting of NIR-spectrometers with the purpose of maximal elimination of shifts on an axis of wavelengths and prenormalization of spectra allow considerably to simplify procedures of calibration models transferring from the master instrument on slaves ones practically without loss of accuracy;

- For the calibration transfer, samples can be selected by the reference values and scores;
- For the transfer of the calibration equations, ten samples selected by the scores or reference data for several characteristics are sufficient. In the experiment, the calibration equations were transferred as per ten samples selected basing on the scores for several characteristics (protein, gluten, glassiness, and moisture);
- Standard Errors of Validation for the master instrument and 14 slave instruments are close to each other, which confirms the validity of the transfer algorithm and the proposed sample selection procedure for the calibration transfer.

REFERENCES

[1] Y. Wang, D. Veltkamp, B.R. Kowalski, *Anal. Chem.*, 63, 2750-2756, (1991).
[2] T. Fearn, *J. Near Infrared Spectrosc.*, 9, 229-244, (2001).
[3] L.A.Rusinov, K.A.Zharinov, E.L.Sulima, and V.A.Zubkov, *This issue,* 204-212
[4] Y. Wang and B. Kowalski, *Appl. Spectrosc.*, 46, 764-771, (1992)

In: Progress in Chemometrics Research
Editor: Alexey L. Pomerantsev, pp. 245-263
ISBN 1-59454-257-0

Chapter 17

SAMPLING IN PRACTICE: A TOS TOOLBOX OF UNIT OPERATIONS

Lars Petersen and Kim H. Esbensen
Aalborg University Esbjerg, Denmark, www.acabs.dk

ABSTRACT

A basic knowledge of the Theory of Sampling (TOS) and a set of only eight sampling unit operations is all the practical sampler needs to ensure *representativeness* of samples extracted from all kinds of lots: production batches, - truckloads, - barrels, sub-division in the laboratory, sampling in nature and in the field (environmental sampling, forestry, geology, biology), from raw materials or manufactory processes etc. We here can only give a brief introduction to the Fundamental Sampling Principle (FSP) and these eight Sampling Unit Operations (SUO's). Always respecting FSP and invoking only the necessary SUO's (dependent on the practical situation) is the only prerequisite needed for eliminating all sampling bias and simultaneously minimizing sampling variance, and this is in addition a sure guarantee for making the final analytical results trustworthy. No reliable conclusions can be made unless the entire sampling process corresponds with TOS.

Keywords: Theory of Sampling, sampling unit operations, Fundamental Sampling Principle

INTRODUCTION

As scientists, engineers and process operators we spend a significant amount of time, money and efforts on optimizing and validating analytical procedures and methods. This of course seems eminently reasonable, getting the analytical error under strict statistical control, ensuring the optimal quality of the analytical results. But what if it is stated that very nearly always *other errors* arise that are in the order of 10, 100 or even 1000 times larger than this analytical error? Surely this should change our perspective! It is fully documented, in fact over more than 50 years, that all primary, secondary and tertiary sampling of heterogeneous materials indeed produces such incredibly larger errors [1], [2]. The Theory of Sampling is more than 50 years old today, but is despite of its relevance little known and/or heeded in both industry and academe.

The situation is not all hopeless however; there are many ways of getting such sampling errors in complete control. But like in almost any other situation nothing is free – we cannot get these dominating sampling errors reduced and/or eliminated without doing the necessary work (indeed knowing what constitutes the work needed is half the game) – the only proper place to do this work is when doing the sampling. There are no possibilities at all for ever correcting or making up for flawed sampling; this is unfortunately not known to everybody: e.g. "bias-corrections" or "sampling robustness" and similarly unfortunately does not remedy erroneous sampling.

But representative sampling is eminently doable, only it does not come about by itself. The extra cost put into what TOS terms "correct sampling" is greatly rewarded in the long run however, because it leads to more representative results and hence more reliable conclusions. Sampling is not gambling – sampling is a science like any other. Proper sampling takes place on many scales, from primary sampling in the field/process to the smallest sub-samplings with a spatula from sample vials in the laboratory of the order of only micrograms, because all materials are inherently heterogeneous. Sampling of heterogeneous materials will always result in sampling errors (sampling variances), because any two samples extracted as close to each other as physically possible will necessarily still make for different analytical results, precisely because one cannot sample the same volume twice, and because there will always exist a local, residual heterogeneity at the scale of the sampling tool volume.

It is simultaneously equally important to realise that *all* sampling procedures and stages by themselves will generate their own inevitable errors.

In this paper we will try to supply the reader with a number of simple tools (conceptual as well as practical) for optimal sampling in practise. After intensive studies of the Theory of Sampling (TOS), developed by the French scientist Pierre Gy, we have formulated the salient aspects of TOS in a small number of simple "Sampling Unit Operations" (SUO) that cover all aspects of practical sampling, ranging from simple procedures such as mixing to more complex heterogeneity characterizations and variographic experiments. Of these unit operations some are intuitive and easy to perform while others are more labour intensive and perhaps not obvious to untutored minds. We here give an introduction to the necessary minimum of theoretical understanding of the principles of TOS, without which the system of SUO would degenerate to rote learning.

This paper gives a short introduction to *material heterogeneity*, an introduction to the ever important *fundamental sampling principle*, and a description of the eight *Sampling Unit Operations* (SUO). But it is important to be aware that the present brief synoptic introduction is not tantamount to a complete sampling course. We also refer the reader to the general literature on sampling [1]-[4], [7].

HETEROGENEITY

The basis for understanding all practical sampling starts with the material to be sampled; in particular we need to be able to describe the phenomenon *of heterogeneity*.

If homogeneous materials existed, sampling would not be necessary...

Often the term "homogeneous" is used to describe a given material (thought to be so). This is *always* a mistake. No naturally existing material is in fact homogeneous. The term homogeneous is in TOS defined as a theoretical limiting term only, describing the extreme, and practically unobtainable, minimum of heterogeneity. Any material is heterogeneous at *some* level with regard to *some* of its parameters. Even a so-called 100% pure particulate chemical substance to be used in high-cost laboratory experiments is not homogeneous: If we keep on magnifying the scale of an "observation module", we will at some point observe differences in particle size, shape and composition: in practise we would of course neither ever encounter substances 100.0000000 % pure etc.

The question for practical sampling rather is whether the material is "sufficiently homogenized" at all scales from the total lot down to scale of the sampling tool volume (that we use to do the actual sampling from the lot). In order to address this issue competently we need to understand the following minimum of the concepts of heterogeneity.

A material must be described by two different kinds of heterogeneity: Constitutional Heterogeneity (CH) and Distributional Heterogeneity (DH). Constitutional heterogeneity, CH, describes the differences in composition, shape, density etc. between the smallest individual and separable constituents of the lot: the *fragments*. Thus CH is concerned with the heterogeneity *between* these fundamental units at the smallest scale level of interest. Distributional heterogeneity, DH, then further describes the part of the overall heterogeneity stemming from spatial distributions at a higher scale level, namely that representing the set of so-called *observation modules*, which in practise corresponds to a sample volume. Thus DH deals with the intermediate to full lot scale heterogeneity aspects stemming from the spatial distribution of the entire set of *virtual samples* together making up the entire lot. It is in fact equal to the heterogeneity contributions *between* these sample-size units of observation at the intermediate scale. The generic term for a potential sample volume to be selected is an "increment".

CH can be affected (reduced) only by processes that alter the physical composition or appearance of the individual particles (crushing, comminution, etc.), whereas DH can be affected (reduced) by mixing, blending and other, similar means of homogenization.

We cannot go into more details on heterogeneity in this introduction, but the interested reader is referred to the extensive literature on this topic [1]-[3].

THE FUNDAMENTAL SAMPLING PRINCIPLE

The overlaying governing rule for all sampling is the Fundamental Sampling Principle (FSP). This can be stated:

> Any individual virtual increment of the lot must have a non-zero and equal probability of being materialised as the final sample.

Respecting the fundamental sampling principle (FSP) ensures *probabilistic sampling*. Any deviation from FSP leads to "incorrect sampling" which will result in an unavoidable sampling bias. Imagine a large stockpile of pebbles to be sampled. A normal procedure would be grapping a few here and there from the surface (the readily available parts of the lot) or at the most within a reasonably accessible depth. This is clearly non-probabilistic sampling and since there are very many parts of the lot (all the interior parts as well as the bottom of the pile), which in this scenario with certainly has a zero probability of being sampled, there is no way the sample can be representative of the whole lot. What would happen if e.g. the pile was built from slowly increasing pebble sizes for example and the samples were to be used for size distribution analysis? Grab sampling is in fact never up to no good [6]!

SAMPLING UNIT OPERATIONS

In order to perform correct sampling in practice, always respecting the principles of TOS, we only have to know about seven *basic* sampling unit operations (SUO) pertaining to normal sampling of 0-D lots and one in addition, that specifically pertain to the special case of sampling 1-D lots (definition of lot dimensionality given below). All these unit operations will be explained in more detail in the following sections.

1. *Structurally correct sampling* is the *only* safeguard against sampling bias
2. *Heterogeneity characterization* of 0-D lots
3. *Homogenization*, mixing, blending
4. *Composite sampling*
5. *Representative mass reduction*
6. *Particle size reduction* (comminution or crushing)
7. *Lot dimensionality transformation* (3D or 2D → 1-D or 0-D)
8. *Variographic characterisation* (1-D lot heterogeneity characterization and process control/optimization)

Armed with these tools anybody can perform correct sampling, all it takes is a little work, which will be repaid generously from the significantly more reliable analytical results.

Structurally Correct Sampling

If handed a particular sample, there is absolutely no way of telling whether it is representative or not – not even by the most refined and expensive analytical techniques in the world, neither by any kind of statistical analysis. Representativeness can only be ensured by performing structurally correct sampling, i.e. representative samples result from performing a representative sampling process. The focus should consequently always be on the sampling *process*. It is outside the scope of the present paper to deal completely with all the principles of structurally correct sampling, but we will give a brief introduction since it is indeed the most important aspect of any sampling process or procedure.

There is first the principal difference between accurate and precise sampling. Precise sampling means that a number of replicated samples will display a low standard deviation, or variance, between them. Accurate sampling means that the replicated samples, regardless of their precision, on average will allow an estimate (the mean) that is closely coincident with the true mean (grade/concentration) of the lot. In other words an accurate sampling process does not posses a significant systematic sampling bias! Performing "structurally correct" sampling then means that all sources of bias have been eliminated from the sampling process.

What are these sources of bias? There exist only three sources, which all luckily can be completely eliminated by respecting the rules and guidelines of TOS specified to counteract three kinds of sampling errors, wiz.

1. Increment Delineation Error (IDE)
2. Increment Extraction Error (IEE)
3. Increment Preparation Error (IPE)

These errors are jointly named Incorrect Sampling Errors (ISE), as they all result from incorrect sampling. A highly appealing fact about ISE is that they can be eliminated completely by careful considerations and efforts during the design, maintenance and use of the sampling procedures and equipment.

Ad 1) To be structurally correct, a sample must be correctly delineated, meaning that it must cover an entire dimension of the sampled material with absolutely parallel sides so that no part of a cross section is represented in higher or lower proportions than any other. This is best illustrated graphically (fig 1):

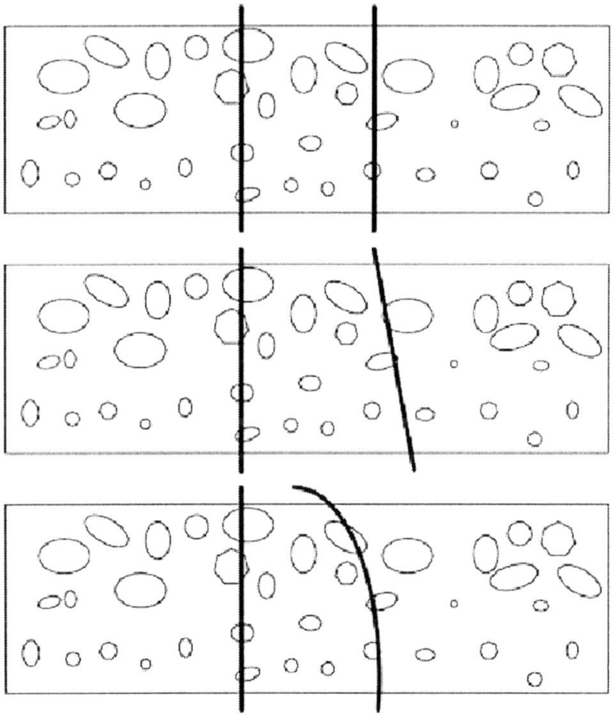

Figure 1. Examples of correct (top) and incorrect increment delineation (middle and bottom). A correctly delineated increment must have fixed parallel sides to ensure equal representation of the whole width of the lot. In the figure one should imagine that the lot is actually moving on a conveyor belt and consequently that only a small portion of the (horizontal) total lot is shown – please notice the stratified distribution of large and small particles emphasizing the need of parallel sides.

Ad 2) After correct delineation it is important that this material is actually the only material extracted. If the geometric delineation "cuts" some particles, they must either enter the sample or be rejected according to the so-called centre of gravity rule. This rule simply states that any particle having its centre of gravity inside the increment should in fact end up in the final sample and vice versa. Even appearing as a triviality this actually has great importance in the physical design of cutting devices, since a falling particle hitting an inclined device wall will just as well bounce in or out of the device if not very carefully designed correctly [1], [2]. This is particularly important if the cutting device is physically *moving* across the process stream, since, for instance, a particle slung into the air above the moving device by the leading edge, which should normally not have ended up in the sample, will in fact end up in there anyway, as the cutter moves underneath it. This matter gains in importance in direct proportion to the average particle size of the material.

Ad 3) The Increment Preparation Error is the only error that is non-statistical in its nature. It consist of a number of potential errors that all occur *after* extraction of the increment or sample. Some examples are:

- Loss of water/moisture due to evaporation, or the opposite (condensates added)
- Unintentional loss of sample material related to materials handling etc.
- Loss of fines, or addition of fines, due to dust
- Contamination from storage devices, equipment-induced cross-contamination
- Deliberate alteration, such as sabotage, or fraud

These errors must be avoided by careful supervision, thorough education of employees and by showing the greatest responsibility towards the sampling procedure. The last point on the above list may perhaps seem provocative to some readers, but history has shown on numerous occasions that since samples don't tell their own stories after extraction, this is also the most likely place for deliberate cheating. No one can in fact ever *prove* that a sample was altered after extraction; we can only assume or have substantiated grounds for suspicions etc.

Common for all the above three errors are that they are the *only* sources of systematic sampling bias, and therefore *must* be eliminated (or at the very least thoroughly minimized).

Sampling is nearly always a multistage process consisting of primary-, secondary-, and perhaps tertiary sampling stages. It is important that errors at all levels are eliminated. But what are the relative error levels at the different stages?

Heterogeneity Characterization

In order to describe quantitatively the constitution of the material to be sampled, we can quantify the heterogeneity in a given lot either empirically (experimentally), or by a model of the typical behaviour of heterogeneous materials (via simplifying real-world estimation of a few salient physical mixture characteristics).

It is always recommended, before design of any sampling procedure, that sufficient efforts are spent on an initial characterization of the heterogeneity of the material to be sampled. In terms of time and money this is always considered to be expensive, but the alternative, which is uncontrolled biased samples and unnecessarily large erroneous analytical results, is always much worse - not only from a scientific point of view but especially also from the economics surrounding the decisions which are to be taken based on such faulty information!

Figure 2. The absolute minimum layout of an experimental design yielding an estimate of the effective errors at different sampling stage levels. Two replications is the ultimate minimum number under strong normality assumptions; however many more replicates, up to 8-10, must in general be recommended. Black dots represent analysis. PSE (Primary Sampling Error); SSE (Secondary Sampling Error) etc; TAE (Total Analytical Error).

An empirical heterogeneity characterisation is always carried out by a multi-stage replication experiment, designed to yield an estimate of the variation at the relevant different scale-levels of interest. An absolute minimum would be two replicates at all levels (fig. 2), which can only be supported under the strong assumption that all sampling errors at all levels are normally distributed – a questionable assumption by itself oftentimes; however up to some 10+ replications is highly recommended. It really does not pay to save the expenditures at this stage of sampling. From experiences with practical application of TOS, we have reached the conclusion that what is needed is far from a prohibitive large number of total replicates – indeed an absolute maximum of 50 would appear to point itself out (and many times a total less than this is sufficient even): this total number of replicates, N^{replic}, needed in order to reach completely reliable estimates of the relative magnitudes for all sampling stages must be distributed over all the relevant (problem-dependent) sampling + analysis stages, for example PSE: 10; SSE: 10; TSE 10; TAE: 10 replicates. Contrast with figure 2.

The summation of all errors is named the Global Estimation Error (GEE) and this is estimated directly as the variance of the uppermost level (replicated primary samplings).

By carefully performing the multi-stage replication experiment as laid out in figure 2, either following the maximum or the minimum layout, one can successively subtract the lower error estimates from the higher, in a bottom-up procedure. The errors may be tested statistically by an F-test, telling whether a particular level of variance is significantly different from the preceding levels or from zero. It will be appreciated how this empirical scheme yields the absolutely most reliable sampling error estimates, because *all* the empirical sampling -, sample preparation - as well as all the analytical procedures are involved. They all contribute to the aggregate errors precisely because of the bottom-up configuration of "ordinary", only replicated samplings delineated. Since there are only empirical errors involved, and since their reliability will be in direct proportion to (square root of) the number of replicates decided upon, this empirical heterogeneity characterisation procedure is greatly popular by all practical samplers: it is easy to understand and is extremely simple to implement. Be aware though: it does not pay to be a miser when settling on the number of replication to be fixed – on the contrary!

Another way of getting information on the lot heterogeneity would be using the "Gy's formula", which derives from the theoretical apparatus of TOS. This simplified formula is designed to give a reliable estimate of the order of magnitude of the *factor of invariant heterogeneity* (which is a material's characteristic irrespective of the total mass present), also called the Heterogeneity Invariant (IH_L), or the variation of the Fundamental Sampling Error (FSE), the error dependent on the material itself that can never be eliminated. Application of this formula requires knowledge of the essential physical properties of the material to be sampled, which is sometimes easy at hand, while at other times not so easily available. Appendix 1 describes the use of the Gy formula in further detail. The simplest expression of "Gy's formula" reads:

$$\sigma^2(\mathrm{FSE}) = \frac{C \cdot d^3}{M_S} \tag{1}$$

where:

C is the sampling constant (actually consisting of four sub-constants, App. 1)
d is the top particle size (upper 95 % average grain size diameter)
M_S is the sample mass

This central formula is rich with information on how to do practical sampling. Once you know three of the four factors involved one can of course very easily calculate the remaining one (more on this below).

Homogenization, Mixing and Blending

In order to minimize contributions to the sampling error by segregation and grouping, we should *always* homogenize the material to be sampled prior to increment extractions. The only allowable exception to this rule is the primary sampling, where this is not always possible for purely practical reasons – but then *composite sampling* comes to the fore with the greatest emphasis, see further below. Sampling effects by segregation and grouping of the lot material is described by TOS as contributing to the Grouping and Segregation Error (GSE). GSE, along with the fundamental sampling error (FSE), make up the "Correct Sampling Errors" (CSE) which can never be fully eliminated, only minimized, by following the guidelines prescribed by TOS. GSE has two contributing parts, as the name suggest, one related to segregation and the other to "grouping". This latter term may not be obvious to the reader, but it simply means contributions to the error stemming from the fact that neighbouring particles often bear *some* compositional resemblance to a higher degree than what concerns particles farther away. Thus simultaneous extraction of more than one particle (a necessary effect of using a sampling tool usually with a volume pertaining to many particles) means more of rather similar particles in the increment. After all has been done with regard to homogenisation, this contribution from grouping can only be further minimized by reducing the size of the extracted increments. In fact it will be totally eliminated if fragments are extracted one at a time, independently, but this is of course very often (completely) impossible in practice.

Heterogeneity contributions from segregation (most often vertical segregation, due to the effect of gravity) can be effectively minimized by homogenization.

Composite Sampling

This sampling unit operation is very closely related to mixing (or homogenization), since it also works on minimizing the contribution to the overall error from GSE. As indicated by the delineations in the above section, it will be appreciated how combined extraction (composite sampling) of many small(er) increments, which are subsequently combined to add up to form the final sample, instead of e.g. extracting just one large increment with the same total mass, M_S, will minimize the contribution from grouping and segregation. This is always the case, and one can also intuitively understand the value of extracting material from many areas/sub-volumes in the lot instead of just one. It is recommended to perform composite sampling whenever at all possible. It is particularly noteworthy that even if this will result in "too-large" primary sample masses, this is no real excuse for not complying with this SUO, as any "over-mass" can if fact always be easily reduced to appropriate size, using the very powerful subsequent SUO: representative mass reduction.

Representative Mass Reduction

When sub-sampling any sample (i.e. reduce the mass of primary, secondary etc. samples), it is of great importance that this procedure is carried out in strict accord with TOS principles; this is in fact the guarantee that SUO Nos 3 & 4 can be effective.

In 2002, a large experimental survey of 17 current devices and methods for reduction of sample mass was carried out (2 kg of a tri-component mixture was reduced to 100 or 125 g respectively). A summary of the central results are shown in figure 3 below, where it is clearly observed that the correct choice of device has tremendous effect on the reliability of the final sample mass. The ordinate axis in figure 3 describes this "representativeness" of the method/device - the smaller the numeric values of this attribute, the better the method. Representativeness is in TOS defined so as to contain measures of both accuracy and precision; it is thus a powerful measure to evaluate and compare such methods. The entire mass reduction reliability survey, together with all pertinent TOS interpretations etc. can be found in [6].

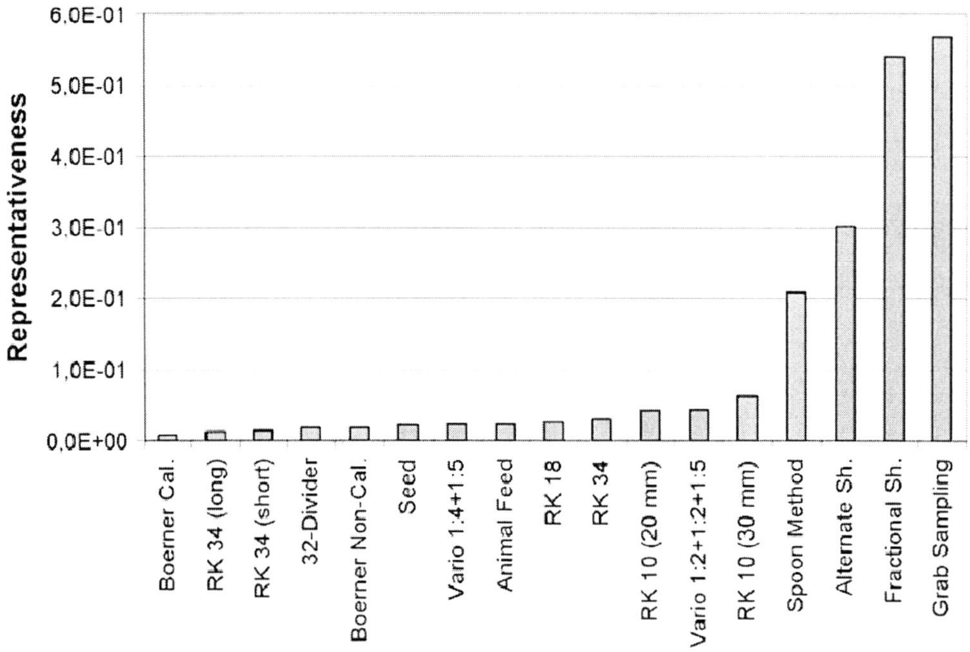

Figure 3. Summary evaluation of 17 widely used methods for mass reduction. A low value on the comparison ordinate axis indicates better representativeness. The representativeness is proportional to the effective total sampling error. Note in particular that the world's most often used sampling technique – grab sampling – is actually the world's poorest performer (further comments in text).

In the figure are summed representativiness contributions from three, very different *types* of constituent elements, representing three different concentration levels: 89.9 %, 10.0 % and 0.1 % by weight, so as to produce information on the mass reduction variations of both abundant elements (90% and 10 % levels) as well as at a typical trace element level (0.1 %, or 1000 ppm).

The methods and devices are fully described elsewhere, *ibid*, and we here only outline that ordinary *grab sampling* (: scooping some material out from an easily accessible part of the lot) is the overall worst performer of all mass reduction techniques, as witnessed by an extremely high value of representativeness. By way of contrast, the principle of *riffle splitting* [1], [2], [6] and its directly equivalents, called *rotational dividers*, all perform very well and are recommended for use in most situations. A word of caution is appropriate here as the choice of device in these categories also has immense influence on the quality of the final results. The following rules-of-thumb are to be obeyed:

- The higher the number of splitting-chutes (or: rotational sectors) the better
- Chute width must be a minimum 3 times the top particle size
- "Correct usage" is vital (amongst other all chutes must be used and fed *evenly*)

Particle Size Reduction

When all other SUO have been invoked etc. but the effective total sampling error is still too high, the only way to minimize the remaining Fundamental Sampling Error (FSE) is to comminute, or to crush, the material particles so as to reduce the average particle size. This is evident from Gy's formula (Eq. 1):

$$\sigma^2(\text{FSE}) = \frac{\text{C} \cdot \text{d}^3}{\text{M}_\text{S}} \tag{2}$$

The particle size is seen to contribute to the error by its third exponent, so it is evident that reducing the particle size greatly reduces the fundamental sampling error. It is also observed that increasing the sample size (M_S) results in a reduced error as well. But since there will always be a maximum physical limit to the size of samples that can be brought back from the field or process, there may be no other option than to reduce the particle size.

The formula can naturally also be rearranged to yield the sample size needed to have a certain error if the particle size is given, or to provide the necessary particle size if sample size and error is fixed etc. (see App. 1).

Lot Dimensionality Transformation

According to TOS it is actually only possible to sample 0-D and 1-D bodies completely correctly. The reader may easily imagine the nature of various 1-D, 2-D and 3-D objects, which correspond to the physical shape of many practical lots. If one dimension is negligible in size compared to the other two, we are essentially dealing with a 2-D object, and if two dimensions are negligible in this sense, we are effectively dealing with a 1-D object. 0-D objects are at first a little harder to understand, as these objects have had this name assigned to them primarily because they are made up of a set of units (virtual increments/samples) without internal correlation. Many 0-D objects in this TOS-meaning, in fact physically corresponds to 3-D objects, but since they display no significant internal spatial autocorrelations they can be viewed and sampled corresponding to the 0-D paradigm. For this lot type, the potential sampling units are not "fixed" in space and in fact they often can be - or have been - mixed prior to being laid up.

Since TOS only deals with sampling of 0-D and 1-D objects one should always consider the possibility of reducing the lot dimensionality prior to sampling. An example will illustrate this.

A huge stockpile of coal (fig. 4) is to be sampled for analytical purposes, e.g. say for sulphur content, prior to being burned in a power plant boiler. It is simply impossible to sample such a large and unmoveable stockpile correctly (according to the fundamental sampling principle) without some extreme manipulation (digging to the centre and the bottom etc.) – and this is clearly not possible in practise. But this does not mean that it is impossible to sample the stockpile correctly and representatively, however!

The solution: When the coal arrives at the power plant (e.g. from a shipload) it is transported via a conveyor belt and laid up in the stockpile by a stacker/reclaiming device (also a conveyor belt). Sampling the coal directly from the conveyor belt, or even better at a transfer point between two conveyor belts (both a 1-D material stream) is by far always the easiest sampling solution.

Figure 4. A large stockpile of coal is for all practical purposes impossible to sample directly in any correct way without enormous, in practise prohibitive, amounts of time and effort. But no stockpile was ever born into this physical form directly however – on the contrary: it was certainly laid up *incrementally*, witness the various conveyor belts present. This simple fact opens up for another approach to completely correct and representative sampling!

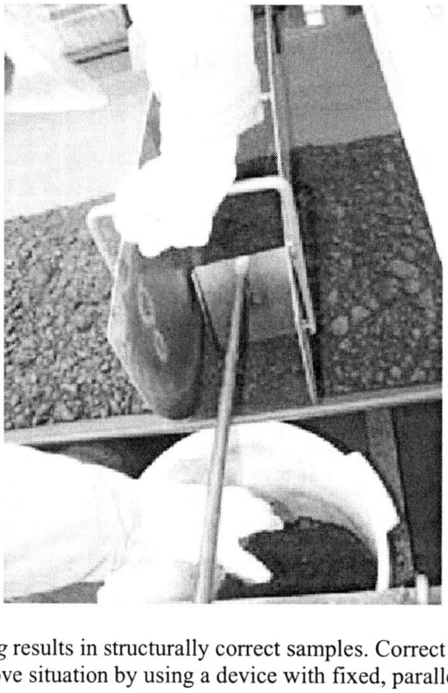

Figure 5. *Stopped belt sampling* results in structurally correct samples. Correct increment delineation and extraction is ensured in the above situation by using a device with fixed, parallel sides and a brush to extract all the fines along the bottom etc.

There are two principal ways to go about this in practise, called the *stopped belt sampling* method (fig. 5) and by use of a so-called automated *cross-stream sampler* (fig. 6). Both will ensure complete, structurally correct sampling, since they both result in correctly delineated and ditto extracted increments with parallel sides etc. The automated approach (the cross-stream sampler) is in fact used at coal power plants in very many countries all around the world today, and is also used extensively in very many other industrial sectors where similar bulk materials handling is on the agenda.

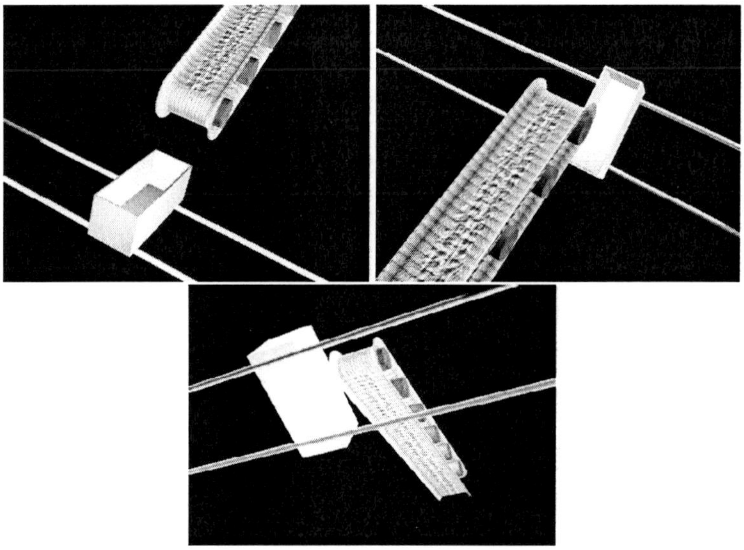

Figure 6. An automated *cross-stream sampler* seen from above-front (left) and above-rear (middle) and below (right). The sample box (white) is moving from side to side – with a *constant* speed of maximum 0.6 m/s (TOS) - and so crosses the stream (fully) twice to collect two increments, combined into one sample, delineated correctly with parallel sides etc. This illustrates an archetype correct sampling device.

Variographic Experiments

The suggested methods for characterization of 0-D lot heterogeneity for e.g. stockpiles etc. are not optimal for 1–D lots however. Instead we use a tool called the *variogram*.

A variogram describes the autocorrelation/variation between individual samples as a function of their "lag" (the distance in time or space between any two pairs of samples) on a 1-D stream, or along a 1-D product lot. The variogram first of all provides an estimate (the so-called "nugget effect") of the Minimum Practical Error (MPE), which includes all the 0-D sampling errors dealt with above, wiz. the Total Analytical Error (TAE), the Fundamental Sampling Error (FSE), the *local* GSE (i.e. the GSE *within* the increment volume) as well as the three incorrect sampling errors (if present). The variogram further provides information on optimal sampling intervals (sampling strategy development/optimization) and the overall variation of the process stream with regard to trends and/or periodic cycles. Below is presented two schematics of a variogram (fig. 7), one showing a "normal" process with autocorrelation between the nearer adjacent increments (or samples) and one for a process showing a clear periodic cyclic behaviour.

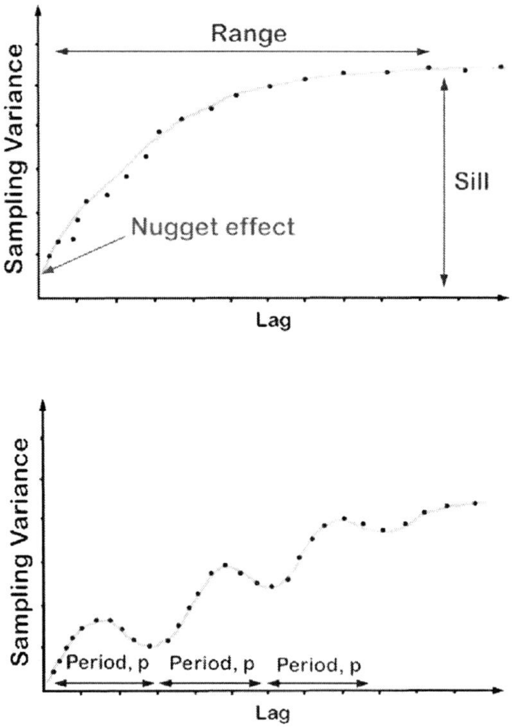

Figure 7. Two schematic variograms. The one on the left is a "normal" variogram with a significant internal autocorrelation for the nearer increments and the one on the right shows a clear periodic cycle. The terms "nugget effect", "range", "sill" and "period" are explained fully in the text.

A variogram can be calculated after performing an experimental investigation of the 1-D heterogeneity (very much like a replication survey), though sometimes two series are necessary, see further below. A number of increments, or samples, (some 40-50 for the highest significance) are extracted equidistantly over the process' running time – or over a period believed to contain possible autocorrelation, trends and cycles. This can be minutes, hours or days, all dependent on the individual process to be characterised. The *heterogeneity contributions* (h_i) from all individual increments - and subsequently the variogram - V(j) - is calculated from:

Individual heterogeneity contributions

$$h_i = \frac{(a_i - a_L)}{a_L} \cdot \frac{M_i}{M_{i,\text{average}}}$$

The variogram

$$V(j) = \frac{1}{2(N_u - j)} \sum_i [h_{i+j} - h_i]^2$$

where

a_i	grade/concentration of increment/sample/unit "i"
a_L	grade/concentration of lot (or best possible estimate/global average)
M_i	mass of increment/sample/unit "i"
$M_{i, average}$	average increment/sample/unit mass
j	"lag" parameter (distance between pairs of increments/samples/units)
N_u	total number of increments/samples/units

It will be appreciated that the variogram expresses much the same ("variability") as the ordinary statistical variance, only the particular V(j) formulation is a distinct function of the (increasing) distances between pairs of measurements. The V(j) vs. lag relationship is very rich with information regarding both the effective 0-D as well as the 1-D variance relationships vs. lag. From the variogram the following interpretations can be made (see fig. 7):

- The "nugget effect" is obtained by *extrapolating* the variogram to intercept the ordinate axis. This value contains MPE (the Minimal Practical Error). In fact the nugget effect describes the difference between two samples taken out at the same time. Since this is not physically possible, extrapolation back to "zero lag" to get the nugget effect is a very clever alternative!
- The "range" is found where the variogram levels off. This can be interpreted as the distance (lag) between samples where there no longer exist any autocorrelation. In some cases the experimental variogram will be found to be essentially flat all over (no autocorrelation at any scale/lags) and there will naturally be no range in these cases.
- The "sill" describes the variation between samples that would be observed if they were extracted at random or with distances larger than the "range".
- The "period" of the cycle is estimated directly as the lag distance between maximums and minimums.

After calculation/interpretation of the overall process variogram it may sometimes also be deemed beneficial to perform a much smaller additional series of extractions to get a better estimate of the details of the "nugget effect" estimation. This is done by extracting samples with the highest possible frequency (seconds or minutes), i.e. for just a very short duration and calculating a new local variogram on this short-term basis. We are in fact only interested in a "blow-up" of the part of the original variogram which is between lags 0 and 1 here. This will provide a much more accurate numerical estimate of MPE and suggest whether a higher sampling frequency (than the one used for the original, overall variogram) is recommendable in general or not.

Discussion and Conclusions

We have briefly illustrated a number of the most central aspects of practical sampling, without going into any deep details of TOS. As a minimum level of understanding, it is critically important that anybody responsible for practical sampling is at least acquainted with the fundamental sampling principle and is familiar with all the seven sampling unit operations. The fundamental requirements are first of all to do everything contained in the sampling process *correctly*, and always to obey FSP. After this we only need invoking the relevant SUO's in the sampling practice as the specific situations call for. The structurally correct sampling will eliminate any sampling bias and also help minimizing the effect of the other sampling errors. This is the only recipe that will result in highly reliable analytical results. There are no shortcuts to this goal, contrary to many myths (e.g. like the presence of a "constant sampling bias" etc.)

The following conclusions can be outlined:

- Only FSP ensures probabilistic sampling.
- Only structurally correct sampling ensures non-biased samples, i.e. elimination of the incorrect sampling errors IDE, IEE and IPE.
- Initial characterization of the lot heterogeneity provides information on optimal sample size/mass, sampling strategy and at which sampling stages most effort should be put into optimizing the sampling procedure.
- Homogenization reduces the error contribution from segregation (part of GSE) and should always be carried out when possible – between all samplings and sub-samplings.
- Composite sampling reduces the error contribution from grouping (part of GSE). A larger number of smaller increments reduce sampling errors very effectively.
- Choice of correct device/method for mass reduction critically minimizes loss of representativeness from this procedure, and can in many cases reduce the time (and hence money) spent on laborious manual methods that does nothing more than decrease the reliability of the final reduced sample masses (with regard to the lot representativeness).
- Comminution reduces the fundamental sampling error (FSE) and is often a worthwhile effort, since a reduction in average particle size influences the sampling variance to the third exponent.
- Lot dimensionality transformation (from 3 or 2-D to 1 or 0-D) brings many sampling problems directly into the practical TOS realm.
- Variographic experiments provide extremely useful information on the variation of production processes and other 1-D lots. It may also disclose hidden cyclic variations and trends over time. If used as a supplement to ordinary X- and R charts in SPC (statistical process control) it will, with proper training, allow the process supervisor to act faster in cases of process changes, since many periodic changes in processes are in fact a product of the process control itself. As an example, the process may often also show periodic behaviour due to an automatic controller somewhere upstream.

INTRODUCTION TO TOS - ONLY

This brief synoptic introduction to the minimum necessary knowledge and experience with TOS is just that – an introduction. The above most certainly is not tantamount to a full course in TOS. The reader is encouraged to delve into the pertinent literature, which is easily available (there are two paradigmatic textbooks at a reasonable university level [1], [2], and at least five additional introductory works [3]-[5], [7] and/or to partake in one of the many expert courses available on the international arena. The literature cited below contains full references to the extensive scientific literature on sampling, especially the overwhelming *oeuvre* from the hand of the originator of TOS himself, Pierre Gy.

APPENDIX 1

"Gy's formula" (Eq 1 and Eq. 2) can provide an order-of-magnitude estimate of the variance of the fundamental sampling error (FSE). The constants c, f, g and β are normally compounded into one "sampling constant" C. The mass of the samples M_S and of the lot M_L are also presumed known or estimated. The full Gy' formula reads like this:

$$\sigma^2(FSE) = cfg\beta d^3 \cdot \left(\frac{1}{M_S} - \frac{1}{M_L} \right) = \frac{C \cdot d^3}{M_S} (if\ M_S << M_L) \tag{3}$$

where M_L is for all practical purposes always much larger than M_S.

The constants c, f, g and β are calculated as follows

- c, the *constitutional parameter*, can be calculated from:

$$c = \frac{\left(1 - \frac{a_L}{\alpha}\right)^2}{\frac{a_L}{\alpha}} \rho_c + \left(1 - \frac{a_L}{\alpha}\right) \rho_m$$

in which

a_L: average concentration of the lot
α: concentration of the critical particles
ρ_c: density of the critical particles
ρ_m: density of gangue material

- f, the *particle shape factor* (dimensionless) describes deviation of particle shape in relation to the ideal of a square. The following standard values of f are defined if nothing else is known:
- square f= 1,
- sphere f= 0.52
- flat disc f =0.1
- g, the *size distribution factor* (dimensionless) describes the particle size distribution:
- If all particles are equal in size then g = 1

- if $1 < d/d_{0.05} < 2$ then $g = 0.75$
- if $2 < d/d_{0.05} < 4$ then $g = 0.5$
- if $d/d_{0.05} > 4$ then $g = 0.25$
- d, the *top particle size*, defined as the square-mesh screen that retains 5% of the material [d must always be expressed in cm].
- β, the *liberation factor* (dimensionless) describes the degree of liberation of the critical component. Liberated particles means $\beta = 1$ and particles incorporating the critical component totally means $\beta = 0$. As a very rough estimate, β can be calculated from the liberation size L:

$$\beta = \sqrt{\frac{L}{d}}$$

It is important to realize that at the time when this simplifying equation was arrived by Pierre Gy (1957), it represented a tremendous leap with regard to the possibilities of even getting a first quantitative handle on the seeming intractable features of heterogeneity – and still does! It will be evident from what follows why this formula only portends to furnish an estimate to within an *order-of-magnitude*, but this is in very many cases more than enough to get a first handle on the relevant sampling issues.

A more detailed *physical meaning* of these summarizing parameters is as follows:

The *constitutional parameter* is an attempt to portray (with a dimensionless, 1-D numerical *index*) the compound density-concentration contracts between the critical analyte material (particles) and the gangue (the "worthless", i.e. non-critical material). This parameter is difficult to envisage since it may vary from unity to several million. Here it really pays off to determine the densities and concentrations as accurately as possible.

The *particle shape factor* simply relates all naturally occurring particle shape deviations from that of a standardized cube; it clearly only embodies a crude average form description (over all particles composed of, or containing, the critical analyte). Again we are forced to deal with a dimensionless, 1-D numerical index.

The *size distribution factor* on the other hand is easy to relate to the material world: This index directly represents the *degree of sorting* of the particle size distribution at hand. Well-calibrated materials have a g-value close to one, while lower and lower values represent progressively irregular, jumbled mixtures.

d, the *top particle size*, is strictly defined: the sieving square-mesh screen-size, which lets 95% of the material through. Therefore this parameter is also often termed d_{95}.

The *liberation factor* purports to describe (with a dimensionless, 1-D numerical *index*) the degree to which the analyte is embedded, "shielded", inside gangue material particles. Only full crushing will liberate totally all the analyte particles in material with a (very) low β.

As should be evident, there will always be some work (field sampling, replication, laboratory determinations etc.) involved in applying Gy's formula, if one is not totally relying only on standard choices for the various parameters (except for d_{95}, which always must be determined experimentally of course). Clearly such "auto-pilot" runs through Gy's formula will not lead to especially detailed and hence reliable estimates, which can only be the result of more careful, individual experimental work for each particular material at hand.

These issues also explain why the direct, empirical replication survey route towards estimates of the effective GEE (or MPE), and its breakdown into PSE, SSE etc, are very popular.

REFERENCES

[1] Gy PM. *Sampling for analytical purposes*. John Wiley and Sons, 1998.

[2] Pitard FF. *Pierre Gy's sampling theory and sampling practice* (2nd ed). CRC Press Ltd., 2001.

[3] Smith PL. *A primer for sampling solids, liquids and gases–Based on the seven sampling errors of Pierre Gy*. ASA SIAM, 2001.

[4] Minkkinen P. *Practical applications of sampling theory*. Submitted for publication in *Chem. Intell. Lab. Syst.*

[5] Petersen L & Esbensen KH. *Representative sampling for reliable data analysis: TOS (Theory of Sampling)*. In prep.

[6] Dahl CK, Petersen L & Esbensen KH. *Representative mass reduction in sampling – A critical survey of techniques and hardware.* Submitted for publication in *Chem. Intell. Lab. Syst.*

[7] Esbensen, KH (Ed.) 50 years of the Theory of Sampling (TOS)–a tribute to Pierre Gy: proceedings from WCSB1 (First World Conference on Sampling and Blending). Chem. Intell. Lab. Syst. Special Issue (in press)

In: Progress in Chemometrics Research
Editor: Alexey L. Pomerantsev, pp. 265-271
ISBN 1-59454-257-0

Chapter 18

PECULIARITIES OF RUBBER-OIL INTERACTION UNDER THE CONDITIONS OF COLD CLIMATE

N. N. Petrova [1]***, A. F. Popova*** [1] ***and O. V. Startsev*** [2]

[1] Institute of Nonmetallic Materials SB RAS, Yakutsk, Russia

[2] Altai State University, Barnaul, Russia

ABSTRACT

It was shown that contact of rubbers with oil causes sequential penetration of hydrocarbons having different composition into the elastomer material (quick diffusion of light hydrocarbons takes place during the first day, longer process of penetration of heavier hydrocarbons) and extraction of a plasticizer, which spoils low-temperature characteristics. Modelling of diffusion processes at room temperature was carried out; coefficients of diffusion were determined and technological approaches, which allow reducing wash-out rate of plasticizers were suggested.

Key words: rubber, climatic stability, plasticizer, frost resistance, diffusion processes.

INTRODUCTION

Elastomers applied as seals, consolidators and collars under the conditions of the North must meet the requirements of frost resistance, oil and benzene stability. In a sense these requirements are alternative, as the combination of low glass-transition temperature, which determines operability of the polymer as a frost resistance material are, for the most well-known polymers, in even inconsistency with their ability to swell in organic solvents and oils. That is why aiming at compromise combination of the said properties, a significant amount of plasticizers – low-molecular weight substances, which reduce intermolecular interaction in the polymer, raise mobility of all structural elements of the system at low temperature, and, as a consequence, improve its frost resistance, is injected into the material. However, interaction of rubbers with oil media (fuel, lubricants) causes diffusion of a plasticizer from the elastomer material into the contacting medium. Extremely low (to –55°C) temperatures in winter, sharp (to 30°C) alternations of temperature passing through zero in spring and autumn, hot summer, which are characteristic of sharply continental climate of Yakutia, strongly impact diffusion processes and changing of rubber properties as a whole.

As a result of carried out investigations of sealing rubbers [1, 2] it was shown that their interaction with oil media under the conditions of cold climate intensifies all diffusion processes in the system "rubber-oil medium". Temperature regime in-cite (seasonal and daily temperature alternations) raises instability of basic operational properties of rubbers. Generally diffusion processes negatively influence such parameters of rubbers as swelling degree and frost resistance. Hereby, after 2-month contact of rubbers based on butadiene-nitrile caoutchouc with oil at ambient temperatures, their frost resistance reduces by 67-94 percent and is from 0.15 to 0, and at temperatures close to that of glass-transition this reduction of low-temperature properties is irreversible. As is well known [3], to provide safe operation of immovable sealing details in the regions with cold climate, allowable threshold of values of the coefficient of frost resistance at -50°C should be set to 0.2, i.e. studied rubbers after short-term exposition in hydrocarbon medium become unserviceable for extreme climatic conditions.

Practically, the most important task is to determine and forecast climatic stability of the elastomer materials, which is evaluated mostly from changes in operating properties. However, it is impossible to predict climatic stability of rubbers, and, hence, operability of the units only from final results of the material aging. It is necessary to carry out detailed analysis of reversible and irreversible processes of the material aging under the impact of low temperatures and aggressive factors, to investigate the mechanism of these processes and their interaction with specific operational properties of rubbers.

Interaction of rubbers with oil causes a number of diffusion processes. As these processes determine the rubber characteristics and its operability under real conditions, it is necessary to study and model the process of rubber swelling in oil, which is a multi-component system. Taking into account complexity of the processes, their dependence on temperature, the first stage of modeling should be carried out at constant temperature. We have chosen swelling degree as a model index for rubber seals, as it is one of the basic ones. It allows controlling the summary process performed at elastomer contacting with the medium and including working medium diffusion and low-molecular weight ingredients washing out. Besides, changes in the rubber swelling degree influences the seal geometry, such as inside diameter, necessary tolerance, etc, and after reaching some definite limits the seal stops executing its functions, operation liquid starts licking [4].

EXPERIMENTAL

Rubber samples were placed into oil from Talakan deposit at room temperature and then studied were the changes in their weight depending on exposition time during 4 months (before reaching equilibrium state). Swelling degree was computed from the formula:

$$Q= ((m-m_0)/m_0)\ 100\% \qquad (1)$$

where m_0 – sample weight at the initial moment; m – sample weight after swelling.

The experiments showed the presence of the initial stage, which was observed during the first day of exposition of samples in oil, whose presence, more likely, could be explained by quick penetration of light hydrocarbons of oil. Mathematic processing of the results showed that it is this stage that initiates the process of the plasticizer – dibutylphtalate – washing out from the rubber volume. Further, simultaneously with the plasticizer washing out, the material swelling occurs, which is ties to diffusion of heavy fractions of oil into the elastomer. To perform mathematical description of diffusion processes in the system "rubber-Talakan deposit oil" we applied a summary equation, suggested by the scientists from SRI of Ecological monitoring SB RAS, which was obtained on the analogy of description of the processes of moisture transtition in the polymer composite materials [5]:

$$Q = Q_\infty \times \left((1 - S_a) + K \times (1 - S_b)\exp(-D_c \times t)\right) \qquad (2)$$

where Q_∞ is yield swelling degree; (1-S_a) is basic part of diffusion, i.e. penetration of heavy hydrocarbons into the rubber; K is a part of the rubber weight change due to diffusion of light oil fractions and trace of reaction products, K sign reflects prevalence of one of such processes; (1-S_b) is a part describing accumulation of light hydrocarbons during the first day; exp(-$D_c \cdot t$) is a multiplier describing the plasticizer extraction from the rubber matrix

$$S_a = \sum_{n=1}^{4} A_n \times \exp(-\alpha_n^{\,2} F_a),$$

where $\alpha_n = \frac{\pi}{2}(2n-1)$; $A_n = \frac{2}{\alpha_n^{\,2}}$; $F_a = \frac{D_a \times t}{h^2}$.

$$S_b = \sum_{n=1}^{4} B_n \times \exp(-\beta_n^{\,2} F_b),$$

where $\beta_n = \frac{\pi}{2}(2n-1)$; $B_n = \frac{2}{\beta_n^{\,2}}$; $F_b = \frac{D_b \times t}{h^2}$

D_a is coefficient of diffusion of heavy oil hydrocarbons;
D_b is coefficient of diffusion of light oil hydrocarbons;
D_c is coefficient of the plasticizer extraction;
t is time of exposition of samples in oil;
h is sample thickness.

Equation (2) was obtained from solution of equations (4), (6) with initial and boundary conditions (5), (7) in the unidimensional area (3). It determines kinetics of distribution of light and heavy oil fractions in rubber proceeding from the assumption that Fick's law works in this system

$$\Omega = \{x | 0 < x < h\} \tag{3}$$

$$\frac{\partial c_1}{\partial t} = D_1 \frac{\partial^2 c_1}{\partial x^2}; \quad x \in \Omega, \ t > 0 \tag{4}$$

$$\begin{aligned} c_1(x,0) &= 0; \quad x \in \Omega \\ c_1(0,t) = c_1(h,t) &= c_{1\infty}; \quad t > 0 \end{aligned} \tag{5}$$

$$\frac{\partial c_2}{\partial t} = D_2 \frac{\partial^2 c_2}{\partial x^2} - kc_2; \quad x \in \Omega, \ t > 0 \tag{6}$$

$$\begin{aligned} c_2(x,0) &= 0; \quad x \in \Omega \\ c_2(0,t) = c_2(h,t) &= c_{2\infty}; \quad t > 0 \end{aligned} \tag{7}$$

where $c_1(x, t)$ – concentration of heavy oil fractions; $c_2(x, t)$ – concentration of light oil fractions in the rubber sample; k – extraction constant; D_1 and D_2 – coefficients of diffusion of heavy and light fractions, respectively; h – sample thickness.

The search of the equation (2) coefficients was performed by means of "Fitter" software. Studied were the peculiarities of swelling in oil for elastomer materials with different chemical composition: rubbers based on butadiene-nitrile caoutchoucs with different amounts of acryl acid nitriles BNKS-18 and BNKS-40, their combinations (BNKS-18 and BNKS-40), as well as those based on propylene-oxide caoutchouc (SKPO).

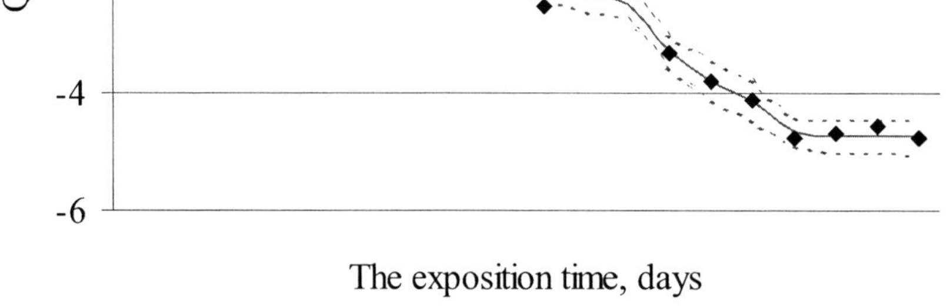

Fig. 1. Kinetics of swelling for rubbers based on BNKS-40 in oil from Talakan deposit at room temperature

Comparison of experimental data and modeling results on swelling in oil for rubbers based on butadiene-nitrile caoutchoucs BNKS-40 (Fig. 1) shows that the suggested model is suitable for description of kinetics of the sample swelling. Computed with "Fitter" values of swelling degree correlate with the experimental data.

The values of calculated values of diffusion coefficients for processes arising while studied rubbers contacted with the hydrocarbon medium are listed in Table 1.

Results of modeling of swelling degree for studied rubbers allow comparing calculated values of diffusion coefficients with their basic operational characteristics. The degree of swelling of an elastomer material in hydrocarbon media is, first of all, determined by presence of polar groups in its composition. For example, rubbers based on BNKS-40 perform higher oil and benzene stability compared to those based on BNKS-18 due to different amounts of nitrile acid acryl in the copolymer (40 and 18 weight percent, respectively). Diffusion of hydrocarbons of oil into the elastomer material based on BNKS-40 will occur slower compared to those based on BNKS-18, which is confirmed by the obtained results: values of diffusion coefficients for light and heavy fractions into rubber based on BNKS-40 are lower (Table 1). Rubbers based on new propylene-oxide caoutchouc perform higher frost resistance (T_c = –74°C), but, due to low polarity degree, their swelling in the hydrocarbon media is stronger compared to rubbers based on butadiene-nitrile rubbers. Rubbers based on SKPO unlike those based on BNKS contacting with oil perform the process of swelling prevailing upon plasticizer washing out (Q_∞=19.67%), which corresponds to obtained higher values of diffusion coefficients (Table 1).

Table 1 Values of parameters of the mathematic model, which describes the process of swelling of rubbers in oil

Rubber Parameter	BNKS-18	BNKS-18+zeolites	BNKS-40	BNKS-18+BNKS-40	SKPO
Q_∞, %	-0.06	4.02	-4.11	2.21	19.67
K	-2.12	-52.73	0.89	-16.2	0.35
D_a, mm^2/day	13.89	0.11	0.73	0.06	19.00
D_b, mm^2/day	12.39	0.03	1.48	0.05	15.30
D_c, mm^2/day	1.11	0.80	0.061	0.97	2.75

Hereby, obtained computational data correspond to theoretical assumptions on structure and properties of rubbers and the experiments carried out at room temperature. There also is a correlation between calculated values of diffusion coefficients and the rubber behavior under the conditions of combined impact of climatic factors and oil. Now one is able to optimize sealing rubber composition at the stage of laboratory development and avoid nature experiments, which will allow saving significant amounts of time, ingredients and efforts. In this case the recipe of the material is developed, its basic properties are investigated at room temperature, modeling is carried out, corrections are made, and the most "successful" materials performing balanced set of properties, are placed to in-cite exposition. Correctness of this method can be conformed by the following examples.

Modification of rubbers based on BNKS-18 by natural adsorbents – zeolites from the Yakut deposits changes greatly operational properties of produced rubbers. Zeolites perform high adsorption capacity, participate in the process of the material vulcanization, which leads to thickening of the surface net of chemical bonds and reduction of swelling degree in the hydrocarbon media. Calculated values of diffusion coefficients D_a, D_b and D_c for modified rubbers are lower compared to unmodified ones (Table 1), so one can make a conclusion that modification of rubbers based on BNKS-18 with zeolites significantly slows down diffusion processes within the system “rubber-oil medium”. The results of in-cite exposition in oil under the conditions of Yakutsk, the rubber modified by natural zeolites performed higher stability of operational parameters after 2 years of exposition compared to initial rubber [1].

Of all the rubbers listed in Table 1 the lowest values of diffusion coefficients were reproduced by rubbers based on the mixture of rubbers BNKS-18 and BNKS-40 in correlation 1:1. Application of polymer mixtures is one of the most perspective ways to produce materials having a new set of properties compared to initial ones. Improvement of properties can be achieved due to development of heterogenous structure of a definite morphology within the material. Application of two rubbers (BNKS-40 and BNKS-18) within a material allows producing a material performing less penetrability for hydrocarbons compared to BNKS-40 rubber based materials. This rubber has already shown the advantage under the conditions of in-cite exposition.

Conclusions

Therefore, studied were the peculiarities of interaction of rubbers of different chemical composition with multi-component hydrocarbon media; using “Fitter” software carried out was the modeling of diffusion processes occurring at contacting of the material and oil; computed were the diffusion for different oil fractures. It was shown that the data obtained in the process of investigation of diffusion processes for the system “rubber-oil” at room temperature correlate with the data obtained while studying rubber properties under complex impact of oil and ambient temperatures. Injection of adsorbents into the rubber composition, application of mixtures of elastomers, one of which has higher oil and benzene stability and prevents quick penetration of hydrocarbon media into the material, allow slowing down diffusion processes and washing out of a plasticizer from the rubber matrix.

References

[1] Petrova N.N., Popova (Fyodorova) A.F. Modification of Butadiene-Nitrile Rubbers with Natural Zeolites as the Means to Improve Stability of Properties under the Conditions of Combined Impact of Hydrocarbon Media and Low Temperatures, in *Polymer Aging at the Cutting Edge,* Eds Zaikov G.E., Bouchachenko A.L., Ivanov V.B., Nova Science Publishers, 2002., 117-124.

[2] Kurlyand S.K., Bukhina M.D. *Frost resistance of Elastomers.*, Moscow, 1989..

[3] Petrova N.N., Popova A.F., Startsev O.B., Filistovich D.V., Krotov A.S. Diffusion of Oil into Rubbers Based on Butadiene-Nitrile Rubbers // *Proceedings of 2nd Int. Conf. “Experimental methods in Physica of Structurally Ununiform Condenced Media”.* – Barnaul., 2001,. 251-256. *Seals and Sealing Technique. Reference Book* / Ed. A.I.Golubev, L.A.Kondakov, 1986. – 483 p.

[4] Startsev O.V., Krotov A.S., Kuznetsov A.A., Anihovskaya L.I., Senatorova O.G. Experimental study and modeling of moisture diffusion in composite materials / Book of abstracts of *12th International Conference on Mechanics of Composite Materials*, June 9-13, 2002, Riga, Latvia. – 2002, 192-193.

In: Progress in Chemometrics Research
Editor: Alexey L. Pomerantsev, pp. 273-282
ISBN 1-59454-257-0

Chapter 19

NIR-SPECTROSCOPY: CHEMOMETRIC ANALYSIS OF BENZINE FRACTIONS AND OILS

R. Z. Safieva*[1], *R. Z. Syunyaev*[1], *S. Purevsuren*[1], *O. G. Chulyukov*[2] *and S. S. Demygin [2]
[1]Gubkin State Oil and Gas University, Moscow, Russia[2], "Lumex-Centrum"Ltd., Moscow, Russia

ABSTRACT

Chemometric analysis of benzine fractions and oils is made by the PLS-method using the laboratory reference values and the NIR spectral data. Benzine distillate fractions (50 samples) are distillated under the atmospheric pressure. These samples were kindly provided by the oil refinery (Orsk, Russia). Also, 25 samples of origin oils are investigated. Spectral data are obtained and processed by analyzer "INFRALUM FT-10" designed by Russian company "LUMEX" with original software. Calibration models with high prediction ability for the properties of benzine fractions (density, fractional composition: initial and final boiling point, the temperature of distillation 10%, 50%, 90%; octane number; contents of resins) as well as for properties of oil samples (density, fractional composition, initial boiling point, output to 200C, 300C, 350C; viscosity at 20C; 30C; 50C; temperature of freezing) are created. The standard error of cross validation (SECV) is calculated Cross validation shows a rather high precision of the calibration models. The predictive ability of the suggested models obtained for oils could give a basis for development of express-method that evaluate density and output of fraction up to 300C. In many cases the repeatability of spectral analysis was higher than repeatability of reference methods.

Keywords: express method, near infrared spectroscopy, benzine fractions, oil, chemometrics, calibration, Standard error of cross validation (SECV)

Introduction

At production of commodity benzines the mixture of various benzine fractions is applied. Therefore, it is necessary to control the ratio of components and their quality characteristics by means of periodic standard tests. The knowledge of quality parameters of benzines (octane number, contents of sulfur, density, fractional composition, contents of aromatic hydrocarbons, etc.) has a special meaning for quality benzine production.

At the stations of mixture it is often occurred production of nonstandard benzine, which requires correction. This happens because of high time consumption of routine analysis. In particular, standard evaluation of octane number requires a long-term motor installation test. As the result, the standard laboratory methods used in oil refining allow to determine the octane number not earlier than in 1-2 hours, and in real operating conditions of factory laboratories, as a rule, in 2-3 hours.

Moreover, periodic laboratory control of oil quality (density, fractional composition, viscosity, *etc.*) is necessary for operative management of technological installations. This is performed by various standard chemical methods. For example, prompt information about quality of oil allows making good and fast decisions for management of the technological parameters during the process of atmospheric oil distillation.

Conventional tests for quality control of benzines and oils are rather long and expensive. On the other hand, the parameters of quality of benzines and oils can be obtained with by chemometric methods using NIR spectra. There are numerous publications about express-methods based on this chemometric approach for the oil products quality control [1-3].

The purpose of this work is application of chemometric method for the express-analysis of some parameters, which characterize quality of distillated benzine fractions and oils. This method utilizes correlation, which exists between reference and spectral data.

Experimental

Experimental Equipment

Benzine fractions and oils spectrum are received by the standard NIR-analyzer "INFRALUM FT-10" with manual sampler. Measurements are performed in the range of 8000 - 14000 cm^{-1}. Quartz cuvettes with optical length of 10 cm are used for benzine fractions and of 30 and 5 mm for oil samples. The analyzer management and data processing is carried out with the help of software developed by "LUMEX" company.

Calibration Models

We applied linear regression analysis to obtain the models, which calibrate spectral and reference values. For this purpose we used the PLS method that is implemented in accordance with ASTM E1655-99. It is assumed that a quality parameter can be presented by equation

$$y_i = a_o + \sum_{j=1}^{k} a_j \cdot x_{ij}$$

where i is a sample number, y_i is a predicted value for the i-th sample; a_j are regression parameters, x_{ij} are transmittance values at wavelength λ_j.

Chemometric approach assumes a standard procedure for data processing. The whole data set is divided into calibration and test sets. Calibration model is constructed using calibration set; and standard error of calibration (SEC) is calculated. After that we predict the response values for the test set. However, in a case of short data set, it is expedient to use cross validation method. The procedure consists of serial ejection of each calibration sample, recalculation of calibration model, and prediction of response for the thrown out sample, and calculation of SECV. In accordance with ASTM E1655-99, it is calculated by a formula:

$$SECV = \sqrt{\frac{\sum_{i=1}^{n} \left(y_i^{lab} - y_i\right)}{d-1}}$$

where d is the number of samples, y_i^{lab} is the result of chemical analysis of some characteristic (reference value) for sample i; and y_i is the corresponding value predicted by the calibration model.

Results and Discussion

Benzine Fractions

Benzine distillate fractions (50 samples) are received at redistillation of wide benzine fraction and are intended as a raw material for reforming. These samples were supplied by Orsk Oil Refinery (Russia). The standard routine methods are used for determination of the following parameters: density, octane number (motor method), fractional composition: the temperature of initial and final boiling points, temperature of distillation of 10%, 50%, 90%, and the contents of resins.

Calibration data set consists of 48 samples; 2 samples are used in the test set to check the calibration model. The examples of spectra of benzine fractions are shown in Fig. 1.

Using correlation between the spectral data and quality characteristics of benzine we construct calibration mode. Fig. 2 demonstrates correlation between predicted and measured values of various quality characteristics of benzine fractions.

The results of calibration models validation for the unknown (test) samples are presented in Tabl. 1. It shows that parameters of quality (density, octane number, fractional composition, contents of resins) of benzine fractions can be predicted using absorption spectra in the NIR area.

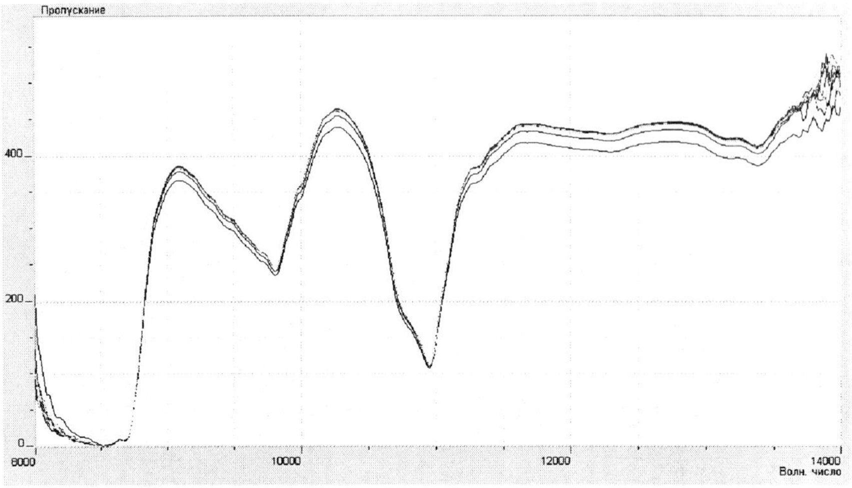

Fig. 1.Examples of distillate benzine fraction spectra

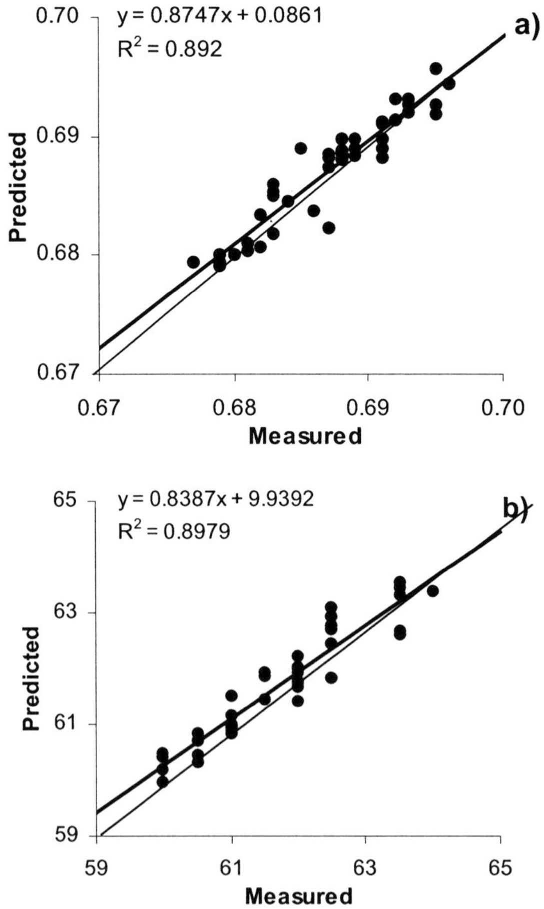

Fig. 2. Predicted vs. measured values of benzine quality characteristics.a) density, b) octane number (Continued on next page)

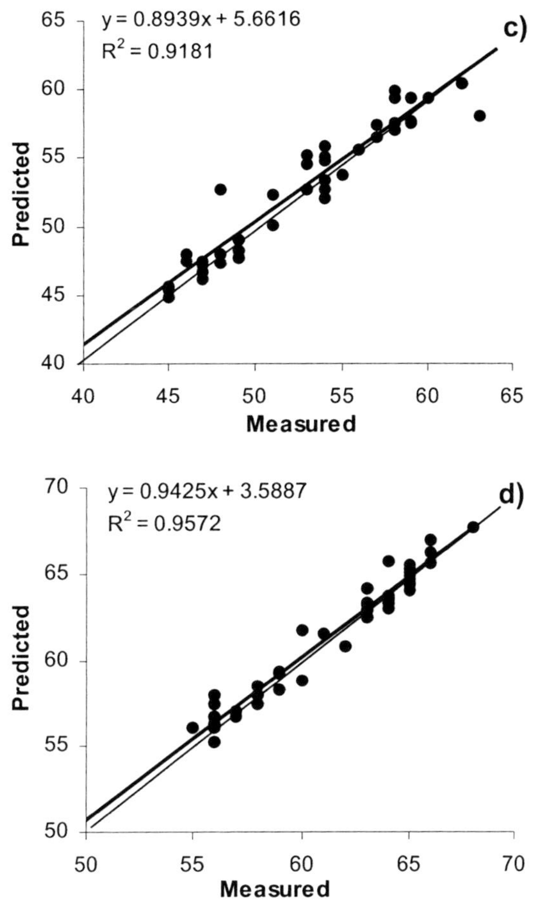

Fig. 2. Predicted vs. measured values of benzine quality characteristics. c) initial boiling temperature, d) temperature of 10% of distillation

Table 1 Results of the test validation for unknown samples

Quality characteristic	Sample 49		Sample 50	
	Measured	Predicted	Measured	Predicted
Density, g/cm^3 at 20°C	0,692	0,691	0,694	0,691
Initial boiling temperature, °C	59	58,64	62	59,9
Temperature of 10 % of boiling, °C	67	66,706	68	67,6
Temperature of 50 % of boiling, °C	75	74,13	75	75,6
Temperature of 90 % of boiling, °C	92	93	93	93,3
End of boiling temperature of, °C	105	108,4	105	107,8
Octane number	60	59,79	60	60,6
Contents of resins, mg/100cm^3	2	2,1	1,6	2,2

Characteristic of calibration and results of cross-validation are given in Table 2.

Table 2 Characteristic calibration models for definition of quality characteristics of benzine fractions (48 samples)

Quality characteristic	State Standard	Characteristic range	Precision of measurement, %	Wavelengths range (cm^{-1})	Precision of measurement, %.	Correla-tion	SEC, %	SECV, %
Density, g/cm^3 at 20°C	3900-47	0,679-0,784	0,7	11412-13000	0,02	0,8920	0,0021	0,0024
Initial boiling temperature, °C	2177-99	42-63	2,5	8764-9596	1,10	0,9101	1,93	2,058
Temperature of 10 % of distillation, °C		55-68	2,0	10918-11462	0,4	0,9572	1,23	1,26
Temperature of 50 % of distillation, °C		67-78	2,0	10994-12598	0,4	0,9076	1,04	1,14
Temperature of 90 % of distillation, °C		87-102	2,3	10900-13232	0,7	0,8549	1,65	1,99
Final boiling temperature, °C		104-123	3,0	10802-11462	1,7	0,8531	2,1	2,285
Octane number	511	58,5-64	0,3	9164-10360	0,02	0,8979	0,51	0,596
Contents of resins, $mg/100cm^3$	8489-85	1,6-2,8	0,4	12610-13232	0,2	0,8767	0,19	3,224

From Table 2 it can be concluded, that calibration models are characterized by the high predictive ability estimated by Standard error of cross validation (SECV).

Oil Samples

Typical NIR-spectra of dehydrated oil samples are shown in Fig.3.

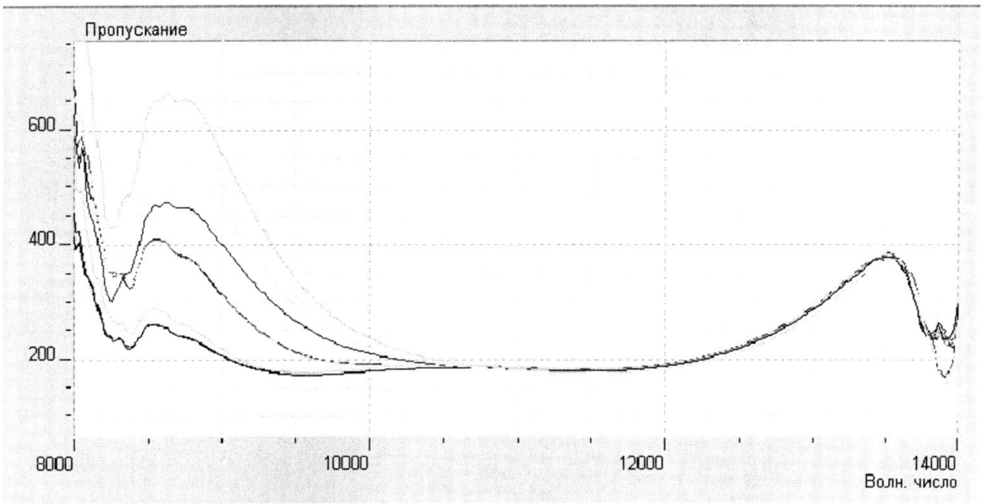

Fig.3. Spectra of some oil samples in the NIR area

Oil samples were gathered from the different oilfields (Russia, Kazakhstan, Mongolia, and Nigeria). The content of water in crude oil samples is changed from 0,5% to 60 % mass, so these water-oil emulsions were preliminary dehydrated in laboratory. Preparation of crude oils samples includes their demulsification, with conditions (temperature, kind and amount of the special demulsifier) that are specially chosen for each sample of crude oil.

Some quality parameters (density, fractional composition: an output of fractions with initial boiling point up to 200C, 300C, 350C; viscosity at 20C; 30C; 50C; temperature of freezing) of oil samples are measured by the standard procedures.

Calibration models are constructed using a set of 25 demulsified oils. In Fig.4 the predicted values are plotted against measured reference values. All results are collected in Table 3.

For some oil samples there are no data on kinematic viscosity at 20C, 30C, and 50C: output of distillated fractions up to 350C and temperature of freezing.

It can be seen that the investigated parameters of oils quality might be predicted using absorption spectra in the NIR region. Therefore, it is possible to use calibration models for prediction if more samples will be collected.

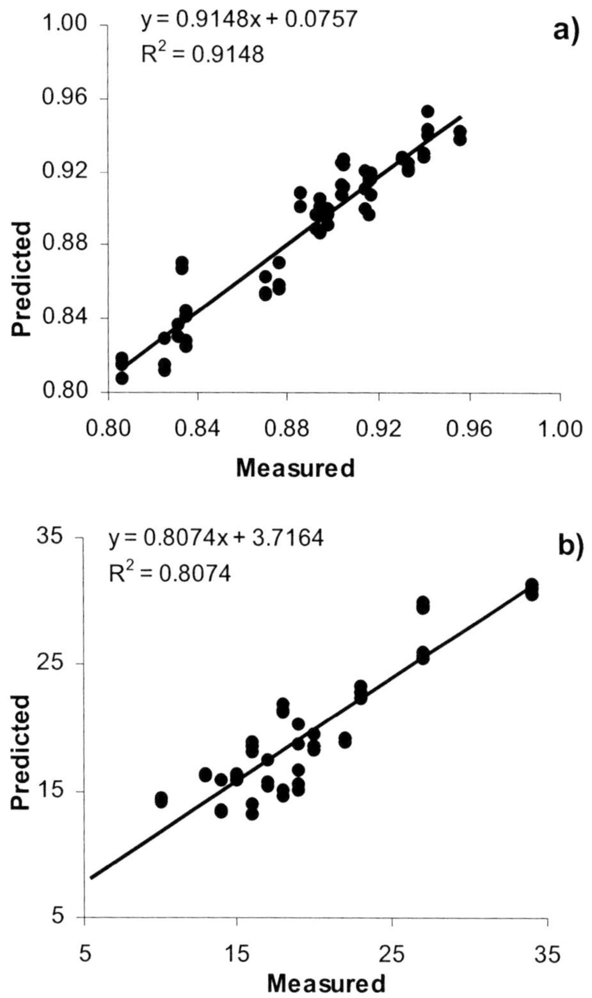

Fig.4 Predicted vs. measured values of oil quality characteristics.a)density, b) output of distillated fractions (up to 200^{0}C)

CONCLUSIONS

Models, which predict quality parameters of direct benzines and oils using spectral data, have been developed. It is shown, that basic quality parameters (density, fractional composition, etc.) of benzine distillate fractions and oils can be predicted with high accuracy using created calibrations.

It is worth mentioning that NIR-spectroscopy method allows to define various quality parameters of benzines and oils in on-line regimes as the time of measurement is just about 3 minutes.

Table 3. . Characteristics of calibration models for definition of quality characteristics of oil samples

Quality characteristic (number of samples)	State Standard	Characteristic range	Precision of measurement, %	Wavelengths range (cm^{-1})	Precision of measurement, %.	Correla -tion	SEC, %	SECV, %
Density at 20 ^{0}C, g/cm^3 (25)	3900-47	0,806-0,956	0,8	8000-13996	0,16	0,915	0,013	0,0017
Kinematic viscosity at 20 ^{0}C, mm^2/s (9)	33-82	3,74-612	0,3	10668-12280	0,006	0,996	6,25	206,2
Kinematic viscosity at 30 ^{0}C, mm^2/s (13)		2,81-228	0,2	10896-12512	0,006	0,986	11,47	85,1
Kinematic viscosity at 50 ^{0}C, mm^2/s (14)		2-17	0,1	8768-10280	0,006	0,898	7,7	29,63
Output of distillated fractions up to 200 ^{0}C, % vol. (24)	2177-99	10-34	2,0	8196-13730	0,35	0,807	2,65	3,67
Output of distillated fractions up to 300 ^{0}C, % vol. (25)		19-66	2,0	8016-13092	0,35	0,857	3,11	4,25
Output of distillated fractions up to 350 ^{0}C, % vol. (14)		29-75	2,0	8768-11024	0,35	0,717	7,9	11,4
Temperature of freezing, ^{0}C (14)	20287-91	-45÷+20	0,1	8796-10662	0,12	0,701	9,16	12,56

References

[1] Bohacs Gy., Ovady Z. *Near Infrared Spectroscopy*, 1998, 6, 341-348
[2] Kjell Arne Ulvund. *Hydrocarbon Technology Int.*, 1998, 19.
[3] Chung H., Vin-Sik Ku. *Applied spectroscopy*, 54, (2000): 239-245.

Part IV. Biology

In: Progress in Chemometrics Research
Editor: Alexey L. Pomerantsev, pp. 285-297
ISBN 1-59454-257-0

Chapter 20

ECOLOGY AND TOXICOLOGY MONITORING PESTICIDES IN AGROSYSTEM: STRUCTURING AND ANALYSIS

Galina Ye. Larina
Institute of Phytopathology, Moscow, Russia, galina_larina@vniif.rosmail.com

ABSTRACT

The monitoring (ecology and toxicology) of pesticides presence in a soil results in a set of heterogeneous information regarding: the number and the kind of weed plant in sowing of different cultures, the term of herbicide presence in agrosystem components, the ecological loading, etc. For pesticides impact assessment of agrosystem offered combined evaluation of scale that include ecological-agrochemical parameters (physical-chemical properties, solubility in water and organic solvents, DT_{50} and etc.) and toxicological-hygienic parameters (LD_{50}, ADI and etc.).

The scheme of the structuring and analysis of data for several pesticides was designed. The initial data are entered in working area of the tables formed in the program Excel and STATISTICA. Then they are subjected to mathematical processing. The basic methods are dispersion and correlation analysis. Then information is imported to a database "Herbicide 99" (been created on the basis of MS ACCESS 97 program). Now, we keep in database information about 304 active ingredients, which are herbicides (162 preparations), fungicides (90 preparations) and insecticides (87 preparations). Database allows (with the help of the integrated tools) to spend additional mathematical processing of data, to query the searches and to receive the reports about necessary information. For example, it is possible to create the report about average values of productivity, increase of a crop and ecological loading on different cultures with respect to the place of application herbicides. This method helps the researcher to apply herbicides in practice, to perform the operative management of agrosystem regarding the greatest degree of safety for an environment. The behaviour of herbicides (metsulfuron-methyl, prosulfuron, rimsulfuron, sulfometuron-methyl, chlorsulfuron, imazapyr, imazethapyr and imazamox) in agrosystem was investigated applying database data as well as the own experimental data (the green house and the laboratory experiments).

Keywords: pesticide, herbicide, database, impact assessment, risk for agrosystem, sulfonylurea, imidazolinone

INTRODUCTION

The application of chemicals for plants protection in agriculture means operative management of agrosystem's components with respect to the greatest possible safe for environment [1-2]. The result depends on a large set of various parameters, which are: the number and the structure of weeds in crops of different cultures, ecological loading of pesticides, persistence of the xenobiotic residuals in the soil, in the plant, and in the water, and other. It is important to structure the data correctly. Monitoring of pesticides is performed with the help of information systems, which are focused on support of decision-making. These systems should satisfy with the proper requirements, which have to be taken into account during their designing.

The accumulation of this knowledge is based on the simplified methods of estimation of the pesticides pressure on environment. For example, researchers use mark scales (gradation "well - poorly"), indicators of steady development, and etc [3].

The purpose of our research is ordering and analysis of monitored information, for example, application of herbicides preparations in agriculture.

EXPERIMENTAL

Materials

In the research we use data [4] about activity of the sample pesticides, which are: the ecological-agrochemical parameters (EAP - physical-chemical properties, solubility in water and organic solvents, DT_{50}, and etc.) and the toxicological-hygienic parameters (THP - LD_{50}, ADI and etc.). In a number of experiments we studied the efficiency and the behaviour of herbicides preparations (active ingredient – a.i.), which are as follows.

A class of derivative of sulfonylurea-metsulfuron-methyl (Laren®) at concentration 9g of a.i. ha^{-1}, prosulfuron (Pik®) at 19 g a.i. ha^{-1}, rimsulfuron (Titus®) at 12 g a.i. ha^{-1}, sulfometuron-methyl (Ankor-85®) at 15 g a.i. ha^{-1}, chlorsulfuron (Kortes®) at 5 g a.i. ha^{-1}.

A class of derivative of imidazolinone-imazapyr (Arsenal®) at concentration 250g of a.i. ha^{-1}, imazethapyr (Pivot®) at 100 g a.i. ha^{-1} and imazamox (Pulsar®) at 40 g a.i. ha^{-1}.

Methods

For the consecutive analysis of data regarding the behaviour of pesticides we use the information modeling tools (program EXCEL and STATISTICA) and database utility ACCESS 97 (table 1).

Table 1 Tasks and software applied

TASKS	FORMS	SOFTWARE	THE TARGET DATA
Reception of the primary information	The working protocols (acts)	DOS, WORD, LEXICON, EXCELL	The situation a field - facilities (circuit of experience, level of a contamination, productivity of culture, parameters of behaviour pesticides)
Ordering and archiving of the data	Spreadsheets of automatic account of parameters	WORD, EXCELL, ACCESS	The situation a region-field - (circuit of experience, level of a contamination, productivity of culture, parameters of behaviour pesticides)
Storage and management of the data (analysis, statistics)	Spreadsheets of automatic account of indexes and integrated factors	ACCESS, STATISTICA, *Deductor (Raw-Data Analyzer)	The situation a land-region- facilities (efficiency and behaviour pesticides)
Modeling of systems, studding of features of pesticides behaviour, forecast	Systems of the analysis of files, trees of the decisions, self-learning systems	**STRATUM 2000, Deductor (SOMap Analyzer, Tree Analyzer, Neural Analyzer)	Search of the latent laws, practical recommendations

* http://www.basegroup.ru (program Deductor company BaseGroup Labs.)
** http://www.pstu.ac.ru (program Stratum Computer 1.3)

Method of Analytical Detection of the Herbicides Residuals in Soil

Analytical indication of the herbicides residuals (chlorsulfuron, metsulfuron-methyl, rim-sulfuron, sulfometuron-methyl, imazapyr, imazethapyr, and etc.) in the soil samples is performed by the following adapted methods: high-pressure liquid or gas chromatography (accordingly HPLC and GC).

The chemical methods of analysis of the herbicides residuals include the following stages: extraction by the organic solvents determined instance from matrix (test-soil); concentration of an extract; branch of analyzed connection from accompanying impurity by a method second- ex-traction; additional clearing of extracts; identification and quantitative definition of analyzed substance in matrix.

All analytical methods are described in detail in Refs [5-7]. Metrological characteristics are given in Table 2.

Table 2 Analytical and bioindication (*) methods of detection of herbicide's residuals in soil

Active ingredient	Method	Limit of detection, mg/kg	Standard error, % (P=0,95)	Standard deviation, %
metsulfuron-methyl	HPLC	0,001	88 ± 5,9 %	5,3
prosulfuron*)	fodder beet (*Beta sp.*)	0,0008	70 ± 14,0	8,0
rimsulfuron	GC	0,001	89 ± 5,0	4,3
sulfometuron-methyl	GC	0,01	75 ± 7,3	5,5
chlorsulfuron	HPLC	0,002	86 ± 5,0	4,1
imazapyr	HPLC	0,2	81 ± 14,7	5,8
imazethapyr	HPLC	0,2	85±13,4	5,4
imazamox	HPLC	0,1	89±4,4	1,8

Bioindication with the Help of the Green Tests-Plants (Total Phytotoxicology of Soil Sample)

The experiments was carried out in the chambers of artificial climate (LAC) [7], under the following conditions: illumination 20 000 Lux within 16 hours, term without illumination is 8 hours; temperature is 25^0C (light periods) and 20C (dark periods); humidity of air is 70 %; daily moistened is 60 % of the total moisture capacity (TMC).

Soddy-podzolic soil (pHsol 5,4, humus 2,4 %) were taken, dried to the air-dry condition and sieved through a bolter with 5 mm mesh. Then, the soil sample (with total dry weight 3,0 kg) was sprayed with water solution of herbicide (at the rate of a working solution 200 l/ha) at different rates descreasing the biomass of tests-plants aerial by 10-90 %. The soil samples were kept in the paper cups by volume of 600 $см^3$ (without a drainage). A top soil layer (1-2 cm) was removed before seeding, the soil remaining in the pot was slightly compacted; seeds were put on a soil surface, covered with removed soil and the soil was compacted again. Experiments are repeated in five replicates. At the same time, a check experiment was carried out with pure soil that was not treated with herbicides. Soil moisture was maintained at the level of 60 % of the TMC due to a daily irrigation with tap water.

The herbicide activity is estimated after 28 days of soil processing and the change of biomass of test-plants is calculated:

$$I_o = 100 \cdot \left(1 - \frac{m_o}{m_к}\right) \tag{1}$$

where I_o is an indicator of crop biomass growth inhibition, % from the control; m_o and $m_к$ are weights of the crop green part in grams in the experimental and the control versions, accordingly.

The dependence of the total soil phytotoxicity of the sulfonylurea and the imidazolinones preparations versus applied doze is described by the S-shaped rational function (Fig. 1). To compare the activity of various herbicides we used parameter ED_{50} (effective doze of pesticide, what causing the 50 % decrease of biomass of test-plants in comparison with the control test). Parameter ED_{50} is calculated by the probit-analysis method at *P-level* of 5 % [8].

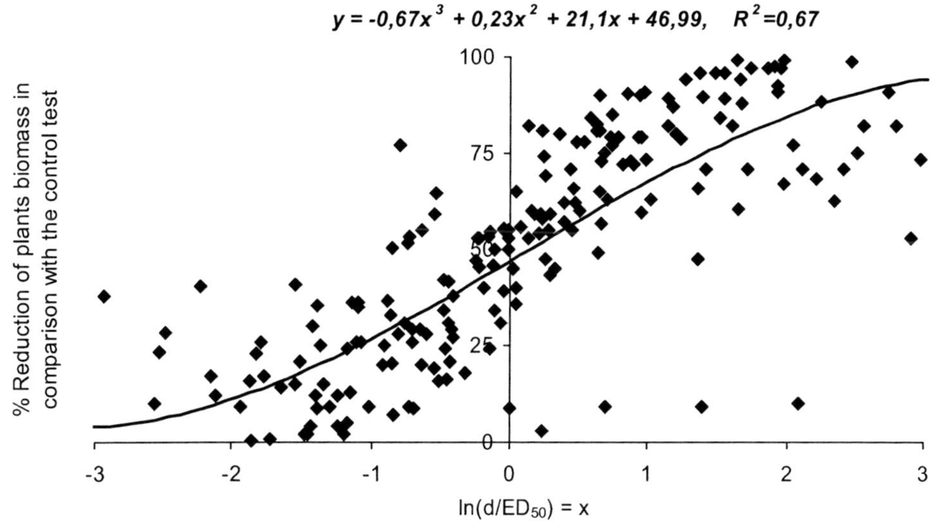

Fig. 1. Dependence of the tests -plants biomass reduction versus parameter $\ln(d/ED_{50})$

Experiments

The first aim of the work is to present the specialized database for pesticides used in an agriculture and choice of characteristic parameters. Our second aim is to investigate the herbicides behaviour in soil using the information from the database and our own experimental data.

Explanatory analysis of the ecological and toxicological pesticides monitoring data is performed using EXCEL workbooks and STATISTICA program. Further information is imported to a database "Herbicide 99" (been created on the basis of MS ACCESS 97 program).

The following characteristics are used for pesticide impact assessment on objects of agrosystem (table 3):- *Ip* is a complex index that presents the pesticide impact assessment of agrosystem. It is calculated as the sum of dimensionless values from the scale of risk classes –

$Ip = EAP + THP$;

$EL_{\text{soil-plant}}$ is a factor of ecological loading for system a soil-plant:

$$EL_{soil-plant} = \frac{DT_{50}(soil) \cdot D_{prep(\text{мах})} \cdot \frac{Da.i.}{100}}{PDK_{ph}}, \tag{2}$$

Table 3. Combined scale of pesticides impact assessment to agrosystem

Parameter	Class of risk	Characteristic of class
Ecological - agrochemical (EAP)		
Solubility in water (S water, mg/l)	1	<10
	2	10 – 1000
	3	>1000
Volatile (P_v, mPa)	1	$<1{,}33\cdot10^{-5}$
	2	$1{,}33\cdot10^{-5} - 1{,}33\cdot10^{-2}$
	3	$>1{,}33\cdot10^{-2} - 1{,}33$
Persistent in soil (DT_{50}), days	1	<15
	2	15–60
	3	60–180
	4	180–360
	5	>360
Logarithmic factor of distribution coefficient between n-octanol and waterwater (log Kow)	1	>3,8
	2	3,8 – 2,4
	3	2,4 – 1,4
	4	1,4 – 0,08
	5	<0,08
Soil parameter - distribution coefficient between organic substance of soil (recalculation on carbon) and water of soil (K_{oc}, sm^3/g)	1	>4000
	2	500 – 4000
	3	75 – 499
	4	16 –74
	5	<15
Toxic-hygienic (THP)		
Limiting and admitted concentration of substance in soil (PDK_{soil}, mg/kg)	1	>0,1
	2	0,01 – 0,1
	3	0 (no permit) or <0,01
The maximal allowable level of the contents of substance in production (MDU_{min}, mg/kg)	1	>1
	2	1 – 0,1
	3	<0,1 – 0,01
	4	<0,01
	5	0 (no permit)
Dose required to kill 50 % of test animals (in our case for rats) at oral introduction of substance ($LD_{50(min)}$, mg/kg)	1	>1000
	2	>200 – 1000
	3	>50 – 200
	4	<50
Acceptable day intake (ADI, mg/kg weights of a body)	1	<0,01
	2	0,01 – 0,1
	3	>0,1
Effective doze of pesticide, what causing 50 % decrease of growth of a test-plant in comparison with the control (ED_{50}, g/ha)	1	<1
	2	1 – 10
	3	10 – 100
	4	>100

Note:
class 1 - not dangerous,
class 2 - weakly dangerous,
class 3 - average dangerous,
class 4- dangerous,
class 5 - strongly dangerous
where

$D_{prep(max)}$ is the maximum recommended doze of herbicides preparation, g/ha;

$D_{a.i.}$ is the contents of active ingredients in a preparation (in %);

PDK_{ph} is the concentration resulting in 10% decrease of the biomass of test-plant in comparison with control grown on the same soil without herbicide treatment, g/ha (for fodder beet ED_{10} = 0,07-0,5 g/ha of active ingredient).

T_{PDK} is the waiting period or period that is necessary for decreasing of the active ingredient content in soil down to the concentration equal to PDK^*:

$$T_{PDK} = \frac{\ln\left(\frac{PDK^*}{C_{i(soil)}}\right)}{k} \quad (3)$$

where

$Ci_{(soil)}$ is the residual concentration (g/ha) of herbicide in a soil at time T_i after application of a preparation, k is the rate constant of herbicide dilution (day^{-1});

PDK^* is the concentration of active ingredient in soil that is equal to PDK_{soi} , which is the upper level of the ecologically safe contents of herbicide in soil or PDK_{ph} (g/ha).

Index GUS (*groundwater ubiquity score*) is a characteristic of substance mobility that also determines the potential risk of pollution by the herbicides residuals at the deep soil horizons and the earth waters [9].

$$GUS=\log DT_{50}(4-\log K_{oc}) \quad (4)$$

Results and Discussion

Database "HERBICIDE_99" systematizes experimental data and conditions of experiments. It can evaluate parameters of preparation efficiency and helps perform quantitative and graphic analysis. The main features of the database are shown in Table 4.

The behaviour of herbicides in agrosystem is determined by the nature of active ingredients as well as by the external factors (soil quality, climate, etc.). For example, if a substance has good water solubility, it can be actively moved with groundwater to the large distances (conditional "*pedomigration*"). Substances with good solubility in the organic solvents are able to penetrate through the biological membranes (conditional "*biomigration*"), so they can move with the food chains.

Fig. 2 demonstrates distribution of 339 active ingredients (from database "HERBICIDE_99") by the various criteria. Substances are divided in two groups: of 1-3 risk classes and of 4-5 risk classes. A special attention should be paid to the active ingredients in the second group, which have the following main characteristics DT_{50} soil is 16,4 %; LD_{50} rat is 10,3 %; Koc min is 31,7 %; S water is 14,8 %; log Kow is 11,1 % and PDKsoil is 48,4 %.

Database "HERBICIDE_99" contains data about 304 active ingredients, which are herbicides (162 preparations), fungicides (90 preparations), and insecticides (87 preparations). Fig. 3 represents distribution of active ingredients that belong to 4-5 risk classes for various characteristics.

Table 4. Structure of HERBICIDE_99 database

REFERENCE DATA		EXPERIMENTAL DATA
Culture		Technical efficiency
Harmful organisms		Economic efficiency
Pesticide (herbicide and etc.)	**Application**	Parameters of pesticides behaviour (constant of disappearance velocity, application dose, DT_{50} and other)
Soil properties i.e. field characteristics		Associations of weed plants
Climate (temperature, sum of deposits)		Phytotoxic level of soil for plant and microorganisms
Description region - area - facilities		Effect of pesticide action in the remote time
Agroengineering		
	Scientific researches	
	Practical recommendations	

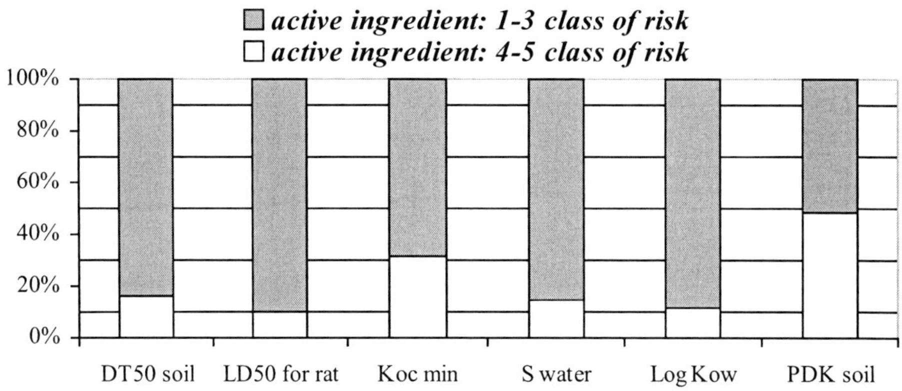

Fig. 2. Distribution of active ingredients by their characteristics

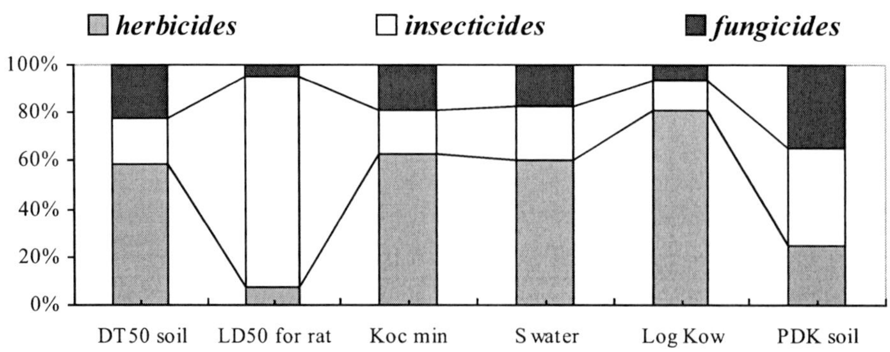

Fig. 3. Distribution of chemicals of 4-5 risk classes by their characteristics

Is was shown that the most stable in soil (DT_{50} > 180 days) are herbicides based on chlorbromuron, acifluorfen, diflufenican, terbacil, sulfentrazone, terbumeton, ethofumesate; fungicides based on triadimenol, penconazole, triticonazole, fenarimol, and insecticides based on pirimicarb and heptachlor.

The processes that determine the behaviour of substance in a soil are interconnected. It is known, that the longer active ingredient keeps the activity, the more probability of residuals migration in agrosystem. The analysis of the pesticides distribution with respect to various characteristics allows to claim the high mobility some 15 % of active ingredients in the assortment. They have S-water > 1000 mg/l and Koc < 74 l/kg. Among them prevail herbicides, which residuals are capable to penetrate through the biological membranes ($-3{,}3 \leq \log K_{ow} \leq 0{,}08$). However, herbicides are not dangerously toxic for the warm-blooded animals, while 38 % of insecticides have characteristic $LD_{50} \leq 100$ mg/kg (i.e. they are highly toxic). This fact is well demonstrated by a toxic-hygienic parameter PDK water, which in most cases specifies the presence of residuals of fungicides and insecticides in water.

Active ingredients of different chemical nature have been classified regarding the EAP parameters (physical - chemical properties, solubility in water and organic solvents, DT_{50}, and etc.) and regarding the THP parameters (LD_{50}, ADI, and et.). The Russian State standard (17.4.1.02-83) "Classifications of chemical means by a degree of their risk" establishes the three groups of environmental risk from herbicides (table 5).

Table 5 Classification of herbicides regarding classes of risk

Class of risk	Parameter (*Ip= EAP+ THP*)	Active ingredient
I(low)	≤19	vernolate, tri-allate, cycloate, imazamethabenz-methyl, chlorotoluron, trifluralin
II(medium)	20 – 24	bentazone, glyphosate, glyphosate-trimesium, glyphosate-ammonium, atrazine, prometryn, propazine, simazine, butylate, molinate, pebulate, dalapon, imazaquin, imazamox, imazapyr, imazethapyr, alachlor, acetochlor, metazachlor, metholachlor, amidosulfuron, bensulfuron-methyl, dichlorprop-P, ioxynil, diuron, metsulfuron-methyl, nicosulfuron, pyrazosulfuron-ethyl, rimsulfuron, sulfometuron-methyl, thifensulfuron-methyl, triasulfuron, tribenuron-methyl, triflusulfuron-methyl, chlorimuron, 2,4-D, picloram, chloramben, bromoxynil, isoproturon, fenuron, fluometuron, chlorbromuron, dimethenamid, clomazone, pendimethalin
III(high)	≥25	clopyralid, dicamba, monolinuron, tebuthiuron, propachlor, chlorsulfuron

A special attention should be paid to the highly dangerous substances (group III). It is necessary to standardize these objects in environment and to supervise the normality of their application. For example, the active ingredients of group III are characterized by the high volatility, which is more than 1,3 mPa (exception is chlorsulfuron that has $Pv=3 \cdot 10^{-6}$ mPa); by the water solubility, which varies from Swater=580 – 735 mg/l for propachlor and monolinuron up to Swater =31800 – 143000 mg/l for chlorsulfuron and clopyralid; by parameter LD_{50} (determined in experiments with rats), which is equal 500 – 2000 mg/kg, and thus they are classified as medium dangerous (excepting chlorsulfuron with LD_{50}=5545 mg/kg). With respect to parameter MDU, the active ingredients of group III are characterized as dangerous and their presence in foodstuff is forbidden. The analysis shows that monolinuron, tebuthiuron, propachlor are dangerous with respect to the toxic-hygienic parameters, while clopyralid, dicamba, chlorsulfuron are dangerous with respect to the ecological-agrochemical parameters. In result, monolinuron, tebuthiuron, propachlor are forbidden to application, and clopyralid, dicamba, chlorsulfuron are recommended to use mainly in the mixtures with others pesticides.

The distribution diagram together with the limits of index *Ip* value for different pesticides are shown in Fig. 4.

Picture demonstrates that pesticides based on tiocarbamate and imidazolinones are the most safe (*Ip* ≤22). Parameter *Ip* equals 18 – 30 for derivatives of carbonate acids, phenylurea and sulfonylurea. Why there is a large difference between bottom and top border of *Ip* in this case? Probably, it depends on behaviour of active ingredients in environment; for example, the rate constant of herbicide in soil could vary in dependence of pH, in dependence of the contents of organic substance, and in dependence of the hydrothermal characteristics, etc.

In table 6 and 7 we demonstrate the comparative analysis of individual active herbicides ingredients using several derivative of sulfonilurea and imidazolinone.

a)

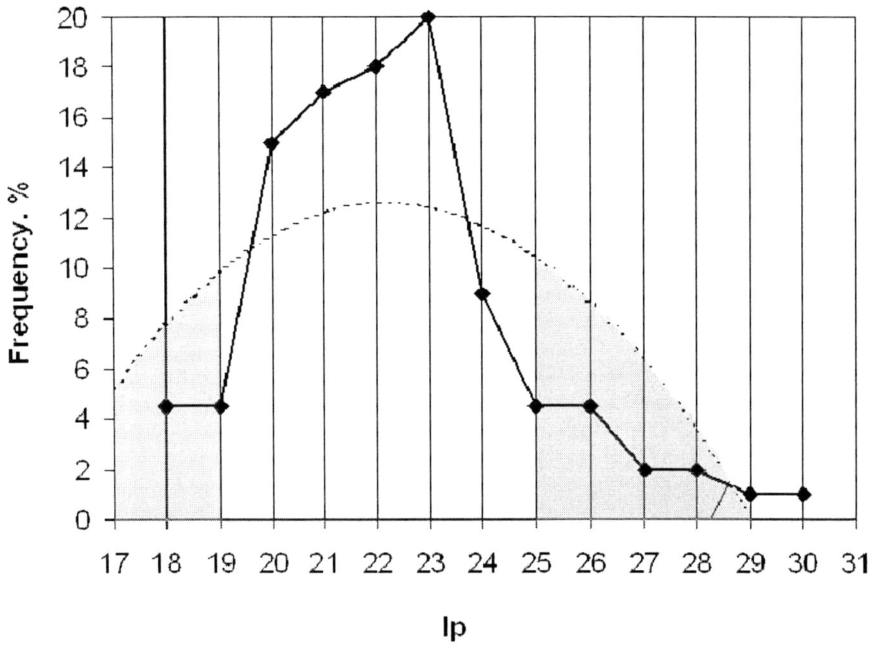

Fig. 4. Distribution (a) of index *Ip* for different pesticides (Continued on next page)

b)

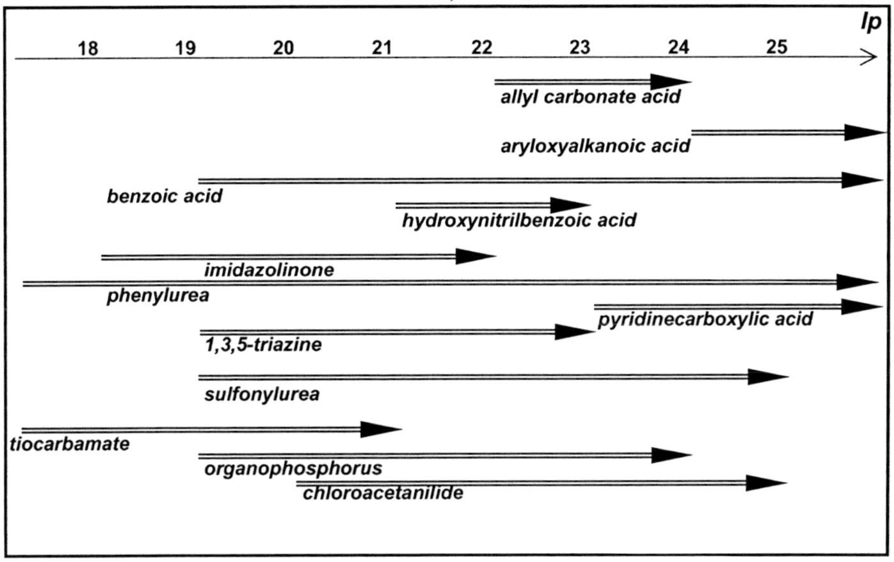

Fig. 4. Value limits (b) of index *Ip* for different pesticides

Table 6. Characteristics of herbicides behaviour in podzolic soil

Active ingredient (a.i.)	Doze, g/ha a.i.	k, day^{-1}	DT_{50}, day	* ED_{10} (fodder beet), g/ha a.i.	PDKsoil, g/ha a.i.
		sulfonylurea			
metsulfuron-methyl	9,0	0,0593	11	0,09	0,0012**
prosulfuron	18,75	0,0299	23	0,08	0,001
rimsulfuron	12,5	0,0596	11	0,1	0,036
sulfometuron-methyl	15	0,0501	15	0,5	0,048
chlorsulfuron	5,25	0,0154	45	0,3	0,0024**
		imidazolinone			
imazapyr	250	0,0307	23	0,9	0,2**
imazethapyr	100	0,0448	16	2,4	0,10
imazamox	40	0,0289	24	3,9	0,1**

* - bioindication in conditions of podzolic soil (pHsol 5,4, humus 2,4 %),

** - PDKsoil of this active ingredient does not s present in literature; therefore we use the mean limit of detection by the analytical method (HPLC or GC)

The value of $EL_{soil\text{-}plant}$ parameter changes in the large range: for podzolic soil from 0,02 up to 0,27, and for chernozem and dark-chestnut in the range 0,04-0,59. The tendency of decreasing $EL_{soil\text{-}plant}$ value in podzolic soil (or the minimum of pesticide impact on the agrosystem) could be marked. It specifies the active degradation of the active ingredient in the acid soil. It is possible to relate parameter GUS with the medium mobile ingredients, which are metsulfuron-methyl, rimsulfuron, and sulfometuron-methyl (2,36 – 2,62), and with the high mobile ingredients. which are chlorsulfuron, prosulfuron, imazapyr, imazethapyr and imazamox (3,27 – 6,73).

Table 7. Integrated parameters of behaviour of herbicides in agrosystem

Active ingredient	$EL_{soil\text{-}plant}$ for different soils Podzolic (pHwater 5,9, humus 2,8 %)	Chernozem (pHwater 6,7, humus 5,0 %)	Dark-chestnut (pHwater 8,3, humus 4,5 %)	*GUS*	$T(PDK_{ph})$ days	$T(PDK_{soil})$ days
sulfonylurea						
metsulfuron-methyl	0,07	0,41	0,55		150	78
prosulfuron	0,27	0,59	---	3,27	330	183
rimsulfuron	0,02	0,04	0,03	2,55	98	81
sulfometuron methyl	0,26	0,28	0,28	3,96	115	68
chlorsulfuron	0,11	0,28	0,27	2,36	499	186
imidazolinone						
imazapyr	0,58	---	---	6,73	210	81
imazethapyr	0,07	0,08	0,02	6,23	176	77
imazamox	0,02	0,02	0,03	6,31	127	22

Characteristic $T(PDK_{ph})$ characterizes the level of chlorsulfuron and prosulfuron residuals that is above the permitted hygienic specification PDK_{ph} in podzolic soil at the end of green season. However, at the moment of sowing ($T(PDK_{soil})$) the contents of residuals of investigated herbicides is more than achieved level of *PDKsoil*. Thus, the set of characteristic parameters (Swater, LD_{50}, DT_{50}, and etc.) together with a set of integrate indices ($EL_{soil\text{-}plant}$, GUS, and others) help to estimate a level of pesticide persistence or to compare ability of soil to independent clearing.

CONCLUSIONS

The step-by-step procedure for testing of assortment pesticides allows estimating the efficiency and the safety of preparation (active ingredient), and also promotes unification of works, facilitates comparison of experimental data. The minimal set of the parameters received by the researcher (*Ip*, $EL_{soil\text{-}plant}$, T_{PDK}, GUS, etc.) depends on designed tasks. They allow to estimate the ecological situation in agrosystem at the moment of herbicide application, and also to predict the change of it contents in view of concrete natural conditions.

ACKNOWLEDGMENTS

Special thanks are to employees of department of herbology and to the chief Yu.Ya. Spiri-donov for technical assistance.

REFERENCES

[1] Guidelines to the use of the WHO recommended classification of pesticides by hazard, *FAO plant protection bulletin*. 28. 1-24, (1980)

[2] M.S.Sokolov, O.A.Monasturskii, E.A.Pikushova *Ecological protection of plants*. Puschino, 1994. 462 p. (in Russian)

[3] Ecoinformatica. *Theory. Practice. Methods and systems*. S-Peterburg, 1992. 520 p. (in Russian)

[4] *The Pesticide manual*. The British Crop Protection Council 11-th. Ed. UK. By ed. C.D.S.Tomlin, 1997. 1606 p.

[5] Wells M., Michael J. *J. Chromatogr. Sci*. 25. 345-350, (1987).

[6] Menne H.J., Janowitz K., Berger B.M. *J. of AOAC International*. 82. 1534-1541, (1999).

[7] Larina G.E., Spiridonov Yu.Ya., Zakharov S.A., Zakharova T.V. Indication of the rests herbicides of imidazolinone in objects of agrocenoses. In *Problems of ecological security in agriculture*. Moscow, Sergiev Posad. 5. 33-40, (2001).

[8] Finney D.J. *International Statistical Review*. 4. 1-12, (1979).

[9] Gustafson D.I. *Envir. Toxicol. And Chemistry*. 8. 339-357, (1989)

In: Progress in Chemometrics Research
Editor: Alexey L. Pomerantsev, pp. 299-311
ISBN 1-59454-257-0

Chapter 21

TEXTURAL AND MINERALOGICAL CLASSIFICATION OF SOILS IN NATIONAL PARK "PLESCHEEVO LAKE"

Yu. N. Zborischuk and V.E. Slyusarenko
Lomonosov Moscow State Uniersity, Russia, slusarenko@laincom.chem.msu.ru

ABSTRACT

The investigation of texture and mineralogical soils' composition in the national park "Plescheevo Lake" have been carried out for different relief forms within of the lake hollow.

Keywords: soil, granular and mineralogical composition

INTRODUCTION

The top-soils of national park "Plescheevo lake" were generated from different glacial, fluvial-glacial and ancient-alluvial accumulations in the dismembered relief conditions. The main soils of investigated region are sod-podzolic. For the most part these soils have pronounced unusual peculiarities. In all investigated soils of Lake Plescheevo region the soil forming podzol process is observed but it not forms a whitish horizons as usually. From more than a hundred of soil profiles only a few have whitish podzol horizons [1]. Different transition horizons are formed in the soil profiles within the lake hollow and surrounding hills.

It is necessary to elucidate a reasons of this phenomenon so as it is important for diagnostics of the soils, determination of their genesis depending on the parent materials origin, their positions in relief, a biological and climatic factors, affecting the soil forming processes.

The soil's diversity is the main reason for the soil formation investigations in this region. Within the lake Plescheevo region clays, loams, sandy loams and sands are the parent materials. They have different texture from sand up to clay. Often parent materials are the binomial: loams are hushed up by sands and sands are hushed up by cover loams.

The parent materials with different structures are distributed mosaically in relief. The biological and climatic features are strengthening the variety of soil forming processes. As a result there is a great deviation of sandy podzols and podzolic soils' properties within small Plescheevo Lake water basin.

This hampers a discovery of general laws in the soil-forming processes but indicates that soil cover of the near lake landscape is in an active stage of development, which is an interesting object for investigators.

To obtain sufficiently authentic information it is necessary to carry out systematic quantitative investigations of large number of soil profiles' properties and their causes. In this work authors attempt to determine a correlation between topographic position of soils and their granular and mineralogical compositions

Objects and Methods of Investigation

Seven soil profiles were under investigation within the national park "Plescheevo Lake" water basin along geomorphologic profile, which has about 900 m length.. The overall height difference along geomorphologic profile was 35 m. Pits 122,135 and 136 are disposed on steep slope (150 m) and pits 127, 132 138 and 139 — on gentle slope. Pit 132 is disposed at the bottom of ravine (Fig. 1).

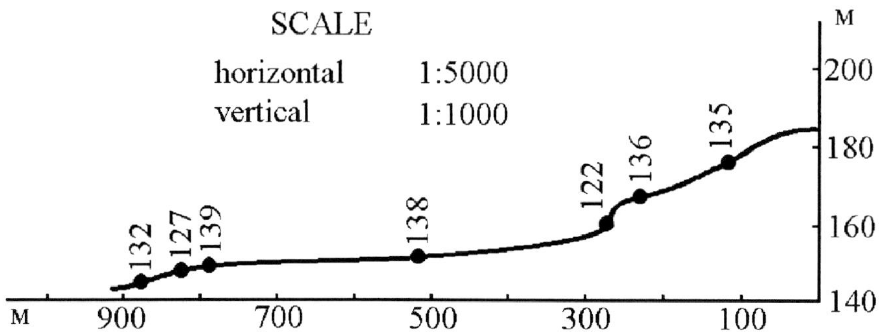

Fig. 1 Soil-geomorphological profile

Granular composition of the soils was determined by Kachinsky [2]. Mineralogical analysis was carried out, after minerals' division into light and heavy parts, in the bromoform solution by microscopic method [3]. Some clay minerals' contents were determined by Kornblum [4] with small alterations. Due to mobile railing minerals' compression the principle of 10 Å reflex intensity estimation and registration was used when the samples get tempered. This method was developed by Japan scientists [5]. The correction of LP-factor has been carried out also [5]. A silt fraction was investigated in two profiles: pit 122 (horizons A, EB, BCg) and pit 139 (horizons AB, B, C).

Granular Composition of Soils

Analysis of soils' granular composition (Table 1) indicates that the investigated soils have a glacial origin. A low concentration of coarse sand (> 1 mm) and predominance of mid and fine sand practically in all horizons are characterizing morainal accumulations. These accumulations consist of minerals, more stable against weathering, than cover loams. A coarse dust fraction is prevailing over sand in strongly weathered cover loams and is accumulated in upper horizons. According Kachinsky classification all investigated soils are sandy loams with predominance of mid and fine sands.

On steep slopes the soils (pits 122, 135, 136) are formed on thick ancient-alluvial sands' accumulations and have practically an equal granular composition in all horizons. The soils (pits 127, 138, 139) in the sloping part are formed on the lamellar (binomial) parent materials. Upper soil horizons are formed on fluvial-glacial sands, which have the same composition as ancient-alluvial sands. Lower horizons of the soils are formed from morine loams. These horizons contain much more physical clay than cover sands in upper layers.

Pit 132 distinguishes from other pits very much. This pit lies at a small distance from lake and is disposing in the ravine. Lower quantities of coarse sand and physical loam in upper horizons show that upper horizons in this profile have a secondary origin by washing down mid and fine sands from the higher parts of slopes. In the investigated soil profile coarse sand is relatively accumulating due to the dust and silt partial taking out to the lake hollow by lateral flow. Lower horizons are formed on more ancient sand accumulations. This sand accumulations are enriched by the coarse sand, what is explaining a decrease of the cryogenic crushing in illuvial horizons and an intensification of chemical decomposition for fine and silt fractions under extra moistening.

Dust coefficient Kf [6], which is a ratio of coarse dust content to sand particles quantity (>1 mm), have been calculated (Table 1). This coefficient usually used for loams nature determination: Kf >70 for covering loams; Kf < 50 for morainal loams. In our case this coefficient is not informative: Kf is the same for sands and morainal loams. Apparently that crushing processes are not principal for investigated soils.

There can be no doubts that the coarse dust accumulation in soils plays an important role in the soil-forming processes and in soil's structure.

Each soil is a mixture of particles with different dimensions, which are packing up in space by chance. Standard division of the soils particles in fractions of sand and dust lead to very big differences between an average size of sand particles and average size of dust particles. We can represent all soils as a packed structures consisting of big sand particles and a holes filled by dust. If the particles have a spherical form then the sand – dust ratio is equal to the ratio of spheres and holes volumes (Kcp) in the spheres' filled space (1).

$$Kcp = V_{sand} / V_{dust} \quad (1)$$

Table 1. Granular soils composition

Horizon, depth, cm	> 1,00 % from soil	1,00-0,25	0,25-0,05	0,05-0,01	< 0,01	Kf	Kcp
		% from the sum of fraction < 1.00					
Pit 132 (H = 142 м)							
A 3-28	2	25	48	13	14	15	2.78
B_2 28-41	1	22	44	18	16	21	1.97
B_3 41-60	6	28	42	21	9	23	2.55
BC 60-105	4	27	46	17	10	19	2.86
C 105-160							
Pit 127 (H = 144)							
A 5-14	4	32	42	18	8	20	3.01
AE 14-25	2	31	40	20	9	22	2.52
EB 25-33	3	25	42	23	10	26	2.12
B_1 33-60	2	24	41	24	11	27	1.91
B_2 60-80	3	22	37	25	16	30	1.51
B_3 80-150	1	20	42	18	20	23	1.66
C > 150							
Pit 139 (H = 146)							
A 5-9							
AB 9-24	3	21	42	25	12	28	1.79
B 24-42	2	15	51	20	14	23	2.00
BC_g 42-76	3	18	46	10	26	14	1.86
C > 76	1	14	36	15	35	23	1.02
Pit 138 (H = 152)							
AE 6-15	5	27	40	19	14	22	2.19
B_1 15-36	4	26	35	23	16	27	1.67
B_2 36-88	4	24	39	27	10	30	1.82
BC_g 88-125	2	20	36	14	30	20	1.32
Cg > 125	4	17	32	16	37	25	0.97

Table 1. Granular soils composition (Continued)

Horizon, depth, cm	> 1,00 % from soil	1,00-0,25	0,25-0,05	0,05-0,01	< 0,01	Kf	Kcp
		% from the sum of fraction < 1.00					
Pit 122 (H = 160)							
A 9-30	3	30	39	21	10	23	2.33
AE 30-49	4	27	40	21	12	24	2.16
B_{1f} 49-100	3	25	41	21	13	24	2.03
B_2 100-125	2	21	45	23	11	26	2.00
BC_f 125-140	1	27	39	21	13	24	1.97
C > 140	2	28	40	23	9	25	2.19
Pit 136 (H = 166)							
AE 6-15	3	20	43	21	16	25	1.79
B_1 15-35	2	18	41	23	18	28	1.49
B_2 35-56	2	26	42	20	12	23	2.19
B_3 56-81	4	27	44	18	11	20	2.59
BC 81-123							
Pit 135 (H = 175)							
A 6-14	4	24	44	17	15	20	2.26
B_1 14-72	5	28	42	20	10	22	2.51
B_2 72-91	2	23	45	20	12	23	2.19
BC 91-124	3	20	46	19	15	22	2.03
C > 124	2	19	45	22	14	26	1.83

This ratio does not depend on spheres' radius. There are three more close-packed structures: hexagonal and face-centered structures with Kcp=2.85 and base-centered structure with Kcp=2.13. First kind of structures is characterized by strong connections between the spheres, second is by weak connections. When Kcp is changing from 2.13 to 2.85 we have an ideal close-packed soil, which is stable to such phenomena as the drying, super moistening, small seismic actions and others (see Fig. 2).

The soils with Kcp < 1.00 are clays and heavy loams. The soils with Kcp > 4.00 are sands. Such soils are unstable on steep slopes and often were eroded. More actively soil-forming processes are developing in loamy and sandy parent materials with Kcp near ideal interval, what is leading to more stable soils' state.

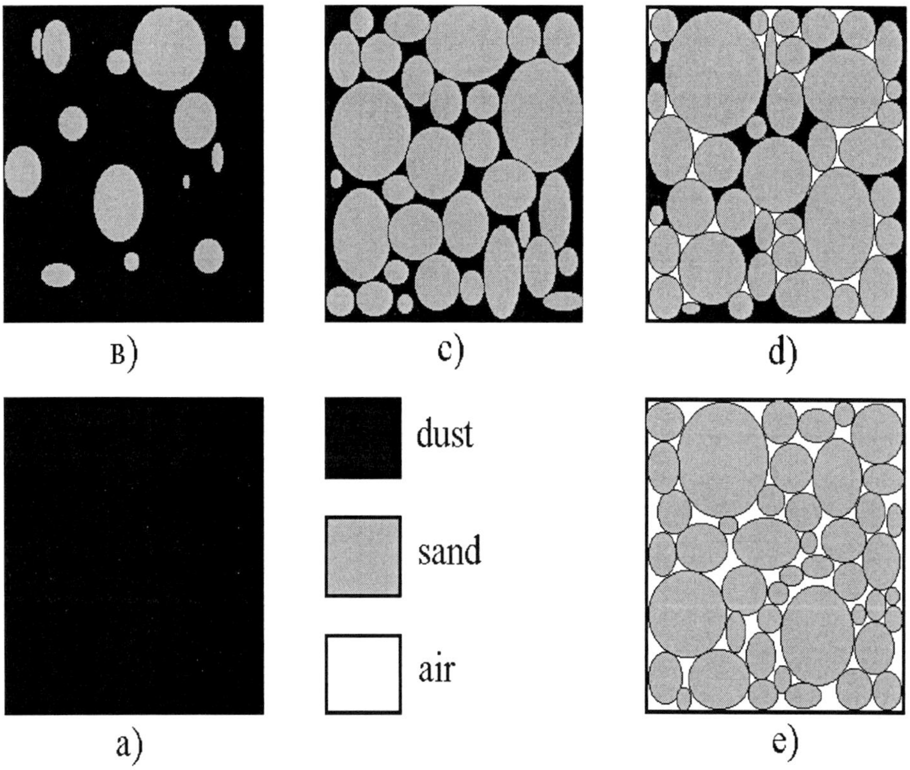

Fig. 2. Soil classification by coefficient Kcp: a) clays (Kcp<1); b) loams (1<Kcp<2); c) close- packed soils (2<Kcp<3); d) sandy loams (3<Kcp<4); e) sands (Kcp>4);

Standard division of dust particles into a coarse and fine fractions leads to the same result since a middle coarse dust radius and is bigger than the one for fine dust. As result we obtain an ideal granular composition for close-packed soils (2).

$$(C_{sand})\% : (C_{coarse\ dust})\% : (C_{fine\ dust})\% = 67\% : 22\% : 11\% \qquad (2)$$

A coefficient Kcp is calculated for all investigated horizons as shown in Table 1.The proposed coefficient Kcp allows to classify any soils by granular composition more exactly. For example, coefficient Kcp indicates that illuvial horizons in pit 127 on the sloping part of relief are light loams but from granular composition data it is not obvious. At the same time the horizon BCg in pit 139 is containing more physical clay has near close-packed structure. Practically all investigated horizons are in close-packed state or near that state.

The carried out experiments and the obtained data allow to suggest authentic hypotheses about parent materials' formation, which determinate subsequent its genesis and development under the action of podzol processes, biological, geological and climatic factors.

Investigated region of lake Plescheevo catchment was covered by Nerl's blade of Moscow's glaciation [7]. Nerl's blade has relatively simple structure: small ridges (10-20 m) consist of sands, boulder loams and gravel. These accumulations are fixing the position of edge glacial tongues depositions and are divided by great number of ice sheds, which are hollows of drain.

In the region of lake Plescheevo the glacier thickness was bigger than in the brims of glacial blades. As a result on steep slopes of ice sheds have been obtained a mighty sands accumulations non-strengthening by vegetation. Such accumulations are gravitationally unstable and in a super-moistening state they are slipping down in the hollows of drain to the foot of slopes as viscous liquids. Therefore doubtful that on the steep slopes sands have a pre-glacial origin. It is more likely that they are ancient redeposited alluvial sands intermixed with fluvial-glacial sands. As a result, more mighty accumulations are formed in the hollows, holes and etc (pits 122, 132.). The granular composition of such sands should be homogeneous what we observe in our investigations (Table 1).

On the gently sloping parts of soil-geomorphological profile parent materials are formed in the lake shallow. Fine sand and clay fractions of aeolian and glacial origin were deposited on the bottom glacial accumulations. After baring of lake bottom parent materials were changed under the action of different factors: washing down of dust fractions from upper horizons of soil-forming materials into lake; filtration of flow through the lower sandy layers; washing down sands and cover loams from steep slopes with laminar structures formation, etc.

Different geomorphologic and climatic nature a soil-forming material is a major reason of soil-forming processes variety in investigated region only. The base factors for subsequent reforming of parent materials into soils are: material digestion; material drifting; stability to weathering, material stability under the actions of different chemical and biological active substances and the like. In this case mineralogical composition of parent material is a principal factor.

Mineralogical Composition of Soils

Pit 122 with the podzol horizon is disposed on the steep slope foot and pit 139 with transitional horizons on the sloping part, were under investigation. Pit 139 has a pronounced laminar structure (a morainal loam covered by fluvial-glacial sand).

Mineralogical composition analysis was carried out for fractions of fine sand and coarse dust. The fraction of fine sand (0,25-0,05 mm) was preliminarily divided into a fraction (0,25-0,1 mm) and a fraction (0,1-0,05 mm). Such division was carried out for a calculation of cryogenic contrast coefficient (CCC) [6]. CCC allows to determine some soil-forming processes using a granular composition of the soils. Coefficient CCC is calculating as a ratio between the concentrations of quartz and feldspars in two fractions of fine sand in accordance with the equation (3)

$$CCC = C_{quartz}/C_{feldspar}\,(\text{fraction } 0{,}1\text{-}0{,}05 \text{ mm}) \,/\, C_{quartz}/C_{feldspar}\,(\text{fraction } 0{,}05\text{-}0{,}01 \text{ mm}) \quad (2)$$

Fraction of Light Minerals

The concentrations of light minerals (mainly quarts and feldspars) in investigated fractions are shown in Table 2. In both fine sand fractions the concentrations of quarts and feldspars are equal for pits 122 and 139 and coincide with early-obtained results [1]. Particles of quarts are dominating. It is necessary to note that in podzol horizon of pit 122 and in horizon B of pit 139 the accumulation of feldspars is visible. Most likely it is connected with more intensive crushing and dissolving of feldspar particles in upper horizons and taking away in lower horizons.

Yu.V.Bystrov, S.N. Sedov, and S.A. Shoba [7] using a feldspar grains in loamy podzol soils found that corrosion of the parent materials particles' surface particles is the reason of intrasoil withering in upper horizons. Process starts by etching holes' formation on the surface of particles and such particles are then destroyed.

The ratio of quarts and feldspars concentrations in a dust fraction is strongly distinguished in both sections. The concentration of easily weathered feldspars is sharply increasing. In pit 122' horizons the feldspars' concentration is lower than quartz concentration and is regularly changing as in the fine sand fractions. In horizons of pit 139, having a distinct laminar structure, the concentration of dust fraction is distinguishing. The concentration of quartz prevails above the concentration of feldspars only in upper horizon AB, but feldspars prevail in other horizons. In illuvial horizons, which are forming on morainal loam, this phenomenon explains a lithological origin. The accumulation of feldspars in horizon B is connecting with dissolution of fine particle of quarts in podzol process.

It is interesting that the concentration of carbonates in dust fractions of both pits 122 and 139 is strongly distinguishe. In pit 122 carbonates are determined in a small concentration only in illuvial horizon C. In pit 139 carbonates are determined in all horizons. Absence of carbonates in sand fraction and their little concentration in dust fraction shows that their formation is secondary process. Pit 139 is disposed near lake where takes place a periodic saturation of all horizons by hard subsoil waters obtaining carbonate from carbonate morainal loams. After the subsoil water level is lowered and soil is drying, carbonates are concentrating by decomposition of supersaturated subsoil waters bicarbonates. Bicarbonates and carbonates react with an active chemical and biological products produced by decomposition of humus and a vital activity of microorganisms. It will restrain a podzol process in soils with small concentration of humus by insoluble salts' formation.

Table 2. Mineralogical composition the fraction of light mineral

Horizon, thickness, cm	Quarts	Feldspar	Carbonate
Granular fraction (0,25-0,1 mm)			
Pit 122			
A (9-30)	80	20	-
AE (30-49)	75	25	-
B_{lf} (49-100)	82	18	-
C (> 140)	79	21	-
Pit 139			
AB (9-24)	79	21	-
B (24-42)	70	30	-
BCg (42-76)	84	16	-
C_g (> 76)	86	14	-
Granular fraction (0,1-0,05 mm)			
Pit 122			
A (9-30)	74	26	-
AE (30-49)	64	36	-
B_{lf} (49-100)	77	23	-
C (> 140)	71	29	-
Pit 139			
AB (9-24)	76	24	-
B (24-42)	65	35	-
BC_g (42-76)	79	21	-
C_g (> 76	80	20	-
Granular fraction (0,05-0,01 mm)			
Pit 122			
A (9-30)	58	42	-
AE (30-49)	55	46	-
B_{lf} (49-100)	67	33	-
C (> 140)	65	33	2
Pit 139			
AB (9-24)	62	32	7
B (24-42)	41	51	8
BCg (42-76)	39	55	6
C_g (> 76	40	48	12

Fraction of Heavy Minerals

Spatial changes of sand accumulations mineralogical composition in moving glaciers has been investigated by V.D. Tonkonogov [8]. The more the sand shift way with fluvial and fluvial-glacial flows and the older deposits, the stronger are a destroying and weathering of any minerals. On the brinks of sand accumulations a part of weathered minerals is decreased. Sand accumulations' composition also dependent on minerals' specific weight: more heavy particles are deposited in the beginning of the way. Total concentration of minerals, which unstable in zone of hypergenesis, are less in more ancient sands.

Sand accumulations in Lake Plescheevo catchment are very poor in heavy minerals (1-2%) what is in accordance with their fluvial-glacial origin. Data of mineralogical analysis for fine sands and coarse dust fractions in investigated pits (122 and 139) shown in Table 3.

Table 3. Mineralogical composition of hard mineral fraction (% from sum of particles

Horizon, thickness, cm	Epidote	Granite	Hornblende	Minerals ore	Sphene	Turmaline	Sillimanite	Biotite	Rutile	Non. denermined
Granular fraction 0,25-0,10 мм										
Pit 122										
A (9-30)	67	15	16	2	-	-	-	-	-	-
AE (30-49)	60	10	10	15	3	1	1	-	-	-
B_{1f} (49-100)	64	18	6	10	1	-	-	-	-	-
C (> 140)	62	28	5	4	<1	<1	-	-	-	-
Pit 139										
AB (9-24)	70	12	2	15	1	<1	-	--	-	-
B (24-42)	71	7	1	19	2	-	-	-	-	-
BCg (42-76)	72	5	1	20	2	-	-	-	-	<1
C_g (> 76)	67	12	1	20	-	-	-	-	-	-
Granular fraction 0,10-0,05 мм										
Pit 122										
A (9-30)	60	18	11	6	4	1	-	-	-	1
AE (30-49)	55	11	5	20	5	3		-	-	1
B_{1f} (49-100)	62	16	2	16	2	<1	<1	-	-	<1
C (> 140)	57	21	2	16	4	-	<1	-	-	<1
Pit 139										
AB (9-24)	64	12	1	20	3	-	-	-	-	-
B (24-42)	69	9	<1	19	2	<1	<1	-	-	-
BCg (42-76)	66	5	<1	24	3	2	1	-	-	-
C_g (> 76)	70	9	<1	18	2	-	-	-	-	<1
Granular fraction 0,05-0,01 мм										
Pit 122										
A (9-30)	57	-	-	-	14	-	14	14	-	-
AE (30-49)	32	9	23	14	9	-	5	5	-	5
B_{1f} (49-100)	30	11	19	22	-	-	7	11	-	-
C (> 140)	38	17	13	13	-	-	21	-	-	-
Pit 139										
AB (9-24)	21	11	26	21	-	-	-	16	5	-
B (24-42)	20	9	30	14	5	-	2	16	5	-
BCg (42-76)	25	7	25	11	10	-	4	14	11	4
C_g (> 76)	16	5	16	16	-	-	-	37	16	-

It is evident, that in both sand fractions, independently from soil profiles are dominating (80-90%) silicates with framework structure (epidotes, granites, sphenes, turmalines), which are stable in the hypergenes's zone. Epidotes grains' content in pit 139 is slightly higher due to the practical absence of hornblende in fine sand fraction. This indicates to its physical-chemical shattering in the overmoisten conditions. As shown in pit 122 hornblende is found only in fine sand fraction within the eluvial horizons.

Table 4. The changes of cryogenic contrast coefficient (CCC) and dust coefficient (Kf) in the profile of sections

Section	Horizon	CCC	Kf
122	A (9-30)	0,3	23
	AE (30-49)	0,6	24
	B_{1f} (49-100)	0,7	24
	C (> 140)	1,0	25
139	AB (9-24)	0,7	28
	B (24-42)	0,4	23
	BCg (42-76)	0,2	14
	C_g (> 76)	0,2	23

Less stable aluminosilicates' with laminar (sillimanite and biotite) and band structures (hornblende) is noticeably increased (up to 40%) in the coarse dust fraction (0,05-0,01 mm) in both pits (122,139). In the illuvial horizon (pit 139), where the podzol processes are in less degree expressed, there is a biggest content of biotite.

Table 5.The combined changes of cryogenic contrast coefficient (CCC) and dust coefficient (Kf) in the soil profile of sections.

	Increase Kf	Decrease Kf
Increase CCC	Cryogenic crushing of quartz	Dissolution of feldspars
ecrease CCC	Cryogenic crushing of feldspars	Dissolution of quartz in dust fraction.

Data of cryogenic contrast coefficients (CCC) and dust coefficients (Kf) calculations for boths profiles are shown in Table 4. In 122 profile coefficient CCC strongly increases down in profile with simultaneous increase of Kf. It indicates that the primary cryogenic crushing of quartz particles was dominated, as follows from Table 5. In 139 profile we observe a reverse picture. Both coefficients decrease down in profile. It indicates that here the primary process was dissolution of the quartz particles in dust fraction.

Silt Fraction

Middle taiga soil's mineralogical composition of silt, which is represented by kaolinite, illite and mobile silicates, is weakly differentiated. But in some cases it gives the definite results, which are defining the intensity of chemical processes in the parent materials. Soil profiles differ predominantly in mineralogical composition of the light minerals part.

Kaolinite and soil chlorite distribution in pit 122 is even. Illite content is increasing twice along the soil profile, when mobile silicate's content is twice decreasing. In pit 139 soil is formed on moraine. This leads to the predominance of rigid structure's minerals (kaolinite, illite) and to the mobile mineral's decrease. In investigated soils the degradation process, is called chloritization, takes place. Illite during this process is converting in mobile structures with chlorites appearance at the very end.

The main question in soil diagnostic is the clay minerals transformation influence on the texturally differentiated soil profiles during soil-forming process. From the silt fraction mineralogical composition analysis (Table 6) it follows, that crushing of such minerals, as kaolinite and illite up to silt fraction's size is obviously not enough for the texturally differentiated soil profiles formation.

Table 6. Mineralogical composition of silt fraction

Horizon	Concentration, % from sum S			Estimation chlorites			
	Kaolinite + soil chlorite	Illite	Mobile structures	Vermikulite	Soil chlorite	Chlorite	Quartz
Pit 122							
	35	27	38	+	+	-	++
AE	39	23	38	+	+	-	+
B_{1f}	39	45	16	+	-	+	+
Pit 139							
AB	56	19	25	+	+	+	+
C	56	31	13	+	+	+	+
C	42	22	36	+	-	-	-

Conclusions

1. Glacial origin of parent materials is established. In the coastal part of investigated area soils are formed on moraine loams covered again by fluvio-glacial sands, and on the steep slopes soils are formed on ancient alluvial sands.

2. It was suggested to use for soils granular composition classification coefficient Kcp, which is calculating as a ratio of sand and dust concentrations (fraction < 0,05). When Kcp is variable from 2.13 to 2.85 we have an ideal close-packed soil, which is stable to such phenomena as the drying, super moistening, a small seismic actions and the like. The calculations of Kcp for all horizons in investigated profiles show that the soils are close-packed or near to this state. A coefficient Kcp also allows to estimate a degree of sand and loam transformation under the action of the soil-forming processes

3. On the gently sloping part of soil-geomorphological profile soils contain carbonates in light minerals' fractions and big amount of biotite in heavy minerals fractions of surface horizons. This may be the main reason of weak podzol process in investigated soils.

4. Combined use the cryogenic contrast coefficient (CCC) and dust coefficient (Kf) lead to the conclusion about cryogenic crushing of quartz particles on steep slopes of catchment

and dissolution of quartz in dust fraction on the sloping part in the glacial period and afterwards.

REFERENCES

[1] Zborischuk Yu.N., Savel'ev D.V., Ul'yanova T.Yu. Vestnik MSU. Ser.17, *Science soil*, 2, (2002).

[2] Vadunina A.F., Korchagina Z.A. *Metody issledovaniya phyzicheskih svoistv pochv*. M., Nauka, 1986. (In Russian)

[3] Kornblyum E., *et al.*, *Soil science*, 2, (1972).

[4] Sudo T., Opiuma K. Acta Universitatis Carolinae, *Geology*, 1, (1961).

[5] Kitaigorodsky A.I. *Rentgenostructurny analis*. M.-L., Gosizdat, 1950. (In Russian)

[6] Makeev A.O., Makeeva O.V. *Pochvy c teksturno-differenczirovannym profilem osnovnyh cryogennyh arealov Russkoi ravniny*. Puschino, 1989. (In Russian)

[7] Pisarev V.V., Sudakova N.G., Gavryusheva E.A., Lecht E.E. *Klyazminsky lednikovyi potok.Mockovsky lednikovyi potokVostochnoi Evropy*. M., Nauka, 1982 (In Russian)

[8] Konischev V.N. *Formirovanie sostava dispersnyh porod v cryolitosphere*. M., Nauka, 1981. (In Russian)

[9] Bystrov Yu.V., Sedov S.N., Shoba C.A., Vestnik MSU. Ser.17, *Soil science*, 1 (1993)

[10] Tonkonogov V.D. *Soil science*, 2, (1971).

In: Progress in Chemometrics Research
Editor: Alexey L. Pomerantsev, pp. 313-321
ISBN 1-59454-257-0

Chapter 22

EXPLANATORY DATA ANALYSIS OF FISH CULTURE IN SIBERIAN LAKES

***A. S. Semenchenko*[1]*, S. M. Semenchenko*[2]*, O. Ye. Rodionova*[3] *and A. L. Pomerantsev*[3]**

[1]Moscow State University of Engineering Ecology, [2]State Fishery Research and Production Center, Russia
[3]Institute of Chemical Physics, Moscow, Russia,

ABSTRACT

Wide breeding of peled (*Coregonus peled*) began in the lakes of Western Siberia in the 1960-s. A lot of lakes were stocked yearly. Total catch of fish exceeded 5 000 tons during fishery arrangement with the share of peled about 90%. Now the amount of 1,400 tons of peled are planting yearly. We study the data set collected for the last 40 years about peled culture in the lakes of Ural and Western Siberia. Till now, MDA methods have not been applied for data processing of Russian fish culture industry. The goals of the research are: the explanatory analysis of MDA methods feasibility in fish culture and prediction of whitefish breeding effectiveness in the Tobolsk region lakes. Steps of analyses: 1) Construction of the PCR model on the whole data set. The results show essential difference between lakes located in different regions. Distinct clusters may be seen in the score plots. 2) For one cluster (Tobolsk region lakes) the PLS1 model is used for prediction of fishery yield. 3) For the groups of lakes with total square less then 60 sq. km, the PCA model helps to analyze the influence of lake conditions on productivity. Software Unscrambler 8.0 was used for data analysis.

Keywords: Fish-culture, White-fish, Siberian Lakes, Partial Least Squares, Principal Component Analysis and Regression.

INTRODUCTION

First successful steps in white-fish (*Coregonidae*) planting were made during the 30-s in the Ural region. Wide breeding of whitefish began in the lakes of Western Siberia in the

1960-s. A lot of lakes have been stocking yearly. Total catch of fish exceeded 5 000 tons during fishery arrangement with the share of peled (*Coregonus peled*) about 90%. Now the amount of 2 000 tons of peled are planting yearly. The peled planting technology is now a routine practice [1, 2] and large amount of information was collected from the end of 60-s through 2003. Now this data is needed to be systemized and analyzed. The goal of the investigation is to evaluate the influence of abiotic and biotic factors on fishery efficiency. We try to educe weather conditions and lake features, which effect on the fish culture and make an attempt to predict the amount of whitefish culture.

DATA DESCRIPTION

Lake farms are located in the forest-steppe region of South-West Siberia, mainly, in Chelabyinsk, Kurgansk, and Tyumen districts. The one-year technology of peled breeding implies small and rich-in-food lakes, which ordinarily froze out during winter. Usually, such lakes are fishless or inhabited with crucian (*Carassius auratus, C. carassius*). Crucian is not a feed competitor for peled and it is not dangerous for peled fingerling. Lakes are planted with peled larva, which is incubated without additional rearing. Usually, this is executed in the second half of May. During summer, fish grows and eats the natural food. In autumn, peled fingerlings reach commercial weight that is about 180 g and the lakes are fished out with drag seines and trap nets. Taking into account the large number of lakes and the low costs of maintenance, such a technology is rather profitable, especially for years with auspicious weather conditions.

Information about lake locations, conditions, and results of planting was gathered with the help of questionnaires, which were filled in by many specialists. For problem simplification, only information regarding peled planting with catching during autumn-winter period is included in data set.

Main Definitions

For clarity and correct result interpretation we need to introduce the following definitions, which are specific for fish planting.

Planting sample is a data collection regarding one cycle of peled planting in particular lake.

Cumulative temperature during vegetation period is the average temperature for the region where lake situated. This value is a sum of average daily temperatures for days when temperature is higher than 10°C. This data was obtained from corresponding reference books [3-8].

Lake square, average depth, water mineralization, pH and Zooplankton density are obtained as the average values for the vegetation period.

Ichthyo-fauna presence is a dummy value that marks presence or absence of other fishes in a particular lake.

Ice-free period is a period, during which a lake is free of ice.

Planting density is a ratio of the fish placement volume to the lake's square that is averaged for given year.

Average fish weight is a fish weight at the end (September – November) of vegetation period.

Catching amount is a total weight that is fished out during one year (provided by statistical report). Poaching and "delivery loss" are not taking into account.

Fish-production is a ratio of catching amount to the average lake square. It is worth mentioning, that this characteristic does not agree with a characteristic "production" [9] used in hydrobiology. Our term does not count the natural death during vegetation period, poaching, fish that are stay in a lake (such fish usually is lost in winter). So, "fish-production" is not exactly a biological feature, but it has a technological meaning.

Specific catch is a ratio of the caught amount to the planted amount. It shows how many tons are fished from every billion of planted ones.

Yield of fishery is a ratio of caught quantity to the planted amount, expressed in percents. The caught quantity is calculated as a ratio of *Catching amount* to *Average fish weight* (see above). Taking into consideration the way of definition of *Catching amount*, the term *Yield of fishery* is a technological characteristic that indirectly represents peled survivability during the first vegetation period.

Results of peled planting were compared using average mass of grown fish and three indirect indexes: fish-production, specific catch and yield of fishery. The staff of the State Center of Fishery managed to collect information about 249 planting samples in 38 lakes totally. Collected data are very noisy, due to the following obvious reasons. It is impossible to obtain the exact number of grown fish for big lakes. It is very hard to catch out all the fish, especially in lakes with a large square. So, catching amount does not represent the exact number of grown fish. Poaching and uncontrolled fishing greatly decreased the amount of grown fish marked in questionnaire. This circumstance mostly influenced data concerns of small lakes. There are cases of total catching out of lakes by natives. Partial questionnaire filling have a significant influence on correct evaluation of fish-planting effectiveness.

Data Analysis

The inherent correlations between various characteristics and essential noise in data set force us to apply methods of multivariate data analysis (MDA) for data processing. Such an approach was not used before in Russian fish culture industry. Variables that are used in data analysis are presented in Tab.1.

Data Overview

On the first step we use the principal component regression (PCR) [10] for general explanatory analysis. The X matrix consists of all variables that characterize the lakes' location and conditions (variables X1-X10 in Tab.1). The response is vector Y1 (*Average fish weight*). The investigated data set consists of 209 samples and the final model employs two principal components (PC's). Explained variance for Y is only 25.5%, and 50.5% for X. Higher-order PC's do not improve the model. Data contains high level of noise; this means that the model cannot be used for prediction. Nevertheless, we use this model for better understanding of initial data structure and object similarities. It also helps us to make some conclusions for further analysis. In the score plot (Fig.1), we can clearly see data grouping.

One of these clusters (group 1 in Fig. 1) represents lakes of Tobolsk region and it is used below for detailed investigation.

Table 1. 10 predictors and 5 responses that reflect lake conditions and fish culture effectiveness

Predictors		Responses	
X1	Cumulative temperature during vegetation period	Y1	Average fish weight
X2	(Lake) Square	Y2	Catching amount
X3	(Lake) Depth	Y3	Fish-production
X4	Ice-free period	Y4	Specific catch
X5	Mineralization	Y5	Yield of fishery
X6	pH		
X7	Zooplankton density		
X8	Ichthyo-fauna presence		
X9	Planting amount		
X1	Planting density		
0			

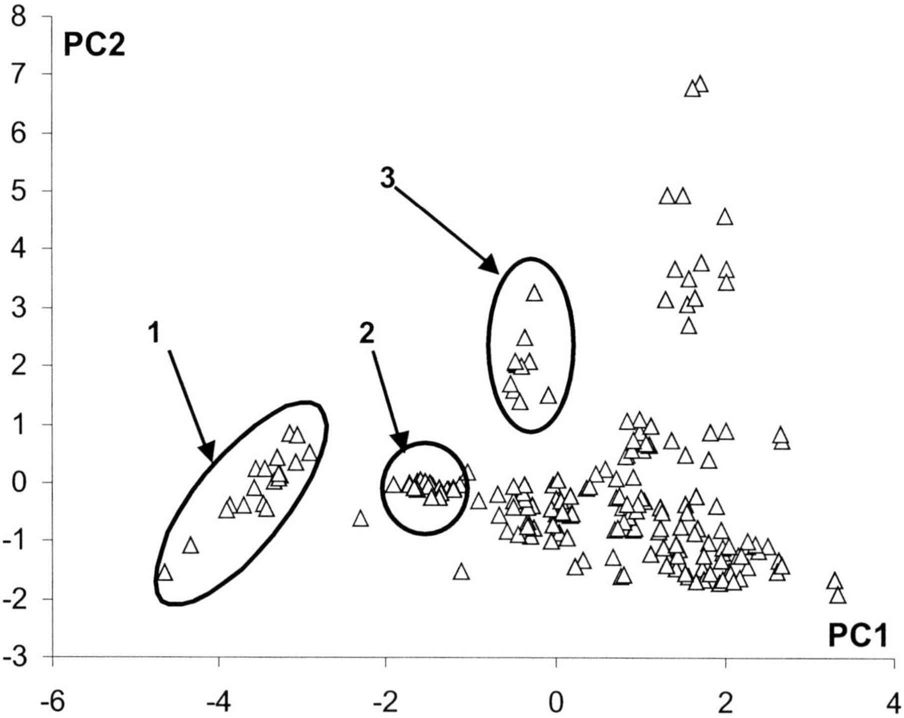

Fig 1. Score plot of the initial data set. Several groups are marked out 1- Tobolsk; 2- Krutinsk; 3 – Karagatinsk.

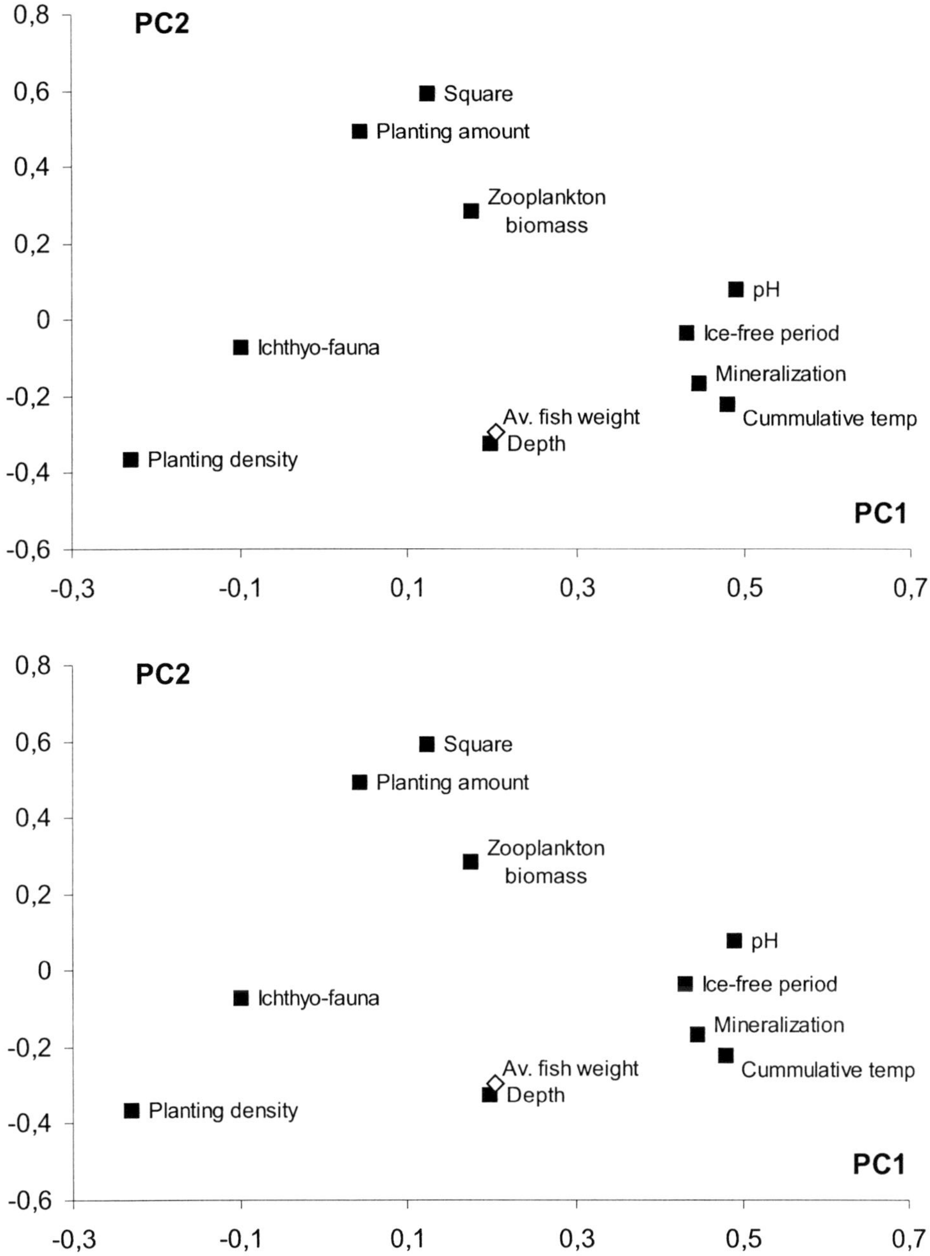

Fig 2. Loading plot for all predictors (■) and the response (◇) that is *Average fish weight*

Simultaneous exploration of score plot (Fig. 1) and loading plot (Fig. 2) provides us with the following information.

1. There are clusters of samples that correspond to geographical regions. Several clusters are marked in the Score plot.

2. Variable *Square* (i.e. X2), has a significant influence on the object variation. So, we decide that it makes a sense to analyze the small lakes (S < 60 sq. km) separately from the large ones.
3. The investigated region of Western Siberia contains different geographical zones and this factor also effects in the sample spread in the score plot.
4. Variable *Square* has a positive correlation with variable *Planting amount.* Such relationship shows us that this model correctly reflects the state of affairs.

Lakes of Tobolsk Region

At the second step we perform the more detailed investigation of data that characterize one group, i.e. the lakes of Tobolsk region (one of the clusters that can be seen in Fig. 1). We apply the partial least square regression (PLS1) [10] for model construction. The X-matrix consists of 8 variable, which are all predictors except X4 and X8 (see Tab. 1). The response vector represents *Yield of fishery* variable (Y5 from Tab. 1). Data set contains 23 samples and therefore cross-validation is applied. The final model employs four PLS-components, which explain a total of 50% of X-variance, but about 73% of Y-variance, with RMSEC=4.91, RMSEP=6.34. Fig. 3 shows the final plot that presents predicted versus measured results. Taking into account that initial data are very noisy, we consider the results to be quite satisfactory. Therefore, they may be used for prediction.

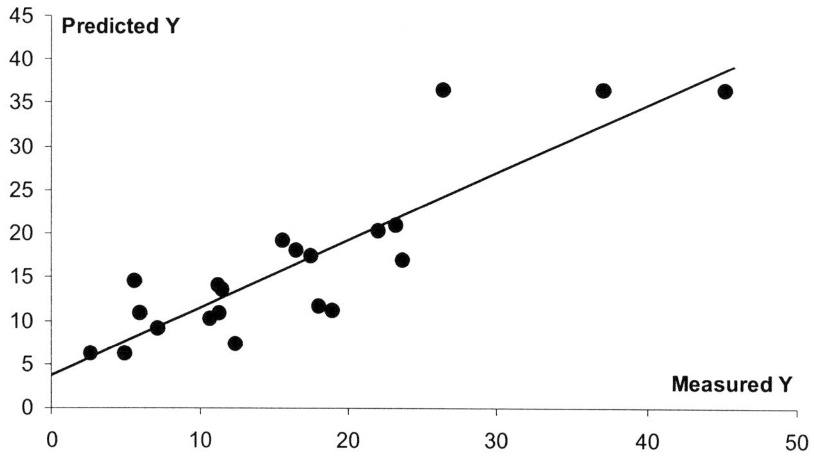

Fig 3. Measured vs. predicted values of *Yield of fishery* response for the lakes of Tobolsk region

Small Lakes (S < 60 km^2)

Detailed investigation of small lakes in order to detect hidden relationships was performed with the help of principal component analysis (PCA) [10]. The data set consists of 167 samples. We combine all predictor and response variables (see Tabl.1) into one X-matrix. So, our data consists of 15 variables. The PCA-model with 10 PCs explains 92% variation in data and helps to understand the interaction between variables. Interpretation of the model is based on loading plots (Fig. 4, Fig. 5). In these plots we can mention the following effects:

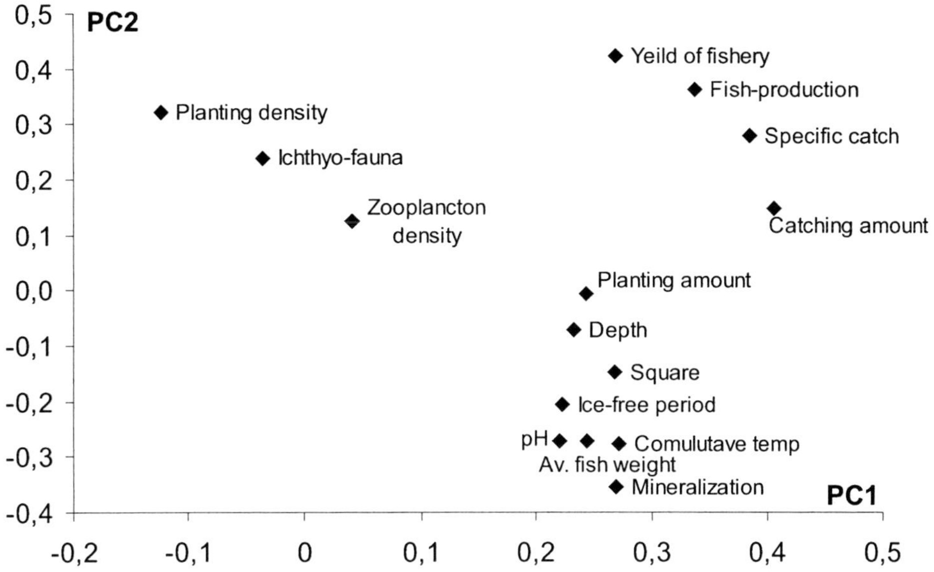

Fig 4. Loading plot for small lakes. PCA model PC1 vs. PC2.

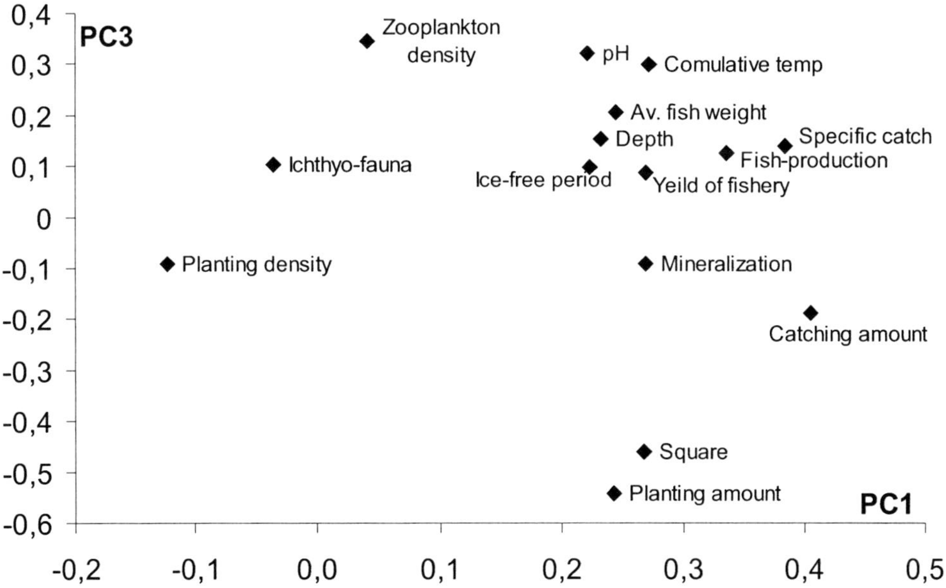

Fig 5. Loading plot for small lakes. PCA model PC1 vs. PC3.

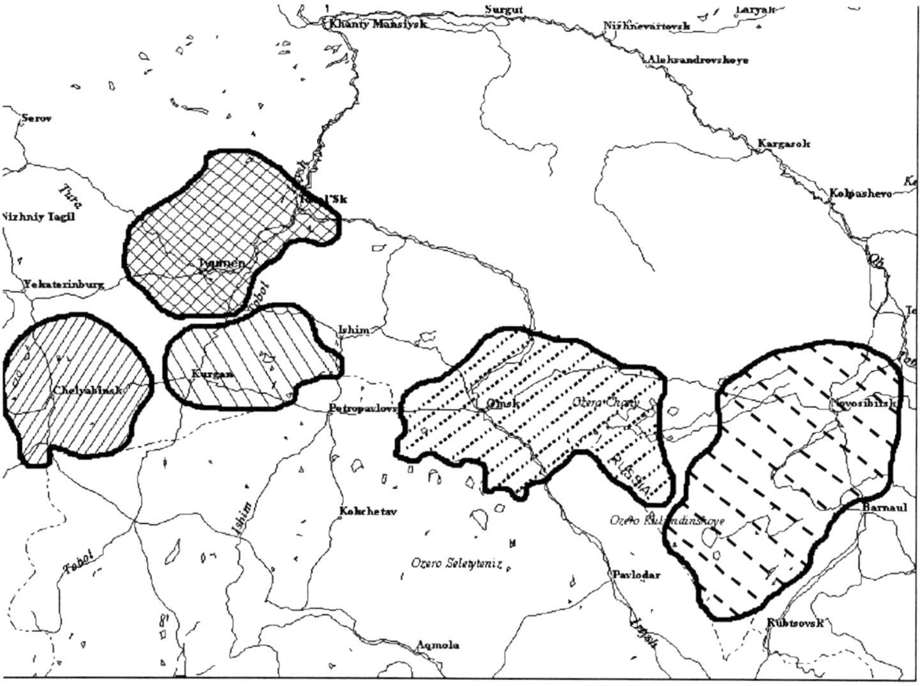

Fig 6. Map of Western Siberia. We marked regions where the lakes with fish farms are located. - Kurgan, - Chelyabinsk, - Tyumen, - Omsk, - Novosibirsk.

1. Variable *Average fish weight* (Y1) has a negative correlation with variable *Catching amount* (Y2) and with other variables that characterize the fish planting (Y3, Y4, and Y5). This means that peled can grow bigger only when *Planting density* is low. In this case *Planting amount* decreases and this circumstance allows peled to grow bigger at the same quantity of fodder.
2. *Planting density* negatively correlates with *Planting amount*. Obviously, the larger the lake the more fish is planted there. But for really huge lakes this variable is rather small. Therefore in all big lakes we have low *Planting density*.
3. *Cumulative temperature* positively correlates with *Ice-free period*. This evident relationship also shows that the model under consideration is reasonable.

CONCLUSIONS

The constructed models show that application of Multivariate Data Analysis is reasonable for exploration of such a complex data. The results are more of qualitative than of numerical nature, and they are mainly useful if the investigator wants to understand the relationships between different factors that characterize peled planting.

1. We have succeeded in construction of prediction model for the lakes of Tobolsk region. Analysis shows that we can predict *Yield of fishery*, but only for each region individually. In future we plan to construct such models for other regions.

2. Gathering of information should be continued. Obviously, methods of data collection need to be improved for decreasing of noise level in the data set.
3. Unfortunately, few or more years will pass before we will be able to collect more "clear" and representative data. Every lake is planted only once a year. Therefore, during one year, number of samples that can be collected are equal to the lakes that are used for peled planting in the definite region.

REFERENCES

[1] Muhachov V. S. *Lake fish culture,* Agropromizdat. 1989, p.161 (in Russian)
[2] Muhachov V. S. *Biotechnics of accelerated peled rearing*, FGU IPP «Tyumen», 2003 , p. 176
[3] *Agro climate resources of Novosibirsk region*, Gidrometeoizdat, 1971, p.155
[4] *Agro climate resources of Omsk region*, Gidrometeoizdat, 1971, p.198
[5] *Agro climate resources of Tyumen region*, Gidrometeoizdat, 1972, p. 150
[6] *Agro climate resources of Chelyabinsk region*. Gidrometeoizdat, 1977, p. 151
[7] *Geographic of Kurgan region* (under ref. of O. G. Zavyalova), Parus-M, 1993, p. 160
[8] *Surface water resources of USSR*, V. 15 №3, Gidrometeoizdat, 1973, p. 244
[9] Alimov A. F. *Introduction in production hydrobiology*, Gidrometeoizdat, 1989, p. 152
[10] K. Esbensen, S. Schönkopf, T. Midtgaard, D. Guyot, *Multivariate Analysis –in practice,* Camo ASA, 1998, p.46-63, 84-92

INDEX

A

B

C

D

E

F

Q

R

S

T

V

W